Astrophysics and Space Science Library

Volume 406

Stella Cottam • Wayne Orchiston

Eclipses, Transits, and Comets of the Nineteenth Century

How America's Perception of the Skies Changed

Springer

Stella Cottam
National Astronomical Research
Institute of Thailand
Chiang Mai, Thailand

Wayne Orchiston
National Astronomical Research
Institute of Thailand
Chiang Mai, Thailand

ISSN 0067-0057 ISSN 2214-7985 (electronic)
ISBN 978-3-319-08340-7 ISBN 978-3-319-08341-4 (eBook)
DOI 10.1007/978-3-319-08341-4
Springer Cham Heidelberg New York Dordrecht London

Library of Congress Control Number: 2014946413

Cover illustration: U.S. Naval Observatory, American Transit of Venus Commission – Courtesy of U.S. Naval Observatory.

Printed on acid-free paper

Springer is part of Springer Science+Business Media (www.springer.com)

Preface

In the middle of the nineteenth century the USA, previously dependent on Europe for general information on astronomy, matured to the point where it also could contribute to this knowledge. No longer would astronomy be considered only a tool for navigation and time-keeping. In this book we will demonstrate how American citizens promoted the participation of their young country in this science. Particular spectacular events and objects such as comets, meteor storms, solar eclipses, and transits of Venus were well covered in newspapers and periodicals, and public lectures on these topics were well attended. Some members of the general public became more involved, and would contribute funds and even participate in astronomical endeavors, as amateur and professional astronomy grew throughout the nation during the second half of the nineteenth century.

Members of the scientific community are most generous and have been willing to share their time, thoughts, and resources with us. It has been our pleasure to communicate by phone, e-mail, and in person with the following individuals, and we would like to express our gratitude to them.

Robert Ariail, of the Antique Telescope Society and co-author of *Alvan Clark & Son – Artists in Optics* (1995), supplied useful information on Clark telescopes, and gave us permission to use his photograph of the Fitz telescope in this book.

Dr Walter Breyer, the Executive Secretary of the Antique Telescope Society, recommended knowledgeable colleagues who proved to be invaluable sources of information.

Ben Burress, Program Content Developer of the Chabot Space and Science Center, provided details on its growing significance since the nineteenth century in bringing astronomy to the public.

Jacqueline Davis, the current Director of the Fort Sam Houston Museum, assisted the first author in locating the historical marker for the American transit expedition site in San Antonio, Texas, and provided some materials that brought the event to life.

Dr Suzanne Debarbat (Paris Observatory) provided some elusive information on the French transit of Venus expedition to St. Augustine, Florida, with the assistance of Virginie Barbet from the Paris Observatory Library. She also had numerous discussions with the second author of this book on the nineteenth century French transit of Venus campaigns.

Dr Steven Dick (ex-U.S. Naval Observatory) brought to our attention records in the National Archives in Washington D.C. relating to public contributions to the Government's 1882 transit of Venus program, and he kindly sent specific information which enabled us to locate these in a timely manner. Meanwhile, on numerous occasions he discussed the American nineteenth century transit of Venus programs with the second author during their collaborative research on these expeditions.

The late Professor Hilmar Duerbeck (University of Münster, Germany) provided information on the German transit of Venus site at Trinity College in Connecticut, and details of other nineteenth century German transit of Venus expeditions.

Bart Fried, a founder of the Antique Telescope Society, kindly discussed the American telescope-maker John Brashear and recommended some useful resources.

Professor Jay Pasachoff from Williams College provided us with useful comments about nineteenth century solar eclipses and transits of Venus.

Dr John Pearson from the National Astronomical Research Institute of Thailand provided us with valuable information on nineteenth century total solar eclipses, and the Lick Observatory's solar eclipse expeditions.

Leslie Proudfit from the Astronomical Society of the Pacific provided information on the growth and accomplishments of the Society during the late nineteenth century.

Carolyn Sanford, Head of Reference and Instruction at the Gould Library at Carleton College, and Eric Hillemann, Gould College Archivist, provided information on their *Sidereal Messenger*.

Professor Chris Sterken (Vrije Universiteit Brussel, Belgium) provided several items of interest and inspiration regarding the 1882 transit of Venus, especially relating to Jean-Charles Houzeau and the Belgian expedition to San Antonio, Texas. He also discussed the accomplishments of the American astronomer David Todd during this event.

John Ventre, the Historian at the Cincinnati Observatory, was always available to verify facts and offer useful suggestions. He also provided the 1870 roster of members of the Observatory.

Associate Professor Alice Walters (St. Francis University, USA) provided a copy of her paper on broadsides and solar eclipses.

Eileen Wild from the Chicago Astronomical Society communicated details regarding the Society, its membership, and its accomplishments.

It is difficult to express adequate gratitude to Dr Tom Williams from Rice University. He provided numerous valuable suggestions that were helpful in creating much of the framework of this book.

In addition, many unnamed individuals were immensely helpful during this project, and we are especially grateful to staff at the Adler Planetarium in Chicago, the Aiken County Historical Society, the University of Cincinnati Archives, the Cincinnati Historical Society, the National Archives in Washington, D.C., and the Shelbyville Historical Society.

Finally the first author would like to thank her family, husband John, and children Emily, Daniel, and Tamara, for their continued support. Meanwhile, the second author is grateful to Professor Boonrucksar Soonthornthum and other staff at the National Astronomical Research Institute of Thailand for their encouragement and support during the completion of this book.

Chiang Mai, Thailand

Stella Cottam
Wayne Orchiston

Contents

1 Introduction 1
1.1 Terminology 5

2 Historical Overview: The United States and Astronomy Until the 1860s 7
2.1 The Leonid Meteor Storm of 1833 8
2.2 The Great Comet of 1843 (C/1843 D1) 17
2.3 American Astronomy in the Early 1800s 25

3 Scientific Overview 43
3.1 Solar Eclipses and the Nature of the Solar Corona 43
3.1.1 Eclipses, Prominences and the Corona 43
3.1.2 Earlier Solar Eclipses of Note 46
3.1.3 Instrumentation 50
3.1.3.1 Photography 50
3.1.3.2 Spectroscopy 52
3.1.3.3 The Polariscope 55
3.1.3.4 The Role of the Telegraph 57
3.1.4 The Solar Eclipses of 1868, 1869 and 1878 57
3.1.4.1 The Eclipse of 18 August 1868 57
3.1.4.2 The Eclipse of 7 August 1869 63
3.1.4.3 The Eclipse of 29 July 1878 75
3.1.5 Evolving Directions for Solar Research 82
3.1.5.1 The Shape of the Corona 82
3.1.5.2 The Chemical Composition of the Corona 82
3.1.5.3 Sunspots and the Corona 84
3.1.5.4 Polarization and the Nature of the Corona 84
3.1.5.5 Intra-Mercurial Planets 85
3.1.6 Conclusions 86

3.2 Transits of Venus and Refinement of the Value of the Astronomical Unit 87
3.2.1 The Astronomical Unit: Its Significance and Early Determinations 87
3.2.2 A Recap of Earlier Recorded Transits of Venus 92
3.2.2.1 The Transit of 6 December 1631 92
3.2.2.2 The Transit of 4 December 1639 93
3.2.2.3 The Transit of 6 June 1761 93
3.2.2.4 The Transit of 3 June 1769 95
3.2.3 Advances in Instrumentation 98
3.2.3.1 The Development of Astronomical Photography 98
3.2.3.2 The Horizontal Photographic Telescope 99
3.2.3.3 The Heliometer 99
3.2.3.4 The Role of Spectroscopy 100
3.2.3.5 The Role of the Telegraph 100
3.2.4 The Transit of Venus of 9 December 1874 102
3.2.4.1 Foreign Expeditions 103
3.2.4.2 The American Expeditions 104
3.2.5 The Transit of Venus of 6 December 1882 110
3.2.5.1 Foreign Expeditions 111
3.2.5.2 Foreign Expeditions to the United States 111
3.2.5.3 The American Expeditions 117

4 Popular Astronomy and the Solar Eclipses of 1868, 1869 and 1878 129
4.1 The Solar Eclipses of 1868, 1869 and 1878 129
4.2 A Sampling of Popular Periodicals 129
4.2.1 General Periodicals 130
4.2.2 Religious Periodicals 153
4.2.3 Scientific Periodicals 175
4.3 *The New York Times* 184
4.3.1 The Solar Eclipse of 18 August 1868 185
4.3.2 The Solar Eclipse of 7 August 1869 185
4.3.3 The Solar Eclipse of 29 July 1878 192

5 Popular Astronomy and the Transits of Venus of 1874 and 1882 201
5.1 The Transit of Venus of December 1874 201
5.1.1 A Sampling of Popular Periodicals 201
5.1.1.1 General Periodicals 201
5.1.1.2 Religious Periodicals 211
5.1.1.3 Scientific Periodicals 218
5.1.2 *The New York Times* 220
5.2 The Transit of Venus of December 1882 231
5.2.1 A Sampling of Popular Periodicals 231
5.2.1.1 General Periodicals 231
5.2.1.2 Religious Periodicals 234
5.2.1.3 Scientific Periodicals 239
5.2.2 *The New York Times* 240

6 Discussion ... 255
6.1 'Amateurs' and 'Professionals' in Astronomy ... 255
6.2 Articles in *The New York Times* ... 263
6.3 Popular Lectures ... 265
6.4 Telescopes ... 267
6.5 Observatories ... 269
6.6 Publications ... 271
6.7 Astronomical Societies ... 273
6.8 Public Participation in Research ... 280

7 Concluding Remarks ... 287

References ... 291

Index ... 329

Chapter 1
Introduction

Of all the sciences, astronomy lends itself best to be a source of entertainment for the amateur. Ormsby MacKnight Mitchel (1810–1862), a lecturer and popularizer of astronomy in the U.S.A. in the nineteenth century, described the wonder in the heavens that commenced thousands of years ago. He speculated that primitive men first recognized the Sun and the Moon, and patterns in the stars. The observant early astronomers would recognize deviations from these patterns caused by the motions of the planets, probably in the order of Venus, Jupiter, Mars, Saturn and finally Mercury (Mitchel 1867, pp. 42–68). Infrequent events, such as the appearances of comets, meteor showers, novae, supernovae and eclipses, would further pique their interest.

As the centuries passed, the practical aspects of this science became apparent for such activities as agriculture and navigation. This led to the development of professional astronomy. Modern professional astronomy has been built on the foundations laid by the earliest amateur astronomers. It fell to the professional to develop theories that both explained and predicted celestial events. Tools were created to facilitate these endeavors. These tools were both of a tangible nature, such as telescopes, and non-tangible, such as mathematical investigations. Many of these were still within the reach of the amateur.

At the beginning of the nineteenth century the United States of America was still a young country experiencing growing pains and challenges on multiple levels. As recently as 1776 the Declaration of Independence was adopted, and the nation was formed from the original 13 British colonies. It was not until 1783 that the Treaty of Paris was signed finally ending hostilities.

Early interest in astronomy as a science beyond its utility was scant in the colonies, although there are instances during these years of those who would make time for scientific pursuits. It is commonly thought that Benjamin Franklin (1706–1790) was America's first scientist but this distinction actually belongs to a fellow Philadelphian and Franklin's mentor, James Logan (1674–1751; Tolles 1956, p. 20). Largely self-taught in mathematics, among other subjects, he published papers on optics and astronomy (Hindle 1956, pp. 21–22). The distinction of being America's first astronomer should go to John Winthrop (1714–1779; Fig. 1.1), the Hollis Professor of

S. Cottam, W. Orchiston, *Eclipses, Transits, and Comets of the Nineteenth Century*, Astrophysics and Space Science Library 406,
DOI 10.1007/978-3-319-08341-4_1

Fig. 1.1 A painting of John Winthrop by John Singleton Copley in about 1773 (courtesy: en.wikipedia.org)

Fig. 1.2 A painting of Bishop James Madison (courtesy: en.wikipedia.org)

Mathematics and Natural Philosophy at Harvard University. Among other observations, he recorded the transit of Mercury on 27 April 1740 (Brasch 1916, p. 159). He presented two lectures on comets after he observed Halley's Comet at its predicted return on 11 April 1759 (Brasch 1916, pp. 162–164). He would be instrumental in organizing the first astronomical expedition of the colonies, to observe the transit of Venus in 1761, and again in 1769 (Brasch 1916, p. 164).

Fig. 1.3 Nathaniel Bowditch (courtesy: en.wikipedia.org)

Another well-known astronomer at this time was Bishop James Madison (1749–1812; Fig. 1.2), who was appointed to the Chair of Natural Philosophy at the College of William and Mary in Williamsburg, Virginia, in 1773 and just 4 years later became the President of the College. As Shy (2002) has documented, Madison taught courses that included astronomy, set up the first institutional observatory in the United States (in 1778) and carried out wide-ranging astronomical observations.

Yet another significant colonial astronomer was Nathaniel Bowditch (1773–1838; Fig. 1.3). Like Logan, he was mostly self-taught (Montgomery 1982, p. 917). Initially a seaman, he distinguished himself as a member of a survey party. He learned to use mathematics to correct navigation tables. He would calculate cometary orbits, and he published a study of a great meteor seen in 1807 (Williams 1984, pp. 64–65). In the *North American Review* of April 1825 it is said that his greatest work was the translation of Laplace's *Mécanique Céleste* into English:

> This work has undoubtedly had the greatest influence on the development of American astronomy, since the proverbial difficulty of the [original] *Mécanique Céleste* renders it almost inaccessible to beginners …

These are isolated examples of citizens who pursued their interest in astronomy in a young country that had other more pressing priorities.

Over the next ten decades the new nation would struggle with defining its system of government, while geographically expanding through the incorporation of other territories obtained by means of purchase, treaty or war. With this growth came the necessity to create a stable infrastructure that would assure future success

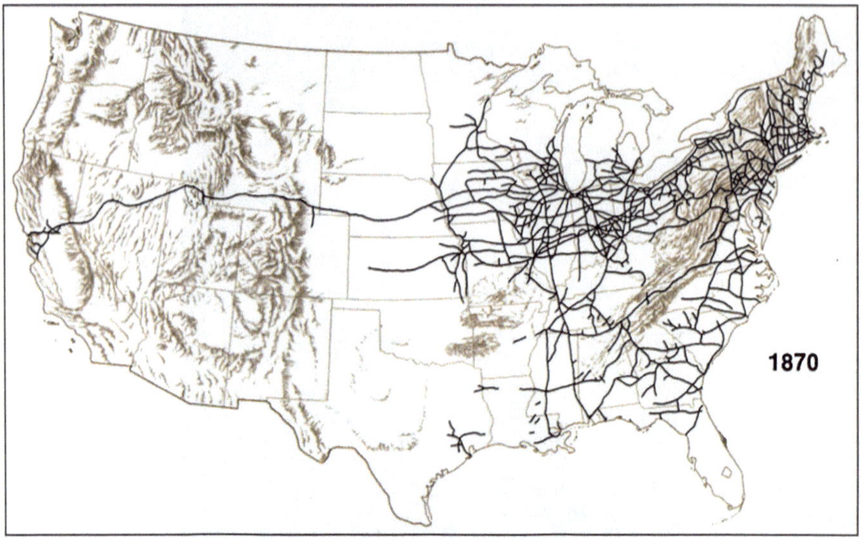

Fig. 1.4 By 1870 the railroad had spread, mainly from Chicago and New York, to cover most of the eastern states, but there was still a conspicuous 'vacuum' in the far west, especially in Texas, and along the Californian coast (after: www.Latinamericanstudies.org/19th-century-transportation.htm)

on economic and social levels. Technological innovations during the nineteenth century, such as the cotton gin, railroad (see Fig. 1.4), steamboat and the telegraph all contributed to a more stable economy. The taming of the expanding frontier, the technology that led to the easing of daily labors and the increase in the availability of printed matter would all contribute to a growth of leisure time for the pursuit of recreational interests. Cultural entertainment in the form of theatrical productions, circuses and scientific exhibitions became more readily available. School systems grew and literacy increased. With more time and resources available American citizens might more intelligently discuss and pursue interests of a scientific nature.

Astronomy did not become 'popular,' that is to say a topic of interest or entertainment among the general population of the United States, until well into the nineteenth century. Then free of the concerns of the Wars of Independence and 1812, Americans could indulge in new-found recreational activities. Practical advances during this time, such as the production and distribution of paper and books, the manufacture of eyeglasses, and the more generalized presence of gas lighting during night-time leisure hours, enabled greater access to, and use of, printed materials. Verbal communication of newly-discovered concepts and entities in parlors, schools and at public meetings enhanced the promotion of the sciences, astronomy in particular, as a popular topic for discussion, education and entertainment. Enthusiastic amateur astronomers would have a variety of outlets for their developing interest in this new democratic society.

Astronomical spectacles that would generate enormous public interest at this time included the expeditions associated with the 1874 and 1882 transits of Venus, and the total solar eclipses of 1868, 1869 and 1878. The second transit and the last two eclipses all were visible from the United States (Clerke 1902, pp. 167, 169, 174).

1.1 Terminology

Transits of Venus are rare events that occur when Venus appears to cross the face of the Sun as viewed from the Earth. Important for understanding its use in the calculation of the Sun's distance from the Earth are the definitions of the 'contacts', and these are shown in Fig. 1.5. First contact (number 1 in Fig. 1.5) occurs when Venus begins its ingress onto the solar disk and is tangential to the outside of the Sun's limb. Second contact (number 2 in Fig. 1.5) occurs when Venus first appears completely internal to the Sun at the limb. Third contact (number 3 in Fig. 1.5) occurs when the planet is internally tangential at the opposite limb as it begins its egress. The fourth contact (number 4 in Fig. 1.5) occurs as egress is just completed and Venus is then externally tangential to the Sun.

In the nineteenth century two such transits occurred. They were on 9 December 1874 and 6 December 1882. The former transit was not visible from the United States, but the latter could be seen throughout much of the country and its territories. However, the young nation was eager to participate in both of these important events, as they would provide an opportunity to pin down the value of that 'fundamental astronomical yardstick', the distance from the Earth to the Sun (Dick et al. 1998). This value, known as the 'astronomical unit', would provide a scale for the Solar System and ultimately the Galaxy.

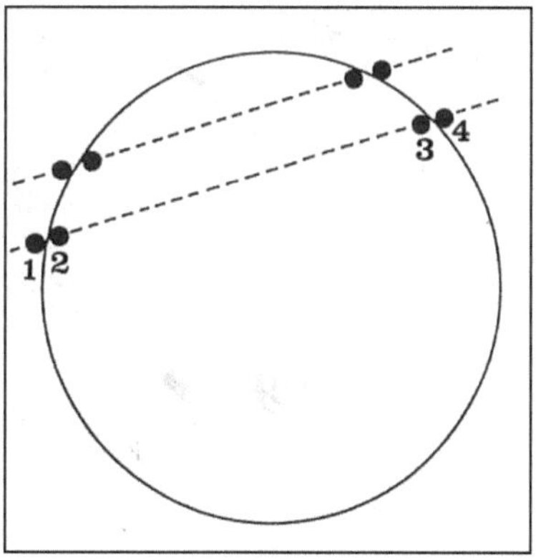

Fig. 1.5 A schematic diagram showing a transit of Venus as viewed from two observing sites widely separated in latitude. In reality the actual separation of the two transects is very much closer than this, but substantial vertical exaggeration has been introduced here to illustrate the concept. The numbers *1* and *2* refer to the two ingress contacts and *3* and *4* to the two egress contacts

Total solar eclipses, when the solar disk is totally obscured by the Moon, happen more frequently. However, they are less easily observed due to their relative short durations and the restricted geographical locations where they are visible. A number of total solar eclipses were visible in parts of the United States during the second half of the nineteenth century. They offered astronomers unique opportunities to investigate the nature of the corona, that mysterious nebulous light surrounding the Sun and only visible at these times, to determine if it belonged to the Sun or the Moon or whether it was in fact an anomalous feature of the Earth's atmosphere (Campbell 1907, pp. 71–72; Cottam, Pearson et al. 2011, p. 340; Pearson et al. 2011, p. 256).

A great deal of published literature is available on the studies of the nineteenth-century transits of Venus and total solar eclipses undertaken by professional astronomers. Academic journals provided preliminary data and later published results obtained by various scientific expeditions. In this book we will describe observations made by amateur and professional astronomers, and examine popular interest in these important astronomical events. An exploration of reports in popular literature, journals and newspapers has been undertaken. The number and nature of such reports before and subsequent to these specific transits and solar eclipses is telling. A review has been made here of periodicals of that time that targeted a variety of demographics and interests. A review has also been made of *The New York Times* (a newspaper that has been in continuous publication since 1851), covering the entire period under consideration (1868–1882). Anecdotal and numerical data are provided that demonstrate a sustained interest in astronomical subjects by the public, as reflected in attendance at popular lectures on scientific topics, non-professional membership in scientific organizations and improved access to and use of astronomical equipment.

Besides the aforementioned opportunities for satisfying their interest, there is evidence of actual public participation in some of these scientific endeavors in response to government requests. Some of these also are described in this book.

Chapter 2
Historical Overview: The United States and Astronomy Until the 1860s

Astronomy as an amateur recreation was entrenched in much of Western Europe by the eighteenth century, where there were the financial means, the knowledge base, the manufacture of tools and the genuine interest among those with the time to engage in such a recreation. It took most of the first half of the nineteenth century for this pastime to become popular in the relatively young United States. This new country had no wealthy aristocracy to indulge in subjects of personal interest. In other words, there was no American Carrington, Common, Cooper, Dawes, De la Rue, Lee, William Herschel, John Herschel, Huggins, Lassell, Lockyer, Nasmyth, Parsons (Third Earl of Rosse), Smyth or South—to mention just Britain alone (see Chapman 1998). Nor was there an established mechanism for the newly-created American Federal Government to finance science education and technology on a national level. With respect to the last, it was felt that much of these matters would, in any case, fall under the purview of the individual states. What little knowledge base was present on the subject of astronomy in America was inherited from Europe and, for most people, that was considered sufficient for the practicalities of everyday life. The value of this science in navigation, surveying and time-keeping was accepted as a given, but most Americans saw no reason to spend money or time to extend knowledge beyond what was needed for these routine tasks. Through much of the first half of the nineteenth century, required astronomical instruments were purchased from Europe rather than manufactured within the United States. It was also some time before American books on the subject were published. The English publications were considered adequate. Americans were satisfied with the situation as it was at the time.

There were documented instances of some early American interest in the science of astronomy for its own sake, but these are rare. A European account of note is cited by Professor W.C. Rufus (1924, pp. 121–122) in his article on astronomical observatories in the United States:

> Mr. Thomas Brattle, of Boston in New England, is the annonymus [sic] person alluded to by Newton, in his Principia, as having made such good observations of the comet of 1680. Several of his observations are preserved in the Transactions of the Royal Society of London.

S. Cottam, W. Orchiston, *Eclipses, Transits, and Comets of the Nineteenth Century*, Astrophysics and Space Science Library 406,
DOI 10.1007/978-3-319-08341-4_2

Another colonial observer, the noted New England scientist and physician Thomas Robie (1689–1729), made significant observations of the lunar eclipse of 15 March 1717 as well as the Sun and the planet Mercury (ibid.). In truth, it must be remembered that these colonists were not Americans, but rather English subjects who had relocated across the Atlantic.

In the second quarter of the nineteenth century in particular, a distinct change in attitude developed regarding the value of astronomy. According to Lick Observatory Director Edward Holden (1846–1914), there was an "... intellectual awakening which came about as soon as our young country was relieved from the pressure of the two wars of 1776 and 1812." (Holden 1897, p. 933). Donald Zochert (1974, p. 463) quotes an editorial comment from a popular newspaper, the Milwaukee *Courier* of 9 April 1845, that exemplifies this rising interest: "The works of Nature are so wonderful – so passing strange – that we are often tempted to turn aside from the tedious duties of editorial life to investigate them."

Just what types of celestial objects or phenomena aroused such curiosity at this time in American history? In the second quarter of the nineteenth century, particularly inspirational events included the Leonid meteor shower of 1833 and the advent of the Great Comet of 1843. While the Leonid meteor shower is visible every year, at 33-year intervals a major shower, termed by astronomers a Leonid 'meteor storm', occurs (see Dick 1998). There was one of these 'storms' in 1833.

2.1 The Leonid Meteor Storm of 1833

On the night of 12–13 November 1833 a meteor shower occurred that was so spectacular that it would be a prominent topic of conversation among scientists and non-scientists alike, and noted in professional publications, popular journals, newspapers and personal chronicles.

In his book of 1878 the author R.M. Devens deemed the Leonid meteor shower of 1833 to be among the 100 most significant events of the first century of the United States:

> Extensive and magnificent showers of shooting stars have been known to occur at various places in modern times, but the most universal and wonderful which has ever been recorded is that of the thirteenth of November, 1833, *the whole firmament, over all the United States, being then for hours, in fiery commotion!* No celestial phenomenon has ever occurred in this country, since its first settlement, which was viewed with such intense admiration by one class in the community, or so much dread and alarm by another. It was the all-engrossing theme of conversation and of scientific disquisition, for weeks and months. (Devens 1878, p. 329, his italics).

Reproduced here in Fig. 2.1 is a nineteenth century woodcut depicting the Leonid meteor shower as it appeared on the evening of 13 November 1833 from the U.S.A. This picture is so famous that Professor David Hughes—an authority on meteor showers—wrote a research paper about it and published this in the journal *Earth, Moon, and Planets* in 1995.

It would be Denison Olmsted (1791–1859; Fig. 2.2), Professor of Mathematics and Physics at Yale University, who would take this opportunity to uncover much

Fig. 2.1 A nineteenth century woodcut of the 1833 Leonid meteor shower (courtesy: en.wikipedia.org)

that had been, until then, unknown about meteors. He was primarily responsible for bringing the study of meteors into the realm of astronomy (see Hoffleit 1992, pp. 24–32; Hughes 1982).

On the evening of 12–13 November 1833 Olmsted was awakened by a friend to observe this spectacular meteor shower which was already in progress. His impressions were published on the 13th in the *New Haven Daily Herald*, where he also requested reports of observations made by others. He was pleasantly surprised to

Fig. 2.2 Denison Olmsted (courtesy: en.wikipedia.org)

receive responses from across the country, well beyond the normal purview of the *Herald*. His significant two-part article summarizing what he gleaned from these reports was published the following year in the *American Journal of Science and Arts*. In Part I Olmsted (1834a, pp. 363–411) first reported on his own observations, and he followed this with a summary of the reports received from others. In Part II Olmsted (1834b, pp. 132–174) revisited some earlier Leonid meteor showers, and he then compared and contrasted the data derived from these with observations made in 1833.

Olmsted observed the shower from New Haven, and he received reports from as far south as Georgia and Mississippi, as far west as Missouri and Louisiana, and as far north as Niagara Falls. He also was sent reports from outlying areas further to the east or south, from ships at sea. Alexander Catlin Twining (1801–1884) sent Olmsted the results of reports received in New York from 15 ships that were located in different positions in the Atlantic when the shower occurred. A student of civil engineering at West Point, Twining provided "… the most careful observation and thoughtful analysis …" (Littmann 1998, pp. 17–18). Those further to the east or south of the United States generally gave negative reports, even when there were clear skies. Some closer to the shores of the nation, such as Captain Gideon Parker on the *Junior*, were more successful in their observations. The professions of those who had contributed reports were not always stated in Olmsted's article, but included college professors and physicians, among others. Olmsted also drew on reports printed in local newspapers from non-professional citizens, including ministers and farmers.

At this time, the most common theory on the origin of meteors linked them to the weather, so the study of meteors was not considered to be under the purview of astronomy.

Thus, the most complete reports received by Olmsted would include meteorological data. Olmsted (1834a, p. 385, his italics) summarized:

> Throughout the entire region where the Meteors were observed, there was a sudden and extraordinary *change of weather from warm to cold*, accompanied by an uncommon transparency of the atmosphere ... It is hardly possible to persuade ourselves that two concurrent phenomena, both so remarkable as the change of weather and the falling stars, were independent of each other.

However, he could not explain this possible relationship. Was it that of cause and effect between the two, or was there something else involved that "... gave origin to both the change of weather and the meteors?" (Olmsted 1834a, p. 402).

The meteors were at their most striking in frequency and brilliance between 2:00 a.m. and 5:00 a.m. It was concluded that determination of an accurate number of meteors could not reasonably be made. Some more methodical observers worked in groups and attempted to keep count of specific sections of the sky. Some compared the number of meteors seen in a section of the sky with the number of known stars. Olmsted (1834a, p. 389) recognized a discrepancy in numbers: "... some describing them as 'thousands' at a time, and some even by 'millions'." While it was not mentioned in Olmsted's article, the distinguished French astronomer and Paris Observatory Director François Arago (1786–1853) reported that no less than "... two hundred and forty thousand meteors were at the same time visible above the horizon of Boston." (Devens 1878, p. 300). Professor Joseph Henry (1797–1878), an authority on electromagnetism and then at Princeton University, estimated a rate of 72,000 per hour (Littmann 1998, pp. 14–15).

Olmsted (1834a, p. 389, his italics) described three varieties of meteors, the most common being phosphoric lines. There were also large fireballs, and among these were those "... *luminous bodies that continued for a long time in view*." One of these last was described by an observer in Poland, Ohio, as lasting more than an hour (Devens 1878, p. 331). These last rare bodies could be valuable in that their height might be determined by applying trigonometry to any of these unique entities observed at sites distant from each other. In particular, the large fireball that exploded near the star Capella may have been seen by four individuals: "... by Mr. Barber at Frederic, Maryland, by Mr. Tomlinson at Brookfield, a few miles north west, and by Dr. Lee at New Britain, a few miles north east of New Haven, and by Lieutenant Crane at West Point." (Olmsted 1834a, pp. 403–404). At the time this article was written the appropriate calculations had not been made. Most observers agreed the phosphoric lines were no more than two or three miles above the Earth. Olmsted (1834a, p. 403) felt the large fireballs were undoubtedly at a greater height. Edward Hitchcock (1834, p. 356) would state that even the common phosphoric lines must be at a height greater than that of the clouds as none were observed between the clouds and the Earth.

In the second installment of Olmsted's article (1834a, p. 368) he described meteor showers of different times and venues. He had already recognized the shower observed by the renowned German naturalist Alexander Humboldt (1769–1859) and the French botanist and explorer Aimé Bonpland (1773–1858) on 12 November 1799 from Cumana, Venezuela, as most like that of 1833. In their personal narrative,

Fig. 2.3 Hubert Anson Newton (courtesy: en.wikipedia.org)

Humboldt and Bonpland (1907, p. 352) described a shower of "… extraordinary luminous meteors …" that began a little after the hour of two in the morning. This was witnessed by almost all the inhabitants who had risen for early morning mass. The oldest of these "… remembered that the great earthquake of 1766 was preceded by a similar phenomena." (Humboldt and Bonpland 1907, p. 353) Humboldt and Bonpland (1907, p. 352) described their own observations of the shower of 1799:

> … from the first appearance of the phenomena, there was not in the firmament a space equal in extent to three diameters of the moon, which was not filled every instant with bolides and falling stars … All these meteors left luminous traces from five to ten degrees in length.

It would later be determined, as these dates seem to support, that there was a cycle to the peak of these November showers of about 33 or 34 years. It was the astronomer and Professor of Mathematics at Yale University, Hubert Anson Newton (1830–1896; Fig. 2.3; see Hoffleit 1992, pp. 47–56), who first recognized this (Dick 1998; Littmann 1998). Thus, predictions could be made.

There were consistencies and inconsistencies in these reports that Olmsted would try to sort out. Based on the accumulated data he would look at a number of parameters, draw some conclusions, make some speculations, or in some cases advise on the need for future investigation.

Among his conclusions was that:

> *The meteors consisted of combustible matter, and took fire and were consumed in traversing the atmosphere.* That these bodies underwent combustion, we have the direct evidence of the senses. We saw them glowing with intense light and heat, increasing in size and splendor as they approached the earth. (Olmsted 1834b, p. 151, his italics).

It is now known that Olmsted was only partially correct. This was frictional vaporization, not the conventional burning of combustible material in oxygen (Littmann 1998, p. 21).

One point of disagreement among Olmsted's reports was whether or not there was any sound associated with the shower meteors. There were reports of hissing sounds, slight explosions or popping sounds, or no sound at all. Although Olmsted himself heard nothing, he did not dismiss the possibility, as many of the reports were similar in nature (Olmsted 1834a, pp. 392–393, 404). The New Zealand radar meteor astronomer Professor Colin Keay (e.g. 1980a, b) has since clearly documented that electrophonic meteors—those producing sound effects—do indeed exist, although they are rare. Olmsted (1834a, p. 397) also described another discrepancy: "Phenomena resembling more or less the Aurora Borealis, were visible in some places, although in many other places no appearances of the kind were observed." He felt this issue could be resolved at a future event by means of magnetic observations.

A significant point of agreement was the fact that the meteors seemed to originate from a common point, which is now referred to as the 'radiant':

> The meteors, as seen by most observers, appeared to proceed from a fixed point in the heavens, which some referred to the zenith, and others to a point a little S.E. of the zenith. Those who marked its position among the fixed stars, observed it to be in the constellation Leo … (Olmsted 1834a, p. 394).

In his discussion Olmsted (1834b, p. 163, his italics) would put forth the inquiry: "*What relations did the body which afforded the meteoric shower, sustain to the earth*?" He considered the possibilities of a satellite, a comet or "… a collection of nebulous matter …" He drew no conclusions on the nature of the body but he would incorrectly conclude (Littmann 1998, pp. 25–26) that this meteor shower derived from "… a nebulous body, which revolves around the sun in an orbit interior to that of the earth …" (Olmsted 1834b, p. 172).

The cometary origin of these phenomena would first be suggested by Daniel Kirkwood (1814–1895; Fig. 2.4), Professor of Mathematics at Indiana University: "… the debris of ancient but now disintegrated comets, whose matter has become distributed around their orbits." (*Danville Quarterly Review* 1861). The comet associated with these Leonids was determined to be what is now known as Comet P55/Tempel-Tuttle, as demonstrated independently by the German-American astronomer Carl Friedrich Wilhelm Peters (1844–1894), the noted Austrian astronomer and mathematician Theodor von Oppolzer (1841–1886) and the Italian astronomer Giovanni Schiaparelli (1835–1910), all of whom published papers on this topic in the *Astronomische Nachrichten* in 1867 (Mason 1995; Yeomans 1981).

As the name indicates, this comet was discovered independently by two different astronomers. One was the German-born amateur Ernst Wilhelm Leberecht Tempel (1821–1889), usually known as Wilhelm Tempel, who lived in Marseilles (France) at the time and detected the comet on 19 December 1865 using a custom-made Steinheil 10.8-cm comet-seeker (Bianchi et al. 2010). The other was the American, Horace Parnell Tuttle (1837–1923), who at the time worked at the U.S. Naval Observatory, and made his independent discovery on 5 January 1866 (Kronk 2003). This comet has a period of 33 years, and each time it reaches perihelion material is ejected from the nucleus, which is thought to have a radius of 1.8 ± 0.2 km (Hainaut et al. 1998), a mass of around 1.2×10^{13} kg (Jewitt 2004, p. 672) and a rotation

Fig. 2.4 Daniel Kirkwood (courtesy: findagrave.com)

period of 15.31 ± 0.03 h (Jorda et al. 1998). Although debris is now spread along the entire orbit of the comet and a Leonid meteor shower is observed every year, it is the concentration of material encountered by the Earth every 33 years—as in November 1833—that gives rise to the spectacular Leonid meteor storms.

The November meteor storm of 1833 was a subject of great public interest, as reflected in the local newspapers of the time. Olmsted's original article in the *New Haven Herald* was recapped and shared with a number of other publications, including the *Pittsfield Sun* of 21 November 1833 (Curious phenomenon 1833) and the *New Bedford Mercury* of 22 November 1833 (From the *New Haven Herald* 1833).

In some newspapers, personal reports were on occasion made in letters submitted to the editor. Nathan Daboll described in the *Norwich Courier* of 27 November 1833 the blazing skies, and he offered the theory that:

> … the meteors are generated and produced by exhalation drawn up by the Sun from aqueous, nitrous, sulphureous, bituminous and various other substances, animal and vegetable … This is, however, a matter of opinion. (Remarkable phenomena 1833).

D. Leavitt's letter of 10 December 1833, of his observations in Meredith, New Hampshire, was printed in the *New Hampshire Patriot* of 16 December 1833: "… The whole horizon seemed to be illumined …" He recounted, and drew comparisons with, the experiences of Humboldt and Bonpland in South America, and he provided his own theory on the nature of meteors:

> … the principal materials which compose those meteors, are phosphorus, nitre, and hydrogen gas, of which the first is the basis … If any [one should] ask why there are more meteors in the air at some time than at others, the answer is because the matter, which composes them accumulates slowly, like the cause which produces earthquakes, and the eruptions in most volcanoes … (Remarks on the phenomena and nature of meteors 1833).

Local citizens' reports might also be quoted or described within articles. In the *New Hampshire Sentinel* of 21 November 1833 there was the following statement:

"In Portsmouth, an elderly lady, of 'wiry nerves' was thrown into the most frantic delirium and fainted. She was positive the world was coming to an end!" (Keene 1833). The *New Bedford Mercury* on 22 November 1833 quoted this item from the *Portsmouth Journal* of one witness of the meteors:

> An old sailor said he had been all over the world, he had been on deck at all hours of the night, and in every sea in all weathers, and he had never seen such a light as this since his name was Sam. (From the *New Haven Herald* 1833).

This meteor shower would emerge as a significant event in personal memoirs. In his autobiography, Elder Samuel Rogers (1789–1877) described the sharing of this event with his family and friends. The phenomenon was seen as he and his family were preparing for a move to central Indiana. One child feared: "… the world is surely coming to an end." Another said: "The whole heavens are on fire! All the stars are falling!" Samuel Rogers, himself, was more detailed. He would describe tracks of light: "… visible for several seconds" and some meteors "… as large as the full moon." (Rogers 1880, pp. 133–135). Another short memoir was included in *Turnbo's Tales of the Ozarks*. Herein John Tabor told of his experience with his brother waking to a sky: "… brilliantly illuminated with hundreds and thousands of stars shooting swiftly down toward the earth." He expressed a common fear as he

> … believed the earth and all living creatures would succumb to the wrath of God that night. I was a wicked man then, but after that date of the falling stars, I did not live so sinful toward God. (Kwas 1999, pp. 321–322).

It was evident that readers might be further interested in previous showers. Subsequent to the shower of 1833 several newspapers printed the remarks of Captain Hammond of the ship *Restitution* who witnessed a similar shower in the Red Sea at Mocha on 13 November of the previous year (From the *New Haven Herald* 1833; The recent phenomena of the heavens 1833; A bundle of coincidences 1833). The last-cited article, which was from the *Richmond Enquirer*, stated the coincidence noted in the *Salem Register* that the three greatest meteoric showers on record all took place on the morning of 12 November, i.e. those observed by Humboldt and Bonpland in 1799, Captain Hammond in 1832 and the most recent of 1833.

Newspapers recognized a sustained interest in meteors, in particular those of 1833, and printed relevant articles over years to come. Announcements of lectures would reflect the lingering interest in the recent meteor shower. For example, in the *New-Bedford Mercury* of 26 February 1836 there was an item announcing a lecture that evening on "… brilliant Meteors of November, 1833." (Astronomical lecture 1833). Other newspapers which would continue to provide details of this shower included the *Pittsfield Sun* (Meteoric shower 1838), *The New York Times* (The meteors of 1833; 1884) and the *Advocate* (Meteoric showers 1925). Mention of other meteor showers were included in articles as well. See the *Baltimore Patriot*'s comparison of 1834 and 1833 (e.g. The meteors 1834). A review of the showers over the past 6 years was printed in both the *Rhode Island Republican* (From the *New Haven Herald* 1837) and the *New-Bedford Mercury* (Meteoric shower of November 1837; 1837). Mathematician and astronomer Professor Elias Loomis (1811–1889; Fig. 2.5) of Yale College, wrote a general article on meteor showers for

Fig. 2.5 Elias Loomis (courtesy: en.wikipedia.org)

the *Farmer's Cabinet*, where he remarked on the anticipated periodicity of that of 1833 (Miscellaneous Readings … 1866). The personal account of Parson J.H. Campbell of the 1833 shower was reprinted in *The New York Times* in 1879 (An eye-witness's account of the display of 1833 …, *The New York Times* 1879). The *Afro-American Advocate* wrote a general article on the subject (The November meteors 1892, p. 4), and the Chicago *Broad Axe*, printed a short article on meteors in 1926 (Shower of stars 1926).

As we have already mentioned, Leonid meteor storms (rather than showers) only occur at about 33-year intervals. It is of interest to ask how the 1833 event we have been describing compared with the previous one seen in 1799, and the following one, witnessed in 1866? Yeomans et al. (1996) have addressed this question and shown that the distance of the Earth from the comet when the two orbits crossed determined the intensity of each Leonid meteor storm. They published the plot that we reproduce here as Fig. 2.6, which shows that the storms of 1796 and 1833 were particularly intense, as they lie close to the critical zero (0) line, and prior to this period were mirrored only by the storms of 934, 967, 1238 and 1566. In comparison, the 1866 event was also intense, but not quite as spectacular as these earlier storms.

Consequently, the 1833 and 1866 Leonid meteor storms have a very important place in astronomical history:

> The celestial origins of the Leonids, the determination of their periodic nature, the recognition that they resulted from an orbiting stream of objects, and the identification of this stream with a parent comet, are all landmark events that take on added significance because they represent the origin of the relatively-recent science of meteor studies. Although the 'August meteors' (now known as the Perseids) also played a concurrent role (Littmann 1998), they were not so important as the 'November meteors' (later known as the Leonids), which periodically tended to storm, and thus demanded an immediate explanation. (Dick 1998, p. 2).

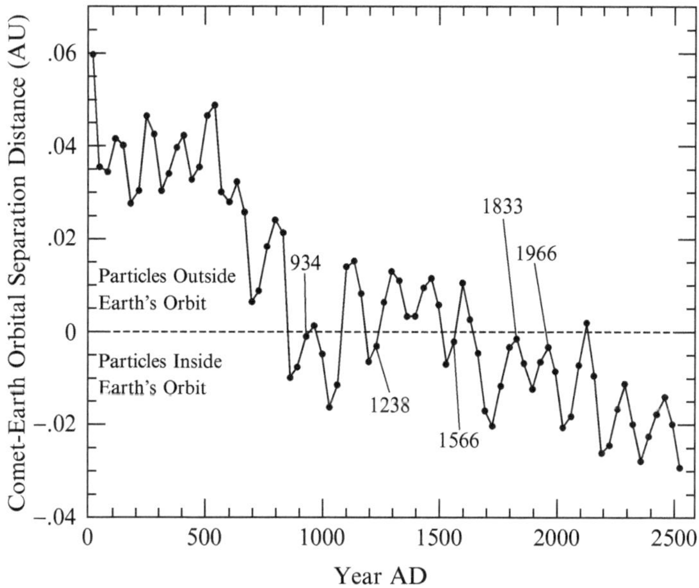

Fig. 2.6 A plot of the minimum distance between Comet 55P/Tempel-Tuttle and the Earth's orbit at the time of the comet's passage through its ascending node (after Yeomans et al. 1996)

Backtracking for a moment to other periodic meteor showers, it was as early as 1834 that one would find in the *Baltimore Patriot & Mercantile Advertiser* a reprint of the *Cincinnati Daily Gazette*'s report by John Locke (1792–1856), a physician, scientist and educator of that city. His observations were of a meteor shower which took place in August of that year. Paling by comparison with the Leonids of the previous November, he approximated there were about 120 meteors visible in 2 h. However, the number was sufficient for him to determine that the 'radiating point' was in Perseus (From the *Cincinnati Daily Gazette* – Meteors 1834). Nowadays the Perseid meteor shower is very popular with amateur astronomers, and is known to be associated with Comet 109P/Swift-Tuttle (e.g. see Brown and Jones 1998; Hughes 1995a; Jenniskens et al. 1998; Jewitt 1996). As we have already seen, this comet played a supporting role in the emergence of meteor astronomy as a scientific discipline (Dick 1998).

2.2 The Great Comet of 1843 (C/1843 D1)

Like meteors, comets were initially deemed to be meteorological rather than astronomical phenomena. Ptolemy's *Almagest* made no mention of comets, while Galileo thought them an optical illusion caused by reflected sunlight (Seargent 2009, pp. 22–23).

Fig. 2.7 The Great Comet of 1843 (after Flammarion 1955, p. 357)

Zochert (1974, p. 449) made the arguable statement:

> Of the many aspects of astronomy, it was comets … which exercised the strongest claim upon the popular mind. The Great Comet of 1843 … generated a popular interest and enthusiasm which persisted throughout its passage.

Marc Rothenberg (1990, p. 45), in his research paper about the Harvard College Observatory, stated: "The Great Comet of 1843 presented Harvard scientists with an opportunity to remind the leaders of Boston that the community leaders were not fulfilling all their responsibilities." David Seargent (2009, p. 233) stated that this spectacular comet, shown here in Fig. 2.7, "… was almost certainly the brightest since 1106 and was probably the most conspicuous … daylight comet on record."

This Great Comet of 1843 was yet another astronomical object noted as among the 100 most significant events in the first century of the United States according to R.M. Devens. He described it as

> … perhaps, the most marvelous of the present age, having been observed in the day-time even before it was visible at night, --- passing very near the sun, --- exhibiting an enormous length of tail, --- and arousing an interest in the public mind as universal and deep as it was unprecedented. (Devens 1878, p. 425).

Camille Flammarion (1842–1925; Fig. 2.8) also noted the initial daylight appearance of this comet: "Cette merveilleuse apparition s'est montrée à tous pour la première fois *en plein jour* le 28 février, a côté du Soleil." (Flammarion 1955, p. 356, his italics).

Fig. 2.8 Camille Flammarion (courtesy: en.wikipedia.org)

Reports of the comet were made from sites around the world, including Rome, Naples and Bologna, in Italy; Pernambuco in Brazil; Trivandrum in India; Van Dieman's Land (now Tasmania) in Australia; Concepcion in Chile; Berlin in Germany; Saint Thomas in the Caribbean; and the Cape of Good Hope in South Africa (Devens 1878, pp. 425–426). The United States also contributed to this: "… permanent scientific history …" largely due to the work of Professor Loomis of Yale College. Loomis stated that the most complete recorded observations were carried out by Walker and Kendall of Philadelphia who followed its course until 10 April of that year (Devens 1878, p. 425). At the time Sears Cook Walker (1805–1853; Fig. 2.9) was an amateur astronomer in Philadelphia, but 2 years later he joined the U.S. Naval Observatory, while his half-brother, Ezra Otis Kendall (1818–1899), was Professor of Theoretical Mathematics and Astronomy at the Central High School in Philadelphia—which had one of the best-equipped observatories in America. Some of their data collected at the Observatory were published in *The American Journal of Science and Arts* (Walker and Kendall 1843), in *Astronomische Nachrichten* (Kendall 1843a), and in the *Monthly Notices of the Royal Astronomical Society* (Kendall 1843b, pp. 304–308; Sabine 1843, pp. 294–295).

Many astronomers computed the comet's orbit and found the perihelion distance from the Sun to be quite small, estimated to be within 78,000 miles (Clerke 1902, p. 104). Loomis' opinion was that this accounted for its great brilliance (Devens 1878, p. 429). The comet's velocity around the Sun was quite large at 366 miles a second, enabling it to escape from the Sun's gravity (Clerke 1902, p. 104). The head

Fig. 2.9 S.C. Walker (courtesy: en.wikipedia.org)

was noted to be small in comparison to the very long tail, which on 21 March reached its greatest visible length (Devens 1878, p. 427).

Numerous professional accounts of visual impressions and observations were published in 1843 in the *Monthly Notices of the Royal Astronomical Society* (Henderson 1843a, p. 266; Henderson 1843b, pp. 267–269; Forster 1843a, pp. 269–270; Nasmyth 1843, pp. 270–271; Littrow 1843, pp. 271–272; Schumacher 1843a, b, pp. 272–275; Close 1843, p. 293; Belam 1843, pp. 293–294; Hopkins 1843, p. 295; Montojo 1843, p. 295; Jacob 1843, pp. 295–296; Forster 1843b, p. 296; Pollock 1843, p. 296; Cowper 1843, p. 296; Smyth 1843, p. 297; Caldecott 1843, pp. 302–304; Kendall 1843a, pp. 304–308). E.C.H. (1843, p. 175) published observations and data in the *American Journal of Science and Arts*.

Also in the *Monthly Notices of the Royal Astronomical Society* were extracts from popular reports found in local newspapers. The *Boston Courier* of 1 April 1843 printed a letter from Professor Benjamin Peirce (1809–1880), who taught mathematics at Harvard University, regarding several observations made by William Cranch Bond (1789–1859; Fig. 2.10), inaugural Director of the Harvard College Observatory during the previous month (Peirce 1843, p. 298). There were also reports from elsewhere around the world including Milan, Italy; Madras, India; and Georgetown, Guyana (Abstract, by the secretary 1843, pp. 298–302). Later that year the *Monthly Notices* would note similar newspaper reports from Port Louis, Mauritius (Abstract from an article in *Le Cernéen* 1843, p. 8); Sydney, New South

Fig. 2.10 William Cranch Bond (courtesy: en.wikipedia.org)

Wales (An article extracted from the *Colonial Observer* 1843, p. 8); and New Zealand (Copy of a letter … 1843, p. 8; cf. Orchiston 2001c).

Examples of personal accounts and letters to the editor may be found in numerous American newspapers. In the *New-Hampshire Statesman and State Journal* of 10 March 1843 a writer commented that there were only three previously-recorded comets that were seen during the day and he was excited to report that: "This day, Feb. 29th, a comet of great brilliancy has been seen, visible through the day, without the aid of a glass, and its brilliancy almost equal to that of Venus." (The Comet of 1843; 1843). On 5 April 1843 Sears Walker sent a letter to the editor of the *Boston Courier* providing additional data on the comet, then disappearing from naked-eye visibility, enabling others to follow it for some days to come. On 18 April 1843 a letter was sent to the *Pennsylvania Inquirer and National Gazette* regarding the aforementioned observations by Walker and Kendall at the Central High School Observatory in Philadelphia (To the editor of the *Pennsylvania Inquirer* 1843). On 16 March 1843 the *Pittsfield Sun* printed a copy of a letter from Princeton's Professor Alexander presumably James Waddel Stephen Alexander, 1804–1859 to Sears Walker dated 7 March, in which details of the comet were brought to the latter's attention. On that same page the *Sun* reprinted several articles from other United States newspapers regarding the comet, including one from a correspondent of the *New Bedford Mercury*, one from 'a Vermont Paper' and one from the *National Intelligencer* (Comets and Commentaries 1843). For the benefit of the public, local newspapers might print some of their own researched informative articles. On 11 March 1843 the *Hudson River Chronicle* printed a recap from a report from the

Fig. 2.11 Dionysius Lardner (courtesy: en.wikipedia.org)

Hydrographical Office on the history of previous comets, as well as specifics on sighting the comet currently in the sky (The "Strange Light" Again 1843).

There were public lectures on the Comet of 1843 and the newspapers would print summary articles that might be of interest and of benefit both to those who could and those who could not attend. On 24 March 1843 the *Mississippi Free Trader and Natchez Daily Gazette* printed a summary of such a public lecture given by the well-known Irish science popularizer and former Professor of Natural Philosophy and Astronomy at University College, London, Dionysius Lardner (1793–1859; Fig. 2.11). A topic of concern was the likelihood of a collision of this comet with the Earth, and Professor Lardner assured his audience that the calculation had been made by the French and the odds of the occurrence of such a calamity were determined to be two hundred and eighty million to one (N.O. Bulletin 1843). The report of another lecture was printed in the *Pennsylvania Inquirer and National Gazette* on 29 March 1843. Professor Peirce speculated on the nature of the comet's tail and described how orbits were determined. He also allayed public fears about a possible collision, stating the same odds as Lardner (The Comet of 1843; 1843).

There were short filler articles to remind and update readers on the current comet. On 31 March 1843 the *Farmer's Cabinet* of New Hampshire, in an article on the political news of the day, had a two-sentence paragraph that the present comet was now receding from view and it had been "… one of the most splendid sights which man has ever been permitted to see." (Practical reform 1843). The *New-Hampshire Patriot* of 13 April 1843 published a two-paragraph summary of the receding comet.

Further calculations to determine the complete orbit would be necessary, but its nearest approach and the inclination of the orbit were stated (Comet of 1843; 1843). It is interesting that recent analyses would suggest an orbital period of 742 years, making this a major (possibly the largest) fragment of Comet X/1106 C1 which fragmented in AD 1106, giving rise to the so-called Kreutz Group of bright Sun-grazing comets (see Sekanina and Chodas 2008).

This comet also would inspire the writing of poetry. On 23 March 1843 the *Daily National Intelligencer* reprinted the poem of eight stanzas, "The Comet," which had been submitted to the *New York American*. A common theme, the final lines recognized the part the poet's God had played:

Father of Light! to thee we bow
We seek not more to know, than Thou
Host in thy wisdom given.

On 14 April 1843 another poem of the same title, submitted by H.J.S., was printed in the *Barre Gazette,* again wondering at the mysterious ways of the Lord:

Whence comest thou? aye tell me whence,
Bright wanderer of the skies.

Lingering interest in this particular comet would be evident in newspaper articles years later. On 7 March 1880 *The New York Times* printed an extract from a current lecture by Professor Peirce on the Comet of 1843 (The Great Comet of 1843. *The New York Times*, 7 March 1880 AND The Comet of 1843 *The New York Times*, 28 August 1881). On 28 August 1881 *The New York Times* reprinted an article from *Frazer's Magazine* which presented some details of its historical appearance (The Comet of 1843, 1881).

The Great Comet of 1843, or Comet C/1843 D1 as it is known to astronomers, was best seen from the Southern Hemisphere (e.g. see Fig. 2.12), and from an historical viewpoint was one of the most impressive celestial visitors of the nineteenth century. It was one of the Kreutz group of Sun-grazing comets, and at perihelion, on 27 February 1843, was a mere 830,000 km from the Sun. Less than 2 weeks later, on 6 March, it was closest to the Earth. At this time it was visible during the day, and was brighter than Venus, but it remained visible to the naked eye at night until 19 April. During and after perihelion passage this comet exhibited a remarkably long tail that reached high into the sky, and we now know to have extended for more than 2 A.U., that is, more than twice the distance from the Sun to the Earth (Kronk 2003).

While the Great Comet of 1843 was certainly one of the most impressive comets to grace the skies of North America following the European settlement of that continent, it was by no means the only one to reach naked eye visibility and create a spectacle. Indeed, during the nineteenth century alone, and prior to 1861, at least 7 comets could be observed without the aid of binoculars or a telescope (see Orchiston 1997a, 2002b), and two of these, seen in 1811 (C/1811 F1) and 1858 (C/1858 L1 Donati, also known colloquially as Donati's Comet), were designated Great Comets (Kronk 2003; Seargent 2009). Just 1 year later, in 1861 they were joined by yet another Great Comet, C/1861 K1 Tebbutt, perhaps the most spectacular of them all (see Orchiston 1998b, c), so we anticipate that in the first six decades of the

Fig. 2.12 Drawing of comet C/1843 D1 by Mary Morton Allport, as viewed from Tasmania (courtesy: en.wikipedia.org)

nineteenth century there must have been few Americans who could not claim to have seen at least one naked eye comet. The Chinese liked to call comets 'broom stars', and we have seen they also succeeded in sweeping up a public fascination for astronomy in the United States.

2.3 American Astronomy in the Early 1800s

American interest in astronomy in the early 1800s was almost always characterized as 'utilitarian', but Zochert (1974, p. 464) refers to an awakening based on *social* utility. More than:

> … the application of science to the needs of production, transportation and material life … [it was for] the social advancement of one's self in terms of wealth, prestige or culture [and] the social advancement of one's class, community or nation.

Indeed there was a new cultural nationalism in the United States driven, in part, by a need for intellectual independence from the Europeans. Initially, however, American scientific achievements would be measured against European standards. Starting in the 1830s many of America's best scientists would travel to Europe to gain a useful knowledge base for their own scientific community.

In the early 1800s astronomy of a practical nature was taught in secondary schools. Students would learn how to calculate figures of the type found in nautical almanacs. At the college level little science of any sort was part of a typical curriculum. Josiah Meigs (1757–1822; Fig. 2.13), the first President of the University of Georgia (1801–1810) had an interest in astronomy and introduced a course on the subject dealing with its mathematical theory only, there being no telescope available for observation (Williams 1996, p. 15). In 1836 Denison Olmsted became Professor of Natural Philosophy and Astronomy at Yale University (Warner 1979). In the years 1846–1847 scientific schools were finally established at both Yale and Harvard Universities. As yet, though, there were still no graduate degree programs available in these subjects (Hubbell and Smith 1992). The pioneering American astronomer Benjamin Apthorp Gould (1824–1896; Fig. 2.14), who among other accomplishments established the *Astronomical Journal*, received his Ph.D. in Germany, and was the first American to do so (Gingerich 1999; Warner 1979).

In Britain and Europe in the 1700s and 1800s public lectures were a popular avocation whereby adults could learn about the sciences (e.g. see Inkster 1980, 1982; Jones 2005). Lectures were attended by women as well as men, and astronomy was a particularly popular subject (Stephenson et al. 2000, p. ii). It took a while for the popularity of the scientific lecture to gain momentum in the United States (e.g. see Inkster 1978), though there were such lectures predating the American Revolution. Sally Gregory Kohlstedt (1990) points out that Benjamin Franklin's interest in electricity was piqued by such a lecture on the subject of chemistry.

Fig. 2.13 Josiah Meigs (courtesy: www.findagrave.com)

Fig. 2.14 Engraving of Benjamin Apthorp Gould (after *Harpers Encyclopaedia* ...: 99, 1905)

Fig. 2.15 Jean Louis Rodolphe Agassiz (courtesy: en.wikipedia.org)

Formalizing the lecture circuit for the education of the American public, associations known as lyceums evolved. In 1829 the Boston Society for the Diffusion of Useful Knowledge was founded (Greene 1958). By 1839, the educational reformer Horace Mann (1796–1859) counted 137 such lyceums in Massachusetts alone. These lyceums would, during the evening hours, convey to the public practical science by means of lectures and demonstrations. At Boston's Lowell Institute there were lectures on religion and science, science being the more popular. Scientists' lectures would need to be repeated to reach all the interested subscribers. Among the popular scientists and educators who lectured at this time were Louis Agassiz (1807–1873; Fig. 2.15) of Harvard University and Ormsby MacKnight Mitchel (1810–1862; Fig. 2.16), who in Cincinnati established an observatory (Fig. 2.17) primarily through donations from the city's citizens (Rossiter 1971).

Among the most influential speakers on the subject of astronomy must be included John Quincy Adams (1767–1848; Fig. 2.18), the sixth President of the United States. Ever an advocate of the study of astronomy, he undertook a number of speaking engagements to promote astronomy and encourage the use of part of the generous bequest from the British chemist and mineralogist James Smithson (1765–1829) to establish a national observatory. In 1839 he answered an invitation by the Lyceum of the Apprentice Mechanics' Association to speak in Quincy and Boston. His lectures were open to the public. Adams described one of his audiences: "The hall was filled to its utmost capacity with two or three women to one man." (Portolano 2000, p. 499).

In its earliest years the mission of Harvard College had been to prepare its students for the ministry. However, in 1642, there were students who defended astronomical theses. As early as 1646, students would be selected to gather astronomical information for publication in the Harvard almanacs. As Genuth (1990,

Fig. 2.16 Ormsby MacKnight Mitchel (courtesy: en.wikipedia.org)

Fig. 2.17 The Cincinnati Observatory, which was founded in 1842 by Professor O.M. Mitchell from Cincinnati College (courtesy: www.cincinnatiobservatory.org)

Fig. 2.18 Details from as portrait of John Quincy Adams, painted by George Caleb Bingham in about 1850 from an original dated 1844. Quincy was a U.S. President with a passion for astronomy (courtesy: en.wikipedia.org)

pp. 9–10) notes, "Almanacs were the most widely distributed item to issue from the colonial press, and prior to the advent of newspapers in the eighteenth century, they were the sole form of periodic literature in New England." These almanacs were printed for the public and served to popularize astronomy with such articles as those on Copernicus, Kepler, comets and telescopic observations.

According to the Boston publication of 1804, *The Catalog of All the Books Printed in the United States*, no more than 20 of the 1,338 titles listed dealt with scientific topics, not counting medical texts (Greene 1958). Kohlstedt (1990) recognized that the early nineteenth century in the United States was a period of increased literacy. Printed matter regarding science in journals, newspapers and books was becoming increasingly desirable and available to the public. Reading became physically easier with the production of eyeglasses and improved lighting. The distribution of books spread from the bookstore and the street hawkers to include mail distribution subsequent to mass advertising. Books also became available through subscription libraries (Zochert 1974).

Most of the earliest astronomical texts in the United States of any scholarly maturity were reprints of English works. Among such valuable works was the first edition of John Herschel's *Treatise on Astronomy*, published in 1824. In his preface to the American edition Sears C. Walker (1842, p. vii) assures the reader that in this volume one would find: "… information nowhere else to be obtained." To read his *Treatise* Herschel (1842, p. 10) recommends only a familiarity with decimal arithmetic, though some knowledge of geometry and/or trigonometry might be helpful for the understanding of the optics of the telescope. According to Deborah Jean Warner (1979) the first such American publication of comparable value was John Gummere's *Elementary Treatise on Astronomy*, first published in 1822. Gummere

Fig. 2.19 John Gummere, the first American to publish a popular textbook book on astronomy (courtesy:www.haverford.edu)

(1784–1845; Fig. 2.19) was a highly-skilled educator who taught at Haverford College, in Haverford, Pennsylvania. However, we could say that Gummere was partially 'trumped' by fellow-American Elijah Burritt, who in 1818 as a talented 24-year old student at William College penned his first book, *Logarithmick Arithmetick*. Despite the off-putting title, about half of this book deals with astronomy. Elijah Hinsdale Burritt (1794–1838) initially taught himself astronomy before continuing his studies more formally at Williams College. Later, while living in New England and running a school equipped with an observatory he wrote *Geography of the Heavens*. This was first published in 1833, and was intended as an astronomy textbook for children. Burritt (1849 edition: preface) also wished it to be considered for "… proper instruction …" for the general public. His book covered the description of the constellations, and some science of celestial phenomena. In his biographical study of Burritt, Brooks (1936, p. 297) says:

> He had a thorough knowledge of the astronomy of that period, and also a gift of writing in such a simple, interesting and attractive way as to hold the attention of the reader. Consequently his books became popular and were used in many schools and colleges.

Women, too, through the printed word, were able to share in the growing popular enthusiasm for astronomy. Magazines intended for women, such as *Godey's Ladies Book* would often contain scientific articles (Kohlstedt 1990). There also were books and pamphlets written specifically for the female audience, for example, Elijah Burritt's 28-page pamphlet titled *Astronomia … Designed for the Amusement and Instruction of Young Ladies and Gentlemen …* (1821) and Denison Olmsted's *Letters on Astronomy Addressed to a Lady* (1840). At the time he wrote *Astronomia …* Burritt was a school teacher in Milledgeville, which was then the capital of the state of Georgia (Brooks 1936). Meanwhile, we have already met Denison Olmsted, the gifted Yale Professor who pioneered meteor studies.

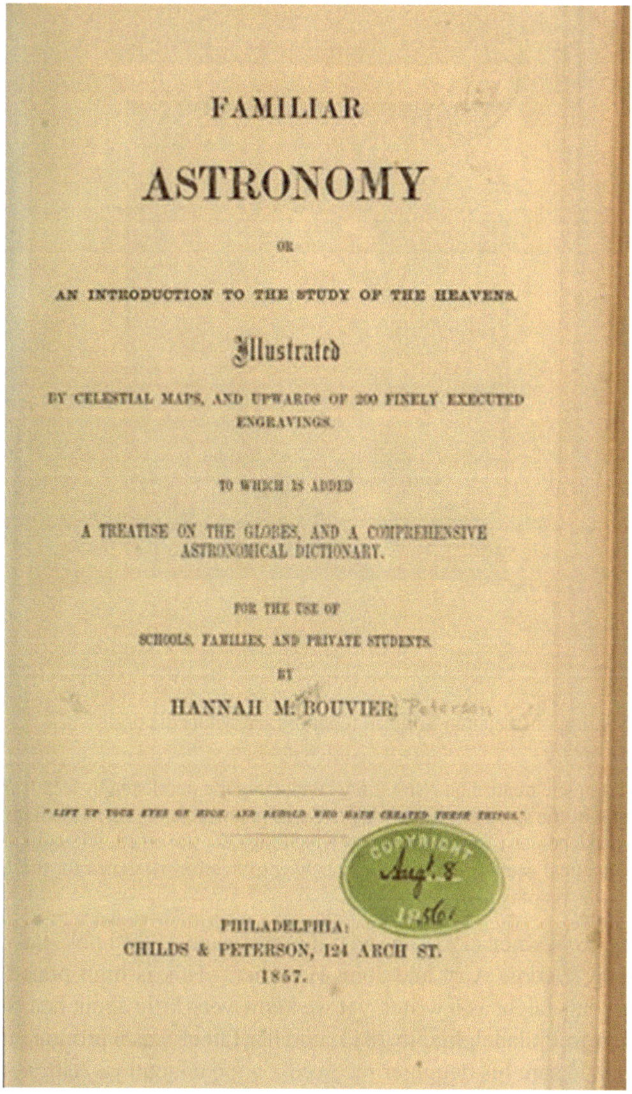

FAMILIAR

ASTRONOMY

OR

AN INTRODUCTION TO THE STUDY OF THE HEAVENS.

Illustrated

BY CELESTIAL MAPS, AND UPWARDS OF 200 FINELY EXECUTED ENGRAVINGS.

TO WHICH IS ADDED

A TREATISE ON THE GLOBES, AND A COMPREHENSIVE ASTRONOMICAL DICTIONARY.

FOR THE USE OF

SCHOOLS, FAMILIES, AND PRIVATE STUDENTS.

BY

HANNAH M. BOUVIER.

"LIFT UP YOUR EYES ON HIGH, AND BEHOLD WHO HATH CREATED THESE THINGS."

PHILADELPHIA:
CHILDS & PETERSON, 124 ARCH ST.
1857.

Fig. 2.20 Front cover of Familiar Astronomy … (courtesy: openlibrary.org.books)

Eventually women would even write books themselves, e.g. Hannah Mary Bouvier Peterson's *Familiar Astronomy …* (see Fig. 2.20), which was first published in 1855 and was reprinted in 1856 and 1857, and then followed by two later editions, both published in 1858 (Reed n.d.). In the preface to *Familiar Astronomy* Peterson (1857) states that her book is intended

Fig. 2.21 A painting of Benjamin Silliman (courtesy: en.wikipedia.org)

> … to be a complete treatise on Astronomy conducting the pupil step by step from the base to the summit of the structure; explaining as far as practicable, by figures and diagrams, all the celestial phenomena, and the laws by which they are governed, without entering into those mathematical details which properly belong to treatises designed for those who propose to make astronomy their chief study.

According to Warner (1978, p. 64), the 499-page 1857 book was: "… highly commended by George Airy and John Herschel." This is high praise indeed and reflects Peterson's talent as a writer, yet we know very little about her. She was born Mary Bouvier, in Philadelphia, in 1811, and her father was a prominent author and jurist so he made sure his daughter received a good education. Astronomy was one of her interests, but it was only after she married the physician Robert Evans Peterson that she had the time, knowledge and inclination to pen *Familiar Astronomy* … She died in Long Branch, New Jersey, in 1870 (Reed n.d.).

American scientific periodical literature was also coming into its own in the United States during these years. Benjamin Silliman (1779–1864; Fig. 2.21) founded the *American Journal of Science* in 1818. Though intended for an audience of scientific men its nature was not very technical and it covered a variety of non-mathematical subjects, including astronomy (Holden 1897).

The publication of a journal, *The Sidereal Messenger* (see Fig. 2.22), by O.M. Mitchel of the Cincinnati Observatory began in July 1846. This was the first astronomical journal in the United States. Of the few astronomical journals in the world, it was the only one of a popular nature. Mitchel planned to have 12 issues per

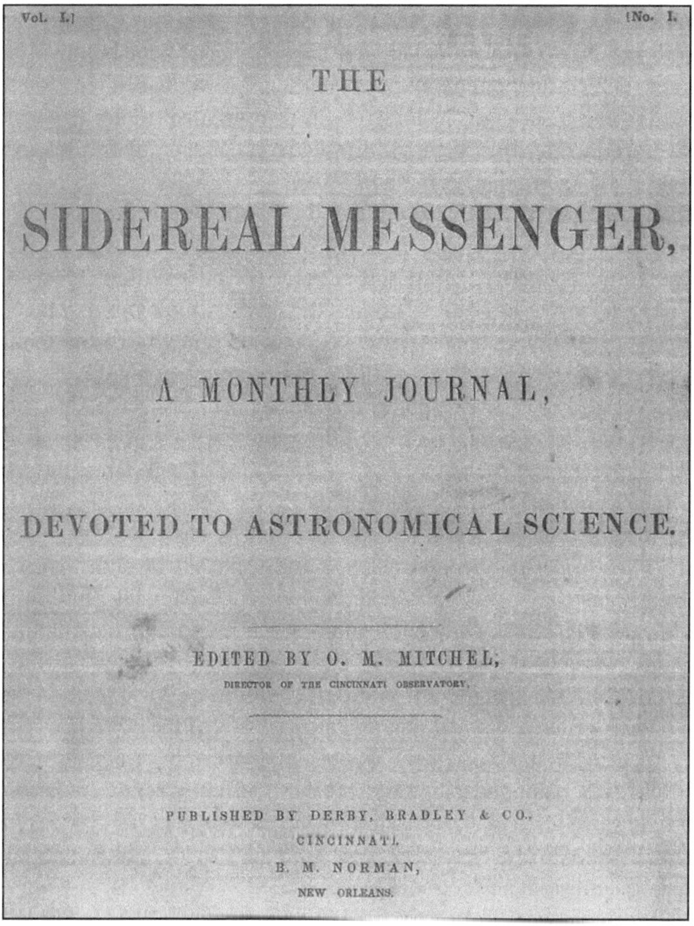
Vol. I.] [No. I.

THE

SIDEREAL MESSENGER,

A MONTHLY JOURNAL,

DEVOTED TO ASTRONOMICAL SCIENCE.

EDITED BY O. M. MITCHEL,

DIRECTOR OF THE CINCINNATI OBSERVATORY.

PUBLISHED BY DERBY, BRADLEY & CO.,

CINCINNATI.

B. M. NORMAN,

NEW ORLEANS.

Fig. 2.22 The *Sidereal Messenger* (courtesy: Cincinnati Historical Society; photograph by the first author)

year with occasional extra issues that would be scientific. He endeavored to keep language simple, to define all scientific terms and to keep the reader current in all significant astronomical research worldwide. The journal was eight pages in length and would include one engraving. He learned to do the engravings himself to save money. Mitchel hoped the journal would be a money-maker to help support the Cincinnati Observatory but he found it increasingly difficult to collect subscription payments. Financial circumstances forced him to end publication after a little more than 2 years (Mitchel 1846–1848).

Broadsides were another form of published matter useful for conveying scientific information to the public. These might provide facts and simple explanations or announce upcoming events. Edmund Halley (1656–1742) published a broadside for the public "Description of the Passage of the Shadow of the Moon over England" describing and illustrating the predicted path for the upcoming solar eclipse of 1715 (Pasachoff 1999). He published a second one for the solar eclipse of 1724 (ibid., Walters 1999). The previously-alluded to presentations of astronomical theses delivered at Harvard were advertised by commencement broadsides (Genuth 1990). Figure 2.23 is an example of one printed in Massachusetts in 1838, advertising and describing a series of lectures on an astronomical topic. Broadsides were inexpensive, frequently illustrated and simple in vocabulary, therefore appealing to the general public (Veron and Tammann 1979).

Kohlstedt (1990, p. 435) pointed out that during the Antebellum period, the years just before the American Civil War, newspaper circulation increased at twice the rate of the population. Zochert (1974, p. 448) quotes the poet William Ellery Channing (1817–1901) who wrote in 1841: "Through the press, discoveries and theories, once the monopoly of the philosophers, have become the property of the multitudes … Science, once the greatest of distinctions is becoming popular." Zochert (1974, p. 449) goes on to state: "… astronomy was clearly the most accessible and appreciated." Of peculiar interest might be the filler material used in newspaper columns relating scientific facts. Zochert (1974, p. 452) shared one from the Milwaukee *Courier* of 1 December 1841 reporting that "… the density of the planet Mercury was equal to that of lead, while that of Saturn was closer to that of cork." *The New York Times* began publication in 1851 and usually was eight pages in length during its early history. It was published daily except on Sunday, until the period of the Civil War when it began daily publication. It has been published continuously up to present day (Mott 1941, pp. 428–429), and it should provide some evidence of the degree of interest in the specific astronomical events to be considered in this study.

Astronomical instruments and equipment available to the American public were frequently imported from abroad in the young nation's history. Available were sundials, globes, orreries, star maps and telescopes, but these were frequently crude conversation pieces, of little value for astronomical studies. James Wilson (1763–1855) of Bradford, Vermont, and later Albany, New York, was America's first commercial globe-maker, producing both terrestrial and celestial globes by 1815 (Warner 1979). Figure 2.24 is a drawing of Wilson at work in Bradford, while the following figure shows one of these globes that is now in the Adler Planetarium in Chicago, Illinois. This was manufactured out of wood, paper, plastic and metal by J. Wilson & Sons in 1831.

Placed next to the celestial globe in Fig. 2.25 is an example of a pocket globe. It was created by the Holbrook Manufacturing Company of Westerfield, Connecticut out of wood and paper in about 1850. Figure 2.26 shows the pocket globe opened. These two items were tools of instruction and popularization, and possibly symbols of class stature. They would be of little or no use to a professional astronomer.

During the Antebellum period more complex apparatus for astronomical demonstrations, such as orreries, were also coming into popularity. The renowned Pennsylvanian self-taught astronomer, surveyor and scientific instrument-maker David Rittenhouse (1732–1796; Fig. 2.27) and the famous Boston clock-maker Joseph Pope (b. ~1750) created 'elegant orreries' for Princeton and Harvard Colleges, respectively (Warner 1979, p. 58). Warner (1979, p. 57) describes how: "… public education for women, practically non-existent in the colonial and early

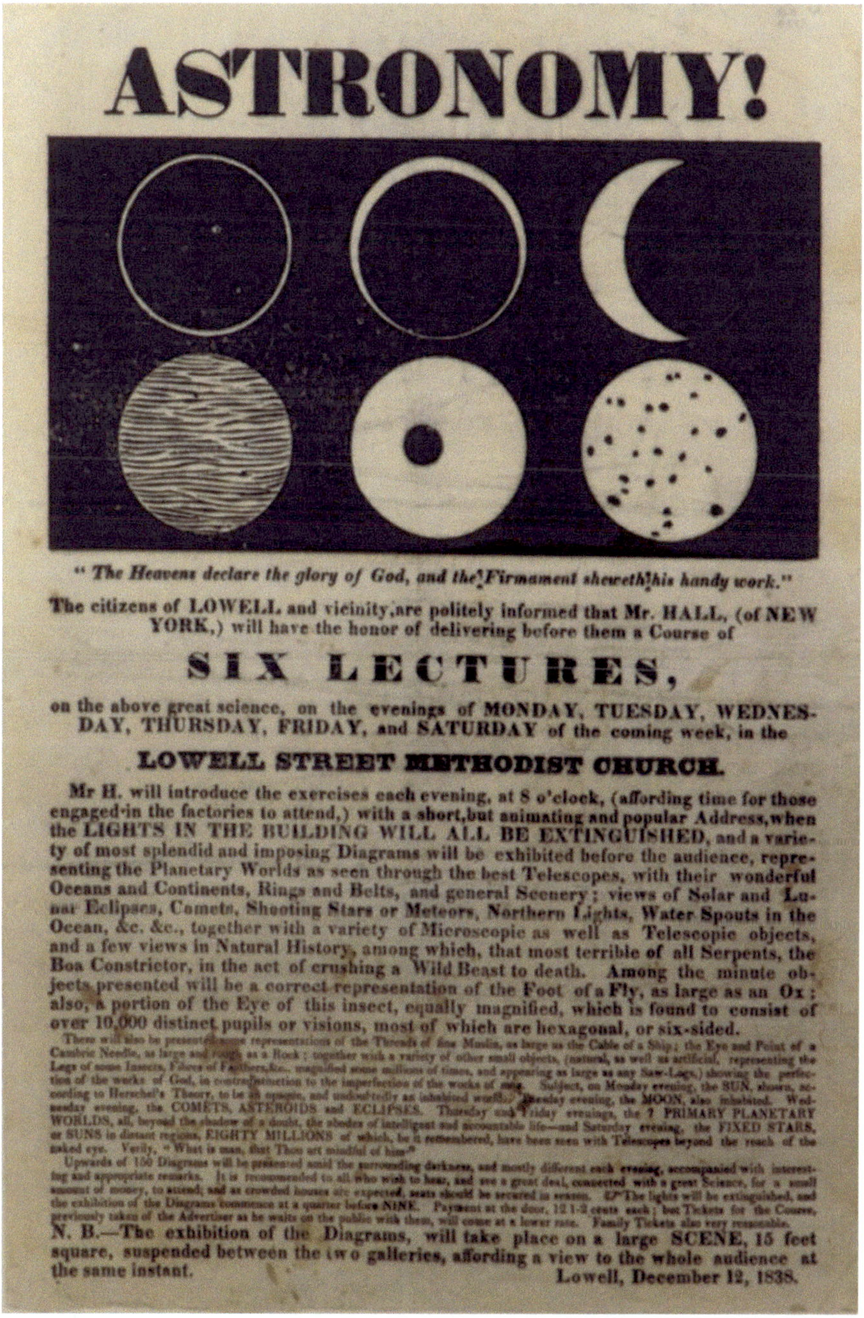

Fig. 2.23 An American broadside (courtesy: Adler Planetarium and Astronomy Museum, Chicago, Illinois)

Fig. 2.24 "James Wilson, the Vermont Globe Maker, Bradford, Vermont, 1810" by Roy Frederick Heinrich (courtesy: www.findagrave.com)

Fig. 2.25 A celestial globe made by J. Wilson & Sons in 1831 (courtesy: Adler Planetarium and Astronomy Museum; Chicago, Illinois, photograph by the first author)

Fig. 2.26 A pocket globe made by the Holbrook Manufacturing Company (courtesy: Adler Planetarium and Astronomy Museum; Chicago, Illinois, photograph by the first author)

Fig. 2.27 Oil painting of David Rittenhouse by Charles Wilson Peale (courtesy: en.wikipedia.org)

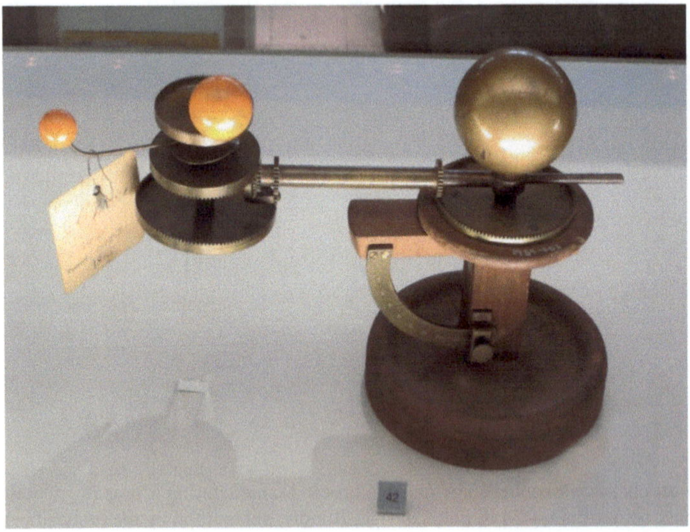

Fig. 2.28 An orrery made by the American John G. Moore, Inventory Number 1989-463 (courtesy: Science Museum, London; photograph by the first author)

republican periods, began to flourish in the 1820s." Female colleges and seminaries evolved and scientific paraphernalia were frequently prime attractors to potential students. The Young Ladies' High School in Boston owned an orrery that it claimed was: "… one of the best instruments of the kind in the country." Warner (ibid) points out that these items served more as public relations devices than as proof of the quality of science education. Figure 2.28 is a photograph of an orrery made by John G. Moore of Philadelphia in 1865. As described where displayed at the Science Museum in London, "This model is an improved design of orrery in which the Earth's axis moves round a small circle to give an elliptical orbit around the Sun."

According to Warner (1979), William Croswell drew the first American-made star chart, published in 1810. It introduced two new uniquely American constellations, the Flying Squirrel and the Bust of Columbus. Though a crude perspective of the sky, it was useful for some basic astronomy instruction (e.g. see Kanas 2009).

A particular kind of star map known as a planisphere was developed in the mid-eighteenth century and became very popular in the nineteenth century (Hughes and Stott 1995; Orchiston 1997a, 2003b). These cardboard devices were based on a circular star map developed for a particular latitude with an overlaying disk that might be adjusted to reveal the particular part of the sky that would be visible for a certain time on a certain date (Hughes and Stott 1995). As stargazing was becoming a popular pastime, planispheres were very helpful in the identification of the constellations. Figure 2.29 is an American-made planisphere dating to about 1828 which is part of the Adler Planetarium collection. Note that decorative planispheres, like this one, with the outlines of the constellation figures, were gradually replaced by the less artistic, but more functional, black-and-white versions over the years.

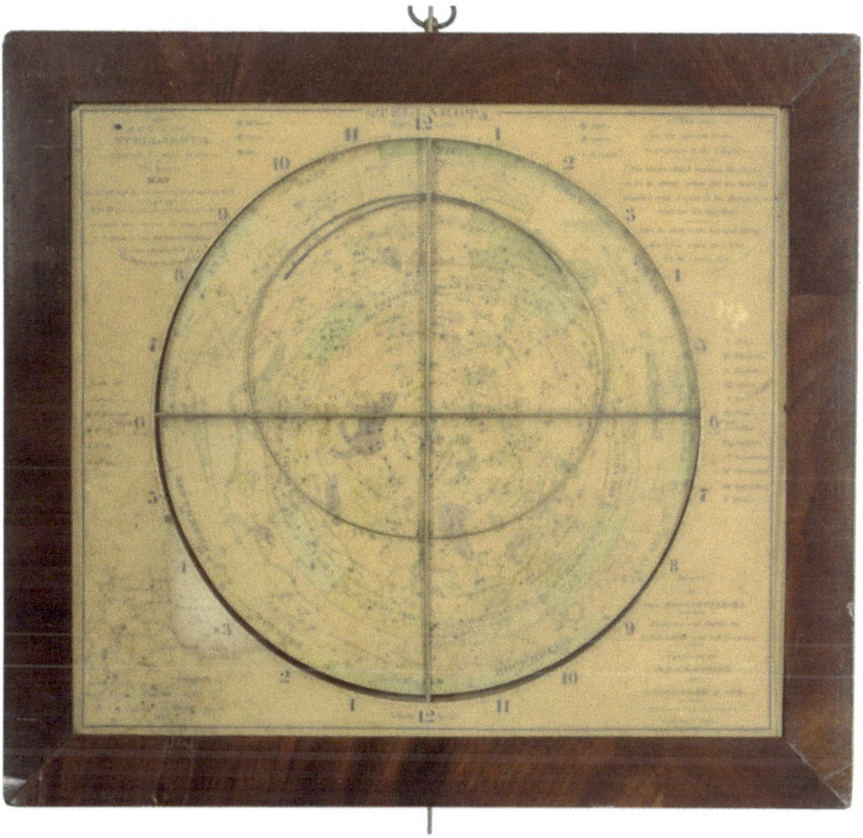

Fig. 2.29 A nineteenth century American-made planisphere (courtesy: Adler Planetarium and Astronomy Museum, Chicago, Illinois)

According to Warner (1979), the most significant advance in astronomy in the Antebellum period of the United States was the new telescope technology. Telescopes were of better quality and they were cheaper, becoming more accessible to more institutions and individuals. At this time refractors were still favored over reflectors, which did not gain popularity until speculum metal mirrors were replaced by silver on glass mirrors later in the century. The newer technology also permitted the manufacture of larger and higher-quality lenses for refracting telescopes. In the 1840s Henry Fitz (1808–1863), the first significant professional telescope-maker in the United States, made refractors that were cheaper than those of his European counterparts (Abrahams 1994), and as a consequence American schools and colleges started buying American telescopes. Thus, Fitz telescopes like the one shown in Fig. 2.30 went to the Elmira Female College, Mount Holyoke Seminary and the Packer Collegiate Institute. Steven Turner (2010) from the National Museum of American History in Washington D.C. notes that between 1840 and 1855 Fitz manufactured about 80 % of the telescopes made

Fig. 2.30 A 7-in. Fitz refractor manufactured in 1849 (courtesy: Robert Ariail)

in America, and these "… were important in increasing American interest [in] and knowledge of astronomy." Later, in the 1850s, Alvan Clark (1804–1887; Fig. 2.31) would make telescopes generally recognized to be equal in quality to those of the Europeans (see Warner 1968).

There was industrial and technological growth in the United States throughout this period, which continued during the turmoil of the Civil War in the early 1860s. After the War this growth was more dramatic. Machine manufacturing was rapidly replacing manual labor, leading to more leisure time for indulgence in entertainment, cultural and scientific pursuits. During this period of growth several significant astronomical events took place. Besides the afore-mentioned meteors and comets, there were the two transits of Venus, in 1874 and 1882. Also of interest were numerous solar eclipses, which would now be better studied and understood due to the new technological advances of recent years. It was during the solar eclipse of 1868, visible from Aden to the Dutch East Indies, that the science of spectroscopy was first applied, but the next total solar eclipse, in 1869, and another in 1878, would be visible over much of the United States.

Fig. 2.31 A portrait of Alvan Clark (after Gerry 1891)

Chapter 3
Scientific Overview

3.1 Solar Eclipses and the Nature of the Solar Corona

3.1.1 Eclipses, Prominences and the Corona

A solar eclipse occurs when the path of the Moon, as seen from the Earth, crosses that of the Sun. In the case of a total solar eclipse the lunar disc completely covers the Sun. Only in this circumstance will the corona, the outermost region of the solar atmosphere, become visible. It is also only during such an eclipse that the smaller 'prominences' (sometimes called 'protuberances'), the red "… masses of great tenuity held in suspension …" (Carrington 1858, p. 177), will be seen, extending beyond the limb of the Moon. The source and nature of the corona and prominences were a topic of intense debate for many years.

The corona has not always been presumed to be a solar feature. At the time of Kepler it was generally accepted to be an illuminated lunar atmosphere. When it was ascertained that the Moon actually had little to no atmosphere, the corona then seemed more likely to be a terrestrial phenomenon (Todd 1894). This seemed logical when one considered the differences in drawings of the corona seen at different observation sites during the same eclipse. According to Lick Observatory Director William Wallace Campbell (1862–1938; Fig. 3.1): "… it was not until the year 1870 that the tide of scientific opinion turned toward the Sun itself as the origin and center of the corona." (Campbell 1907, p. 72). This was further evidenced by observations that showed that the shape of the corona was associated with sunspot activity (Campbell 1907).

In the early nineteenth century means were becoming available to record astronomical phenomena and evaluate other parameters. New instruments would enable the determination of the temperature, polarization, magnetism and chemical composition of celestial objects. With each eclipse more could be revealed regarding the corona by using the latest improvements in instrumentation.

S. Cottam, W. Orchiston, *Eclipses, Transits, and Comets of the Nineteenth Century*, Astrophysics and Space Science Library 406,
DOI 10.1007/978-3-319-08341-4_3

Fig. 3.1 William Wallace Campbell (after MacPherson 1905: facing page 240)

One of Britain's foremost astronomers, William (later Sir William) Huggins (1824–1910; Fig. 3.2; see Becker 2011) delivered a lecture in 1885 about the solar corona (Huggins 1885b, 1900). He listed suggestions for the nature of the corona. It might be a permanent gaseous atmosphere over the solar surface; ejected gases from the Sun; swarms of meteoric particles of sufficient velocity to escape the solar gravity; streams of meteors strongly illuminated about the Sun; debris of comets' tails; or meteor-like friction of solid or liquid particles in the outer solar atmosphere. By then, it was generally accepted that, whatever its source, the corona was indeed a solar phenomenon (see Baxendell 1859; Proctor 1871a).

Britain's leading nineteenth century popularizer of astronomy, Richard Proctor (1837–1888; Fig. 3.3) pointed out that photography, spectroscopy and polariscopy provided the means to a better understanding of the nature of the corona (Proctor 1871d). By then its relationship to the Sun was generally accepted but the question of its origin had yet to be answered. Proctor (1871d, p. 185) had long contended that the solar corona was due mainly to the presence of "... millions of meteoric systems having their perihelion for the most part much closer to the Sun than our Earth's orbit." By this time, however, he admits that, though this theory accounted for some coronal light "... the larger portion remains unaccounted for." (ibid., p. 186). Photographic evidence suggested that "... coronal radiations are phenomena of eruption." (ibid., p. 189). The greater number of prominences evident at the base of a bulging of the corona seemed suggestive of a response to an outwardly-directed eruptive force.

Fig. 3.2 William Huggins, circa 1910 (adapted from *Proceedings of the Royal Society*, 1910: facing page i)

Fig. 3.3 Richard Proctor, about 1870 (courtesy: en.wikipedia.org)

Fig. 3.4 John M. Schaeberle (courtesy: en.wikipedia.org)

Proctor (1873) thus became an advocate of the auroral theory propounded by Professor William Augustus Norton (1810–1883) of Delaware College, Newark (Delaware), in 1845.

Lick Observatory's John Martin Schaeberle (1853–1924; Fig. 3.4) would expand on the theories of the solar origin of the corona in multiple publications in the 1890s with his 'mechanical theory'. He saw the corona as light reflected from ejecta of the Sun, normal to its surface and most active in the area of sunspots. The outward streams of light would curve due to the rotation of the Sun (Schaeberle 1890a, b, c, 1891). However, as early as 1870, Proctor (1870b, p. 223; his italics) appealed to astronomers to focus on a different question during an eclipse: "... I should be glad to see every moment of the short duration of totality devoted both by general observers and spectroscopists to the inquiry *what sort of a solar appendage* the corona might be."

3.1.2 Earlier Solar Eclipses of Note

F. Richard Stephenson (2008) notes the usefulness of eclipses dated as early as 700 BC to AD 1600 in determining changes in the Earth's rotation rate. He remarks that several hundred records of such events exist, originating from countries ranging from Europe to China. While attempts have been made to date earlier eclipses

Fig. 3.5 The total eclipse of the Sun, on 28 May 585 BC (after Todd 1894, p. 95)

(see e.g. De Jong and Van Soldt 1989), the results have been challenged (Pardee and Swerdlow 1993).

A famous eclipse of 585 BC (see Fig. 3.5) has been described by Herodotus as occurring during a war between the Lydians and the Medes, resulting in the cessation of hostilities that had lasted for several years, and enabling the association of a date with this historical event (Todd 1894, p. 96).

The talented Scottish amateur astronomer, Alexander Brown (1814–1893; Fig. 3.6), wrote to the *Astronomical Register* on 12 April 1869 describing his

Fig. 3.6 Alexander Brown (courtesy: www.angus.gov.uk)

calculation of a 14-s solar eclipse of 1598 that was visible in Great Britain. He requested confirmation of his results of this event known at the time as 'Black Monday' (Brown 1869). He also noted that the eclipse of 1652 was a visit of the same eclipse as the recent one of 1868.

The British amateur astronomer T.W. Burr discussed eclipses at a meeting of the North London Naturalist Club on 13 May 1869. He related that the first observation of red prominences was made by a Captain Stanyan at the eclipse of 1706. These phenomena were again observed by the Swedish astronomer Birger Vassenius (1687–1771) in 1733 and by the distinguished British astronomers Francis Baily (1774–1844; Fig. 3.7) and Astronomer Royal George Biddell Airy (1801–1892; Fig. 3.8) in 1842 (The total solar eclipse of 1868, pp. 186–187).

The noted Boston lawyer and amateur astronomer, Robert Treat Paine (1803–1885; Robert Treat Paine 1885), who was to observe the 1869 total solar eclipse in Boonesboro, Iowa, referred to the eclipse of 16 June 1806 as: "… the finest in the United States in the nineteenth century." (Paine 1869a, p. 2). Visible in the north-eastern United States, totality lasted 5 min. Among those who recorded their observations were the Surveyor General of the State of New York, Simeon DeWitt (1756–1834; Fig. 3.9) in Albany, New York; J.J. Ferrer and John Garnett in South Landing, Kinderhook, New York; and Nathaniel Bowditch in Salem, Massachusetts (Lewis 1946).

Baily's beads, glimpses of the solar photosphere shining through lunar valleys, giving the impression of a string of beads along the lunar limb, were first noted by

Fig. 3.7 Francis Baily (courtesy: en.wikipedia.org)

Fig. 3.8 Sketch of George Biddell Airy made in about 1872 (after *Popular Science Monthly*, 1872–1873)

Fig. 3.9 An oil painting of Simeon DeWitt, c. 1804 (courtesy: en.wikipedia.org)

Baily at the beginning of the annular solar eclipse of 1836 (Golub and Pasachoff 1997, p. 27; Todd 1894, p. 12).

These early eclipses up until the mid-nineteenth century were studied with limited instrumentation. The main astronomical equipment available to most astronomers would be a telescope and time-pieces. Records were mainly descriptive, with words or drawings, and limited data referring to time and location. The second half of the century would bring instruments that would enable the scientist to delve deeper into the nature of the corona and to better make permanent and non-subjective records of what was observed.

3.1.3 Instrumentation

3.1.3.1 Photography

Until the mid-nineteenth century the only documentation of the shape of the solar corona observed during an eclipse was from hand-drawn sketches. Accuracy was subject to the judgment, speed and skill of the individual. The corona was a complex entity and varied in extent and internal structure. The best draftsmen would have to be artists as well as scientists. The Royal Astronomical Society provided detailed

Fig. 3.10 Warren De la Rue (courtesy: en.wikipedia.org)

instructions on proper procedures to be maintained during this task. Draftsmen would work beside a telescope. They might be assigned particular quadrants of the eclipse to record. Due to the short duration of totality, final details would be included after the event and would depend on the artist's memory (Pang 1994).

The first photograph of a solar eclipse was a daguerreotype by Berkowski using the Königsberg heliometer in 1851 (Clerke 1902, p. 166; Todd 1894, p. 120). It was with the advent of the 1860 total solar eclipse that practical photographic methods were put into practice. The pioneering British astrophotographer Warren De la Rue (1815–1889; Fig. 3.10) used collodion plates, a wet process faster than the daguerreotype, and more easily printed. He reported the details of his attempts to the Royal Astronomical Society in 1861. Over the next few decades, however, astronomers would convert to a dry plate process. This was more convenient as the dry plates did not require a dark room for their preparation, nor did they require development on the spot, making them more portable (Pang 1994; Staubermann 2004).

It should be noted that, even during this period of emphasis on photography, there was still some appreciation for the talented eclipse artist. William Henry Wesley (1841–1922), Assistant Secretary of the Royal Astronomical Society for 47 years, made his first drawing of a solar corona in 1871:

> The drawing he produced was remarkable, not only for beauty of execution, but for accuracy in delineation … [and] This striking evidence of Mr. Wesley's skill led to him making many drawings of subsequent eclipses, which are well known. (*The Observatory* 1922b, pp. 354–355).

Fig. 3.11 A U.S. Naval Observatory photograph of Asaph Hall taken in 1899 (courtesy: en.wikipedia.org)

His last published eclipse drawing for the Royal Astronomical Society was that of the 1886 eclipse (Pang 1995).

In 1871, the Assistant Astronomer at the U.S. Naval Observatory in Washington, D.C., Professor Asaph Hall (1829–1907; Fig. 3.11) read a paper before the Washington Philosophical Society on the pros and cons of photography. In the case of a clear sky, valuable records may result in good data if properly reduced. However, he felt at this point that the photographic method did not necessarily improve on direct measurement by a micrometer or heliometers for data collection as photographs were subject to a variety of distortions.

3.1.3.2 Spectroscopy

The 1859 publication by the German duo of Robert Wilhelm Eberhard Bunsen (1811–1899; Fig. 3.12) and Gustav Robert Kirchhoff (1824–1887; Fig. 3.13), followed soon after by some laboratory experiments by Kirchhoff, led to the

Fig. 3.12 Robert Bunsen (courtesy: en.wikipedia.org)

Fig. 3.13 Gustav Kirchhoff (courtesy: en.wikipedia.org)

Fig. 3.14 Oil painting of Jules Janssen by his friend, the well-known French artist Jean-Jacques Henner (courtesy: en.wikipedia.org)

recognition of the meaning of the dark lines in the solar spectrum, which were first discovered in 1823 by Joseph von Fraunhofer (1787–1826). These lines corresponded to those in the spectra of specific elements sufficiently heated under laboratory conditions. This realization would enable scientists to identify elements in the solar atmosphere (see Becker 2001; Plotkin 1977).

The year 1868 was significant as it marked the beginning of solar eclipse spectroscopy (see Nath 2013). At the total solar eclipse of that year the British astronomers John Herschel, James Francis Tennant and Norman Robert Pogson, and the French astronomers Georges-Antoine-Pons Rayet and Pierre Jules César Janssen (known simply as Jules Janssen; Fig. 3.14) all succeeded in observing spectra of the solar prominences quite separately from the corona (Orchiston et al. 2006). Conspicuous among the bright lines were those of hydrogen (Janssen 1869a). At the time, Pogson (1868) also noted a bright unrecognized yellow line, which would later to be identified with the yet-to-be-discovered element helium (for details see Kragh 2009; Nath 2013). In 1868 Janssen and the British astronomer Norman Lockyer (1836–1920; Fig. 3.15; Meadows 1972) would speculate independently that it might be possible to make spectroscopic studies of solar prominences even during the absence of an eclipse and both, shortly afterward, succeeded in doing just this (Huggins 1869; Janssen 1869b; Todd 1894, pp. 38–39). The Vatican Observatory

Fig. 3.15 Norman Lockyer in 1909 (courtesy: en.wikipedia.org)

Director, Father Angelo Secchi (1818–1878; Fig. 3.16) would take advantage of this information and make observations of solar prominences without eclipses over extended periods during 1871 and 1872 (Secchi 1872a, b).

It would soon be realized that by applying photography to spectroscopy one could eliminate subjectivity in the interpretation of spectral lines and at the same time obtain a permanent record (Hentschel 1999a, b).

3.1.3.3 The Polariscope

While photography and spectroscopy were useful tools for determining the structure and content of the corona, others would be necessary to determine its dynamics. The German-born British physicist Franz Arthur Friedrich Schuster (1851–1934; Fig. 3.17) from Manchester University listed what questions might be answered about the nature of the solar corona by means of a polariscope (Schuster 1879, p. 36). This device would be used to measure the degree of polarization of the light scattered by solid particles in the corona:

> In what way is the scattering matter distributed in the solar atmosphere?
> What part of the light sent out by the corona is due to scattering matter?
> Is the scattering matter projected outwards from the Sun, or is it falling into the Sun from outside?

Fig. 3.16 Father Angelo Secchi (courtesy: en.wikipedia.org)

Fig. 3.17 Arthur Schuster (after *Physical Laboratories* ..., 1906: Frontispiece)

The degree of polarization measured at different distances from the Sun would give a better picture of this nature of the corona. Over this distance one might determine whether or not there is a progression between solar incandescence and light due to the scattering of small particles (ibid.).

3.1.3.4 The Role of the Telegraph

The telegraph was patented by the American inventor Samuel Morse (1791–1872) in 1837. In a paper published in *The Observatory*, David P. Todd (1881a) described its value in solar eclipse expeditions as an important tool of communication. The duration of totality of any solar eclipse is very brief. But by means of the telegraph, astronomers at one observing station might convey important information to their colleagues at another station, which had not observed totality yet, enabling confirmation or rejection by the latter. Such cooperation would promote more efficient use of limited time (Todd 1894, p. 165).

Besides the afore-mentioned instruments there were others that were found useful by various eclipse expeditions. Some clever devices were developed by members of particular parties for one-time use for their own specific purposes.

3.1.4 The Solar Eclipses of 1868, 1869 and 1878

3.1.4.1 The Eclipse of 18 August 1868

Beginning with the total solar eclipse of 1860, astronomers were able to document what they saw by using photography, but it was only with the eclipse of 1868, visible in India, Siam (present-day Thailand), and the Dutch East Indies (present-day Indonesia), that they were able to first apply spectroscopy to the study of the solar prominences and the corona. With totality lasting as much as 6 min and 50 s in the Gulf of Siam, this proved to be a particularly timely event.

Giovanni Battista Donati (1826–1873; Fig. 3.18), the Director of the Florence Observatory and a highly-regarded pioneer of astronomical spectroscopy, expressed his frustration in the *Astronomical Register* that the Italians were ill-equipped to take advantage of this highly-favorable situation and undertake an eclipse expedition (Donati 1868). No Italian observatory possessed a spectroscope. He hoped that their Government would correct this deficiency before the solar eclipse of 1870, which would be visible in parts of Italy.

The British and French, however, would be ready for the 1868 event (see Nath 2013; Orchiston et al. 2006). Major J.F. Tennant (1829–1915), formerly of the Trigonometrical Survey of India and a temporary short-term Director of the Madras Observatory, requested of the Royal Astronomical Society that two expeditions be sent to India, one to the eastern coast of India, near Masulipatam or Guntoor, and the other more inland, at Hyderabad (Tennant 1867b). Photography, spectroscopy and

Fig. 3.18 Giovanni Battista Donati (courtesy: en.wikipedia.org)

polariscopy would be used (Orchiston et al. 2006). The Royal Astronomical Society agreed to Tennant's proposal, and promptly appointed him Director of one of the expeditions! En route to Calcutta, at which point he would decide on his specific observing site, Tennant stopped off in Aden and communicated with some of the British officers there on how they should study the solar corona during the eclipse. Upon arriving in India, Tennant chose Guntoor as the site for his station and his party arrived there on 3 July (ibid.). He ultimately reported that photographs of the event were underexposed due to a light cloud cover, but he was able to determine that the corona had a continuous spectrum (Tennant 1868, 1869b). He also noted that the solar prominences demonstrated bright lines corresponding to Fraunhofer C, D, and b but time ran out before he could confirm the likely F and G lines. Of these prominences he described in particular the 'Great Horn' (see Fig. 3.19), and when he directed the spectroscope at it, "… suddenly, however, it burst into sight, a glorious brilliant linear spectrum …" (Tennant 1869b, p. 19). One of those in Tennant's party was Captain Branfill from the Trigonometrical Survey of India, and he was able to report that the corona was strongly polarized while the prominences, including the 'Great Horn,' were not at all. This led Tennant (1868) to conclude that the corona was not self-luminous but was reflected sunlight. Photographs of the 'Great Horn' (Tennant 1869b; Plate 5B, Plate 5F) supported the solar origin of the prominences. As Norman Lockyer (1874, p. 112) noted: "… the prominences were eclipsed and uncovered exactly as the sun itself …".

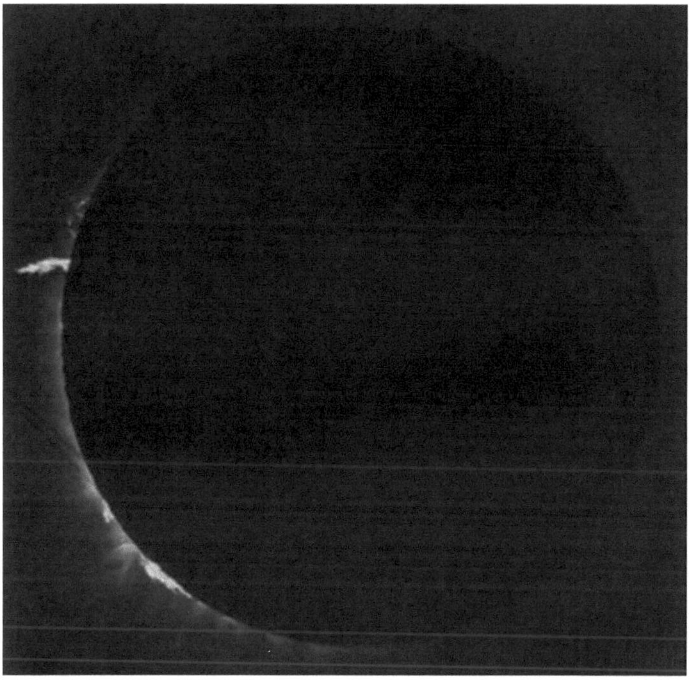

Fig. 3.19 A photograph of the 1868 eclipse showing prominences, including the 'Great Horn' (after Tennant 1869b: Plate 5B)

The second official British expedition was headed by Lieutenant John Herschel from the Trigonometrical Survey of India, the son of Sir John Herschel, and this was based at Jamkandi, inland and further west than the suggested Hyderabad. Herschel encountered more cloud cover during totality but during a period when the clouds parted he was able to make some spectroscopic observations of the "… long finger-like projection …," i.e. the 'Great Horn'. He saw the same three lines as Tennant (Herschel 1868–1869). Polarization observations also led him to conclude that the corona was reflected sunlight (The Eclipse of the Sun, *Astronomical Register* 1868).

A third British party was at Masulipatam, and was led by Norman Pogson (1868; Fig. 3.20), Director of the Madras Observatory and originator of the stellar magnitude scale (see Reddy et al. 2007). Shearman (1913) reported that they successfully observed the spectra of the prominences, and also recorded the bright lines.

This eclipse actually attracted a fourth British party, which was located at Beejapur and was led by Captain Charles Thomas Haig (1834–1907) from the Royal Engineers. Although mostly clouded out, they did succeed in observing the spectra of the prominences, and noticed two bright lines (Haig 1868–1869).

Astrophysics pioneer Professor Jules Janssen (1824–1907) led the French expedition to India. He, like Tennant, was stationed at Guntoor. He was equipped with several large telescopes and a spectroscope. He also was accompanied by a

Fig. 3.20 Normon Pogson (courtesy: dhinakarrajaram.blogspot.com)

draftsman, Jules Lefaucher, to draw the eclipse. When rain and clouds cleared sufficiently he was able to make some spectroscopic observations (Janssen 1869a, c, d). He concluded:

1st. The gaseous nature of the protuberances (bright spectral lines).
2nd. The general similarity of their chemical composition (the spectra corresponding line for line).
3rd. Their constitution (the red and blue lines of their spectra being no other than the lines C and F of the solar spectrum, characteristic as we know, of hydrogen gas). (Janssen 1869a, p. 109).

It was at this eclipse that, having observed the bright prominence lines, Janssen considered the possibility of observing these outside of the event of an eclipse. The very next day he placed the slit of the spectroscope over the same location where he had observed the prominence, and succeeded in again finding the C and F lines (Janssen 1869c).

The well-known Paris Observatory astronomer, Georges Rayet (1839–1906; Fig. 3.21) reported on the second French eclipse expedition, which was based at Wah-Koa, Siam (now Thailand). He also noted very bright lines in the spectrum of the prominences and remarked that the light of the corona was faint in comparison (Rayet 1869; cf. Orchiston and Soonthornthum 2014).

Thus, the British and French collectively were very successful in observing the eclipse, even though the weather was not always fully co-operative. Thanks to the

Fig. 3.21 Georges Rayet (adapted from *The Astrophysical Journal*, 1907: opposite page 53)

imposing presence of the 'Great Horn' all of the observing teams succeeded in recording emission lines, but

> While everyone that day thought that the bright orange line was a signature of sodium in the Sun, Pogson at Masulipatam had some doubts. Unlike Janssen, Tennant or Herschel, he was not so sure about the position of the yellow-orange line … (Nath 2013, p. 155).

Pogson was not convinced that the bright line coincided *exactly* with the position of the well-known sodium D line, and his report (Pogson 1868) "… had a tone of doubt, even if a mild one …" (Nath 2013, p. 156). With the benefit of hindsight we now know that Pogson's suspicions were justified, for the bright line in question was later assigned to the element helium.

Apart from Britain and France, the North German and Austrian Governments also sent out parties to observe this eclipse, and they were stationed at Aden (German observations of the eclipse of the Sun 1864).

Besides these official expeditions, various non-official reports were made by interested individuals. Australia's foremost nineteenth century astronomer, John Tebbutt (1834–1916; Fig. 3.22; see Orchiston 2001b, 2002b, 2004a) reported his visual observations of the eclipse, the Sun being only one-third obscured at his location in Windsor, near Sydney (Tebbutt 1868). Another account was relayed from John Pope Hennessey (1834–1891; Fig. 3.23), the Governor of Labuan in Borneo (Pope Hennessey 1868, pp. 81–90).

Fig. 3.22 John Tebbutt (courtesy: Orchiston Collection)

Fig. 3.23 John Pope Hennessy (courtesy: en.wikipedia.org)

Other accounts were received from observations made at sea. Totality was observed from the *Rangoon* without astronomical instruments, and only a general description of the event was recorded. The location of the *Carnatic* only permitted observations of a partial eclipse. Observers noted:

> … one-sixteenth part of the sun's diameter remained visible [and] the most remarkable appearances … [of] … the irregularity and roughness of the concave portion of the crescent left visible by the moon.

Observations of spectra and polarization were made (Observations of the solar eclipse of 18th August … 1868, p. 1).

The British astronomer Warren De la Rue was not present at any of the sites but nonetheless contributed to the understanding of eclipses as he reviewed photographs of the event. Comparing photographs of the 'Great Horn' taken at Aden and Guntoor, he concluded that the prominence had a very active spiral structure and demonstrated axial rotation (De la Rue 1869a, b; The eclipse of the Sun, *Astronomical Register* 1868). Tennant (1868) also referred to this feature. While the 1868 eclipse was in progress, De la Rue was in England taking his own photographs of the uneclipsed Sun. Comparing these with those taken at the eclipse sites he concluded (1869b) that prominences did not seem to have any direct connection with faculae, the bright regions of the Sun's photosphere best observed near the limb, although they might have been related to them in some way.

3.1.4.2 The Eclipse of 7 August 1869

Total solar eclipses were visible from the United States in 1869 and 1878, thus enabling its citizens to apply new scientific methodologies within their own borders (see Fig. 3.24). Totality of the eclipse of 7 August 1869 at most sites lasted for about 2¾ min, or as long as 4 min near Mt. St. Elias on the Alaska-Canada border (Paine 1869b). It was best visible in Iowa, Illinois, Kentucky and North Carolina.

The nature of the American eclipse expeditions was to be very different from those funded by the European governments, scientific societies or observatories. As Pang (2002, p. 39) has stated:

> American eclipse expeditions, in contrast, were local affairs, planned, funded and outfitted by individual observatories and colleges, or by astronomers borrowing instruments and traveling on small grants from various institutions or agencies.

Although astronomers around the world were interested in the outcomes of the eclipse observations, it would be Americans who would organize the expeditions. In 1869 the event was visible as far west as the Dakota territories and as far east as Newbern, North Carolina, and in the contiguous states and territories (see Fig. 3.25). Observations might also be made further northwest, beyond the American border. For example, the U.S. Naval Observatory sent an expedition led by Asaph Hall to Siberia to carry out magnetic and positional studies (Sands 1869).

Lockyer (1869, p. 15) described the degree of participation of Americans in this event:

> The Government, the Railway and other companies and private persons threw themselves into the work with marvellous earnestness and skill; and the result was that the line of totality was almost one continuous observatory from the Pacific to the Atlantic … There seems to have been scarcely a town of any considerable magnitude along the entire line, which was not garrisoned by observers, having some special astronomical problem in view.

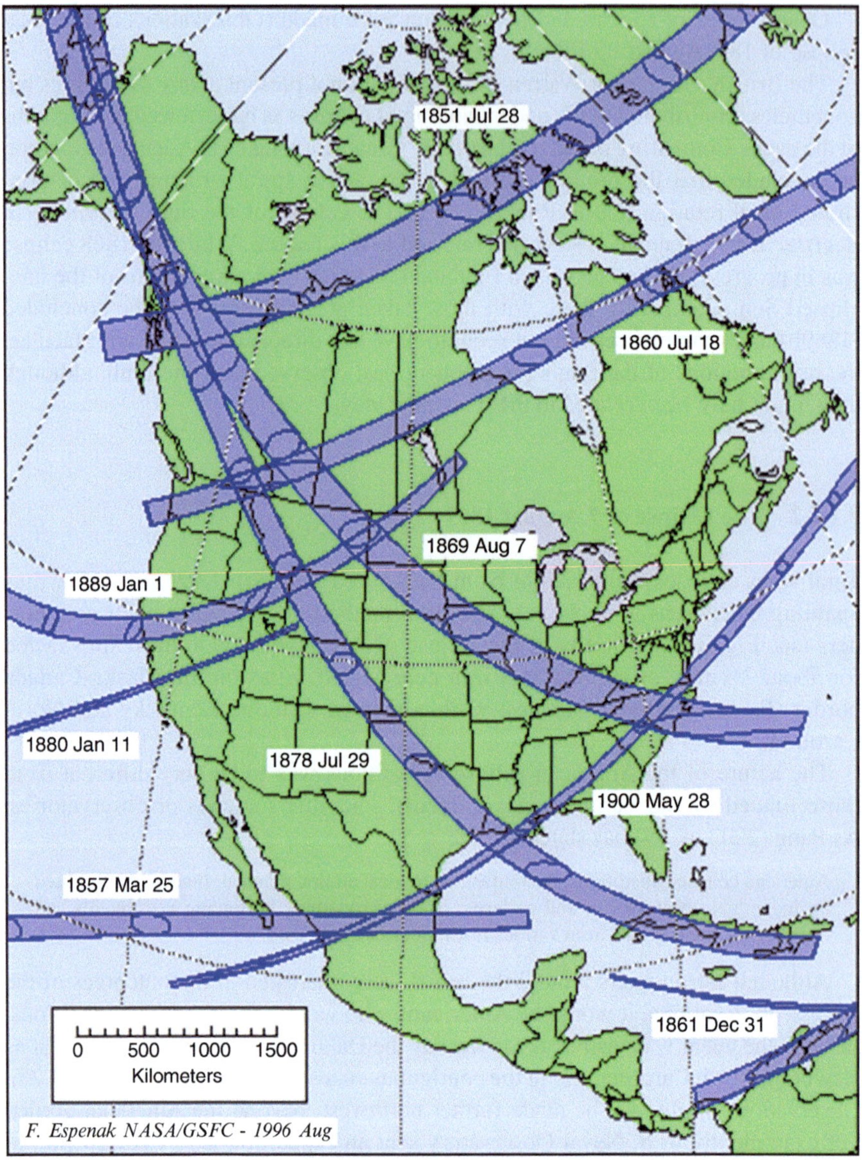

Fig. 3.24 Map showing the paths of totality of solar eclipses visible from North America during the second half of the nineteenth century (courtesy: eclipse.gsfc.nasa.gov)

Many small towns along the "continuous observatory" would share their experiences by reporting to the newspapers. A good number of these were simple descriptions of the event. Noted might be the degree of totality, the weather, effects on animals and/or the human emotions experienced. Newspapers printed such reports from as far east as Boston, Philadelphia and New York, where a partial

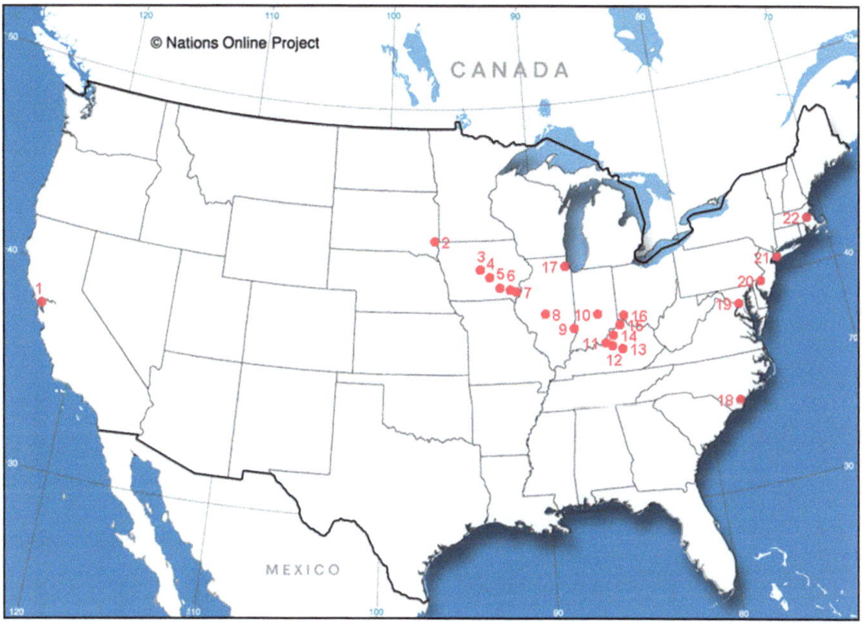

Fig. 3.25 Geographical distribution of U.S. localities mentioned in the text relating to the 1869 eclipse (after Cottam, Pearson et al. 2011). Key: 1 = San Francisco; 2 = Sioux Falls (Fort Dakota); 3 = Jefferson; 4 = Des Moines; 5 = Ottumwa; 6 = Mount Pleasant; 7 = Burlington; 8 = Springfield; 9 = Vincennes; 10 = Martinsville; 11 = Louisville; 12 = Shelbyville; 13 = Nicholasville; 14 = Prestonville; 15 = Big Bone; 16 = Cincinnati; 17 = Chicago; 18 = Wilmington; 19 = Washington, D.C.; 20 = Philadelphia; 21 = New York; 22 = Boston

eclipse was observed by hundreds of thousands, to as far west as San Francisco, where a partial eclipse was experienced on a "… cloudless day." (The eclipse – Darkness upon the face of the Earth 1869). Many smaller towns in between also sent in their reports, including Nicholasville, Big Bone and Prestonville in Kentucky, as well as Vincennes and Martinsville in Indiana (The eclipse – Accounts from the scientific parties 1869),

Significant scientific expeditions that took place during this eclipse included those of Joseph Winlock (1826–1875; Fig. 3.26), Director of the Harvard College Observatory, who would return to his native Kentucky to make his studies, and Cleveland Abbe (1838–1916; Fig. 3.27), Director of the Cincinnati Observatory, who would lead his team of seven to the westernmost site in the Dakota Territory.

Joseph Winlock was born in Shelby County, Kentucky in 1826, graduated from Shelby College in 1845 and went on to become Director of the Harvard Observatory in 1866 (Matthews and Cleveland 2008). As the 1869 eclipse path was to go through Kentucky's largest city of Louisville, Winlock was happy to return to his native Shelbyville, just 33 miles distant (localities 11 and 12 in Fig. 3.25), to lead an expedition as instructed by Benjamin Peirce of the United States Coast and Geodetic Survey.

Fig. 3.26 Joseph Winlock (courtesy: www.kilma-luft.de/steinicke/ngcic/persons/winlock.html)

Fig. 3.27 Cleveland Abbe (after *Popular Science Monthly*, 1887–1888)

Fig. 3.28 The Harvard eclipse expedition party (courtesy: Shelbyville Historical Society)

Eighty photographs were taken, including seven during the totality of 2 min and 29 s (Joseph Winlock 1875). The sons of Alvan Clark, the well-known telescope-maker, George Bassett Clark (1827–1891) and Alvan Graham Clark (1832–1897), assisted Winlock (Warner and Ariail 1995, p. 38). Also present were George Washington Dean (1825–1897) and F. Blake, Jr. from the Coast and Geodetic Survey, who had charge of "… observations of precision …" using the transit instrument, chronograph and astronomical clock; the pioneer photographer John Adam Whipple (1822–1891) from Boston, who took photographs; Professor Robert Tevis from Shelbyville, who was in charge of meteorological observations; and Professor George M. Searle (1839–1918) from New York, was to observe general phenomena and specifically to search for intra-Mercurial planets (see Fig. 3.28). Winlock was in charge of the spectroscopic studies (The eclipse – Darkness upon the face of the Earth 1869).

Other astronomers from across the country also traveled to Shelbyville to set up observing sites for the eclipse (Matthews and Cleveland 2008), while a group of amateurs occupied the observatory atop Shelby College (Fig. 3.29).

Winlock's party succeeded in taking a series of photographs of the eclipse (see Figs. 3.30 and 3.31), and it was from Shelbyville that the first photograph of the diamond-ring effect was made (Golub and Pasachoff 1997, p. 43). It was also here that the first successful non-daguerreotype photograph of the spectrum of the prominences was obtained, and this showed 11 bright lines, 6 of which were seen for the first time (The eclipse – Accounts from the scientific parties 1869). Unfortunately, Winlock would not live to participate in the American eclipse of 1878, as he died in 1875 at the relatively young age of 49.

Fig. 3.29 Shelby College and its observatory (courtesy: Shelbyville Historical Society)

Fig. 3.30 Shelbyville expedition eclipse photographs (courtesy: Shelbyville Historical Society)

Fig. 3.31 A Shelbyville photograph of the 1869 eclipse showing prominences and the corona (courtesy: Shelbyville Historical Society)

Shelby College no longer exists but the site is now occupied by the local elementary Northside School (see Fig. 3.32).

Cleveland Abbe, Director of the Cincinnati Observatory, was the head of an eclipse party at the position "… furthest northwest of any party this side of the Pacific Coast." (Abbe 1869). Abbe, armed with a telescope, would determine the times of contact and note sunspot activity. He was accompanied by W.C. Taylor, photographer, with assistant Longstreet; Professor Alfred G. Compton of New York City Free College who would also note the contact times as well as make spectroscopic studies of solar prominences; James H. Haines, the Astronomical Assistant at the Observatory, who was in charge of photometric observations, the recording of all meteors seen and the search for intra-Mercurial planets; Robert Bowne Warder (1848–1905), then an assistant teacher at Illinois Industrial University (Jensen 1989), who was to record Haines' observations using the photometer, note the contact times, and also was responsible for calibrating the chronometers; and Robert Abbe, Cleveland's brother, who was in charge of meteorological observations and was to render assistance to any other party member as needed. More than 3 min of totality would be experienced at this site (Abbe 1869).

Fig. 3.32 Northside School, Shelbyville (photograph by the first author)

Unique to Abbe's expedition was the actinometer, an invention of Professor Compton and W.C. Taylor. It was described as a

> … disk of cardboard pierced with several fine holes, revolved by clockwork with great uniformity in front of a disk of sensitive paper, the action of the sun's rays combined with the motion of the hole, produced a dark circular line on the sensitive paper, the intensity of whose blackness reveals the varying, photographic power of the sun's rays. (Abbe 1869)

Altogether this device was used for 3 h, including during the total eclipse. Based on the results obtained, Abbe stated: "… the active or photographic power of the atmosphere, as illuminated by the light of the sun's corona and prominences, was very marked." (The eclipse – Scientific observations at Sioux Falls City 1869).

The site of these observations was in the remote Sioux Falls City, then called Fort Dakota, in the Dakota Territory (locality 2 in Fig. 3.25). It was a town of abandoned soldiers' barracks with about half a dozen occupied homes. A small group of local inhabitants gathered around members of the party and these visitors were able to see Baily's beads, some naked-eye prominences, stars and planets (Taylor 1869). Photometric observations demonstrated that at the time of totality the light at the zenith was less than a "… five-hundredth part of that 3 o'clock before the eclipse began …" yet only four planets and the star Regulus were visible. Many successful photographs were taken, seven being of outstanding quality (see Figs. 3.33, 3.34, and 3.35) (The eclipse – Scientific observations at Sioux Falls City 1869).

The time of first contact was reported by four of the party. All observers saw Baily's beads and numerous rosy prominences. Spectroscopic studies confirmed that these latter were gaseous, "… chiefly perhaps, hydrogen, shooting up from the sun's surface, sometimes to a height of 90,000 miles." (Professor Abbe's eclipse observations 1869). Sketches were made of the corona (see Fig. 3.36). As the news-

Fig. 3.33 Some of Taylor's photographs—2¼ inch square—numbered only (courtesy: University of Cincinnati Archives, photograph by the first author)

Fig. 3.34 More Taylor photographs—2¼ inch square—numbered only (courtesy: University of Cincinnati Archives, photograph by the first author)

Fig. 3.35 Some of Taylor photographs—2¼ × 3½ in.—labeled with times (courtesy: University of Cincinnati Archives, photograph by the first author)

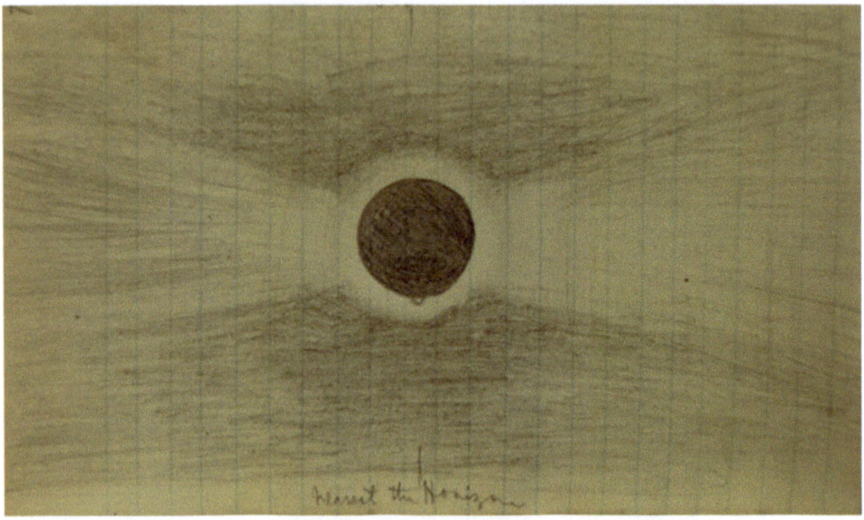

Fig. 3.36 The pencil sketch in Mr Warder's journal (courtesy: University of Cincinnati Archives, photograph by the first author)

paper article further reported: "A month of preparation, two hours of observation, three minutes of swift activity, and the work of the astronomical expedition was over." (ibid.). Abbe would subsequently publish an account of his overall impressions of this, the first eclipse he experienced (Abbe 1955, pp. 264–267).

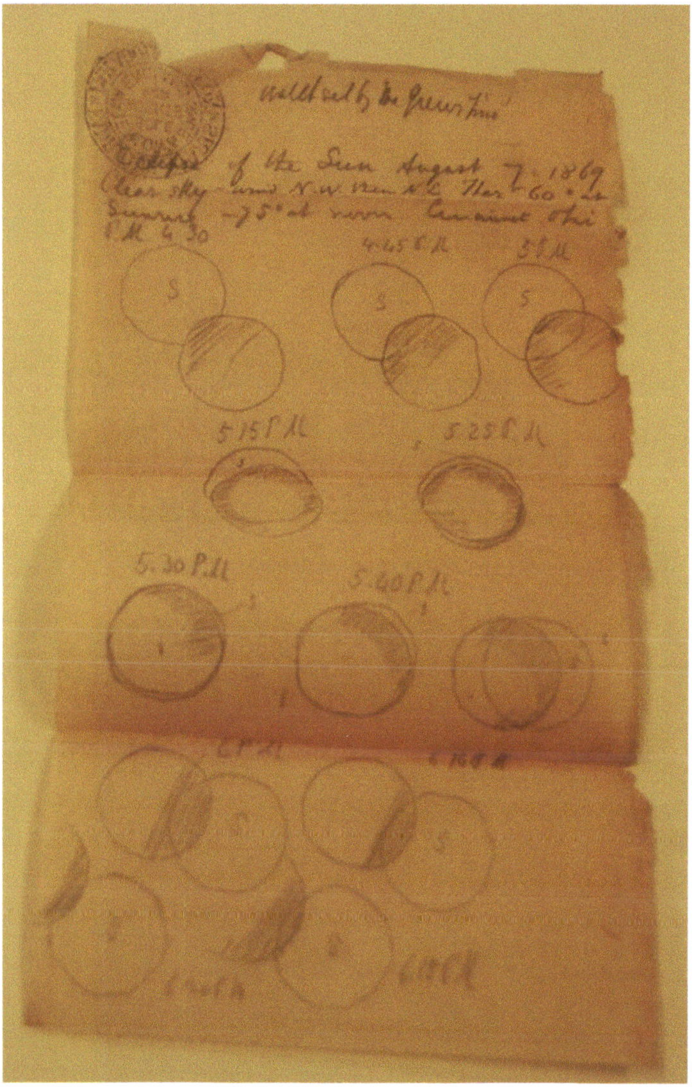

Fig. 3.37 Pencil sketches of the partial eclipse, as seen from the Cincinnati Observatory (courtesy: University of Cincinnati Archives, photograph by the first author)

In Abbe's absence James Russell was left in charge at the Cincinnati Observatory, where D.K. Winder and Henry Twitchell photographed the partial eclipse and made sketches (see Fig. 3.37).

Iowa was the site of some important 1869 eclipse expeditions. Professors William Harkness (1837–1903; see Fig. 3.38), from the U.S. Naval Observatory, and Charles Augustus Young (1834–1908; see Fig. 3.39), from Dartmouth College, were based near Des Moines and Burlington (localities 4 and 7, respectively, in Fig. 3.25), and they both studied spectra of the solar corona. They observed continuous spectra

Fig. 3.38 William Harkness (after *Popular Science Monthly* 1903, p. 86)

Fig. 3.39 Charles A. Young (courtesy: en.wikipedia.org)

traversed by a single bright green line. This line was initially associated with the newly-named element 'coronium' but would later be identified as a spectral line of highly ionized iron (Clerke 1902, p. 170; Golub and Pasachoff 1997, p. 28).

The Franklin Institute of Philadelphia sponsored three parties, authorized by Professor I.H.C. Coffin, U.S.N., Superintendent of the National Nautical Almanac, and they went to Ottumwa, Mount Pleasant and Burlington, all in the state of Iowa (localities 5, 6 and 7 respectively in Fig. 3.25). These were strictly photographic expeditions. Alfred M. Mayer, Professor of Physics and Astronomy at Lehigh University was at the Burlington site and reported that 41 perfect photographs were taken at that venue, five during totality. He sent nine plates spanning partial and total phases to Reverend Thomas William Webb (1807–1885) to present to the Royal Astronomical Society (Webb 1869).

Besides these American parties, there was a Canadian party that published results in the *Monthly Notices of the Royal Astronomical Society*. They were based in Jefferson, Iowa (locality 3 in Fig. 3.25), and witnessed a totality of 3 min and a few seconds. They took photographs but limited financial support precluded spectroscopic studies. In a letter to Warren De la Rue, the leader of the Canadian team, the Director of the Quebec Observatory, Commander Edward David Ashe (1816–1895; Jarrell 1990), wrote: "… I leave to the American astronomers the description of the results obtained by the spectroscope, &c." (Ashe 1869, p. 3). The Canadians succeeded in obtaining several photographs during the partial phase as well as four negatives during totality. The latter were deemed "… full of information …" and were placed in the care of De la Rue (ibid.).

3.1.4.3 The Eclipse of 29 July 1878

This eclipse, visible in the United States, would have a duration of totality of about 2½ min. It was best visible in Wyoming, Colorado and Texas (see Fig. 3.40). Though these less populated regions might lack some of the conveniences of civilization, the clearer skies would be advantageous to the scientific parties. Professor Harkness from the U.S. Navy Observatory published a 30-page document titled "Instructions for Observing the Total Solar Eclipse of July 29th, 1878." It was divided into nine sections covering topics ranging from naked eye and telescopic observations to the use of the various new tools of astrophysics. Several sections were useful to amateurs who wished to contribute valuable observations. Harkness organized the Naval Observatory party, which was at Creston, Wyoming Territory (locality 10 in Fig. 3.40). The telescope-maker Alvan Graham Clark was there, and he and the Observatory's A.N. Skinner (d. 1918) obtained photographs of the event (Warner and Ariail 1995, pp. 38–39).

At some locations there was an unprecedented merging of British and American parties. The railroads providing reduced fares for scientists traveling to expedition sites would extend the same courtesy to European eclipse parties. The British architect and amateur astronomer Francis Cranmer Penrose (1817–1903) published results of his experiences in both *The Observatory* and the *Monthly Notices of the Royal Astronomical Society*. His ultimate site was about 2½ miles to the northwest of Denver, Colorado (locality 6 in Fig. 3.40). During his travels there he would meet with leaders of several other parties, a veritable 'Who's Who' of the American scientific community, including Professors Young and the U.S. Naval Observatory's John Robie Eastman (1836–1914), and Cleveland Abbe, previously of Cincinnati (Penrose 1878a, b). Penrose, himself, succeeded in his main goals of timing the four

Fig. 3.40 Geographical distribution of U.S. localities mentioned in the text relating to the 1878 eclipse (after Cottam, Pearson et al. 2011). Key: 1 = Dallas; 2 = Fort Worth; 3 = La Junta; 4 = Las Animas; 5 = Pike's Peak; 6 = Denver & Cherry Creek; 7 = Central City; 8 = Rawlins; 9 = Separation Point; 10 = Creston

contacts and making a colored drawing of what he saw. He commented: "I must not conclude without a word of recognition of the kindness we received from the American astronomers." (Penrose 1878a, p. 51).

Arthur Schuster, accompanied by his friend and fellow Briton Mr Haskins, observed the eclipse from West Las Animas, South Colorado (locality 4 in Fig. 3.40). He was equipped to take advantage of the newer methodologies of spectroscopy and polariscopy. Haskins was mainly at the spectroscope and at totality noted a continuous spectrum for the corona (Schuster 1878a). Schuster (1878b) concerned himself mainly with the polarization studies. He noted an initial increase in this effect with distance from the Sun, but after reaching a maximum there was a rapid decrease.

A British correspondent for the *Daily News* reported on the work of several parties in Colorado. He noted that Professor Thorpe and Dr Schuster (both from England) were guests of Professor Hall at Las Animas. Fellow Englishman, Norman Lockyer, received multiple invitations including those from General Myer at Pike's Peak, the American astronomical photography pioneer Dr Henry Draper (1837–1882; Fig. 3.41) at Rawlins, Professor Newcomb at Separation Point and Professor F.E. Wright (1836–1915) at Las Animas. He opted for Rawlins where he and Draper took photographs of the coronal spectrum and determined that it was continuous.

Fig. 3.41 Henry Draper (courtesy: en.wikipedia.org)

This same correspondent was impressed by the presence of the American inventor Thomas Edison (1847–1931; Fig. 3.42) who used his invention, the tasimeter (Fig. 3.43), during the eclipse to determine the "… presence of heat waves in the radiation of the corona." (The eclipse of the Sun 1878, pp. 430–432; Lockyer 1878c). What it did was measure infrared radiation from the eclipsed Sun.

William Henry Pickering (1858–1938; Fig. 3.44) observed at Cherry Creek, 2¼ miles southeast of Denver (locality 6 in Fig. 3.40). He used polariscopy. At totality he determined the corona to be radially polarized (Pickering 1878). Lockyer (1878b) had drawn the same conclusion of radial polarization in 1871 and recommended that for future expeditions a means be created to obtain permanent photographic records of this effect.

Maria Mitchell (1818–1889; Fig. 3.45; Booker 2007) of Vassar College also participated in an expedition party. It consisted of six women who would observe from Denver, Colorado (locality 6 in Fig. 3.40). They were equipped with chronometers and telescopes. Notes were taken of visual impressions. Mitchell stated that though there were no significant streamers or prominences "… the corona itself was of great glory." The special artist who accompanied their party captured this image (Kendall 1896).

Fig. 3.42 Thomas Edison much later in life, circa 1922 (courtesy: en.wikipedia.org)

Fig. 3.43 Edison's tasimeter (courtesy: en.wikipedia.org)

Other successful parties included that at Fort Worth, Texas consisting of Leonard Waldo (1853–1929) and Robert Wheeler Willson (1853–1922) from Harvard University, Professor J.K. Riles and W.H. Pulsifer from Washington University in St. Louis, Missouri, and the keen amateur astronomer Frank Evans Seagrave (1860–1934) from Providence, Rhode Island. Besides the afore-mentioned expedition of

Fig. 3.44 W.H. Pickering (courtesy: en.wikipedia.org)

Fig. 3.45 A portrait of Maria Mitchell painted by H. Dassell in 1851 (courtesy: en.wikipedia.org)

Fig. 3.46 Edward S. Holden (courtesy: en.wikipedia.org)

Harkness at Creston, Wyoming Territory (locality 10 in Fig. 3.40) there were other U.S. Naval Observatory expeditions led by the following Naval professors: Asaph Hall at La Junta, Colorado (locality 3 in Fig. 3.40); Edward Singleton Holden (1846–1914; Fig. 3.46) at Central City, Colorado (locality 7 in Fig. 3.40); Samuel Pierpont Langley (1834–1906; Fig. 3.47) at Pike's Peak, Colorado (locality 5 in Fig. 3.40); and David Peck Todd (1855–1939; Fig. 3.48) at Dallas, Texas (locality 1 in Fig. 3.40) (Notes 1878).

Cleveland Abbe, previously at the Cincinnati Observatory, but now working for the United States Weather Bureau in Washington, D.C. led the solar eclipse expedition to Pike's Peak, Colorado (locality 5 in Fig. 3.40). This time he would miss the event. Pike's Peak is more than 14,000 ft high, and Abbe suffered from altitude sickness and had to be carried down the mountain on a stretcher before the eclipse began (Eclipse of the Sun 1878, p. 430). In 1889, however, he would have a third opportunity to observe a total solar eclipse when he would be sent, again by the United States Weather Bureau, to Cape Ledo in West Africa (Abbe 1955, p. 219).

In addition to the afore-mentioned eclipse expeditions, there also were expeditions to non-USA sites like Havana (Cuba) and Quebec (Canada), where positive reports also were obtained (Our astronomical column 1868).

Fig. 3.47 S.P. Langley (courtesy: en.wikipedia.org)

Fig. 3.48 David P. Todd (courtesy: en.wikipedia.org)

3.1.5 Evolving Directions for Solar Research

By the mid-nineteenth century new technology would lay old questions to rest and present the possibility of answering new ones.

Photography, spectroscopy and polariscopy were tools that might help determine the origin and chemical nature of the corona and solar prominences, the existence or non-existence of intra-Mercurial planets and the relation, if any, of sunspots to eclipse phenomena.

3.1.5.1 The Shape of the Corona

Proctor (1870a) initially thought the presence of millions of meteors, some revolving around the Sun, some falling into it, might account for the appearance of the corona (cf. Campbell 1907). Cleveland Abbe supported this theory, feeling the long rays of the corona were streams of meteors rushing inwards or outwards from the Sun (Clerke 1902, p. 177).

Professor Schaeberle from the Lick Observatory proposed a 'mechanical theory' as an explanation for the shape of the solar corona. He suggested that streamers of matter ejected normally from the Sun would emit and reflect light (Campbell 1907). Schaeberle predicted the shape of the corona based on the premise that normal forces are most active in the areas of sunspots. However, the example of a sketch of the predicted shape of the corona in 1893 was not supported by photographs taken during the total solar eclipse of that year (Todd 1894, pp. 74–75).

An electromagnetic theory was proposed by Professor Frank Hagar Bigelow (1851–1924) from the United States Weather Bureau in Washington. He noted that the corona streams occupied "… positions corresponding to the lines of force of in the Sun's magnetic field …" (Campbell 1907, p. 76).

The Swedish scientist and Director of the Nobel Institute Svante August Arrhenius (1859–1927; Fig. 3.49) thought that coronal shape might be the result of the pressure of outward radiation surrounding the Sun (Campbell 1907).

Among other theories there were those that included the influence of gravity and solar rotation (Todd 1894, pp. 73–74).

Campbell (1907, p. 77) thought that aspects of all of these theories might contribute to the appearance of the corona.

3.1.5.2 The Chemical Composition of the Corona

The use of spectroscopy allowed scientists to uncover the nature of the constituents of the Sun and the corona. Harkness and Young discovered the presence of what would ultimately be identified as an emission line of iron in the spectrum of the corona (Golub and Pasachoff 1997, p. 28). The Palermo Observatory Director Pietro Tacchini (1838–1905; Fig. 3.50) reported to the Paris Academy the presence of magnesium in certain regions of the Sun (Tacchini 1873).

Fig. 3.49 Svante Arrhenius (courtesy: en.wikipedia.org)

Fig. 3.50 Pietro Tacchini (after MacPherson 1905: facing page 77)

Fig. 3.51 Edward Maunder (after MacPherson 1905: facing page 192)

Controversial would be Henry Draper's discovery of bright lines of oxygen in the solar spectrum. The British astronomer, Edward Walter Maunder (1851–1928; Fig. 3.51) from the Royal Observatory, Greenwich concluded that what was interpreted as bright emission lines may in fact have been relatively bright lines between the absorption lines (Maunder 1870). The presence of oxygen would actually be confirmed in some dark lines of the Fraunhofer spectrum rather than the bright lines in emission (Golub and Pasachoff 1997, p. 49).

3.1.5.3 Sunspots and the Corona

It would take accumulated data over a number of eclipses to determine if there was a relationship between the shape of the corona and sunspot activity. Eventually, it was established that at sunspot minimum the corona would be elongated while at sunspot maximum it would be circular (Campbell 1907).

3.1.5.4 Polarization and the Nature of the Corona

Arthur Schuster would summarize some conclusions that might be drawn on the nature of the solar corona based on polariscopy. Polarimetric results indicated that the polarization of the corona changed radially with increasing distance from the Sun until a maximum was reached at about 20 min of arc from the solar limb, followed

Fig. 3.52 Sketch of James Craig Watson by Alexander Winchell (after *Popular Science Monthly*, 1880–1881)

by a rapid diminution until it was undetectable, but still within the corona. Polarization was an indicator of light scattering by solid particles. Schuster (1878b, 1879) concluded that at the greater distances, the particles were too large to polarize light, but when the particles broke up as they were attracted by the Sun's gravity they became small enough to cause this effect.

3.1.5.5 Intra-Mercurial Planets

A perturbation in Mercury's orbit led astronomers to suspect the presence of a planet between Mercury and the Sun, which even was assigned a name, Vulcan (see Baum and Sheehan 2003). The search for Vulcan might best be conducted during a total solar eclipse when the Moon blocking out the glare of the Sun would permit the visibility of small faint objects close to the Sun. This search for Vulcan would continue into the twentieth century, and in published articles and letters anticipating upcoming eclipses astronomers were reminded to look for such a body (Barkas 1868; Proctor 1870a).

During the eclipse of 1878 Professor James Craig Watson (1838–1880; Fig. 3.52), from the Ann Arbor Observatory in Michigan, was at Separation Point, Wyoming (locality 9 in Fig. 3.40) when he believed that he had found an intra-Mercurial planet. Similarly, Lewis A. Swift (1820–1913; Fig. 3.53), a successful amateur astronomer from Rochester, New York, believed that he had observed the same planet while based just outside Denver, Colorado (locality 6 in Fig. 3.40), during the same eclipse (Swift 1878). Astronomer Royal Airy (1878) speculated that what was actually seen was the star θ Cancri.

Fig. 3.53 Lewis Swift (courtesy: en.wikipedia.org)

Other extensive and negative searches were conducted into the twentieth century (see Baum and Sheehan 2003), then in 1915 Albert Einstein's General Theory of Relativity would provide a new theory for the perturbations observed in Mercury's orbit (Campbell and Trumpler 1923), and reveal that the existence of Vulcan was no longer necessary.

3.1.6 Conclusions

William Huggins made a long-term effort, late in his life, to photograph the solar corona without an eclipse. He believed he had achieved some success and endeavored to involve others in these attempts. William Pickering tried to duplicate Huggins' results and reported: "… very corona-like effects were certainly produced … unfortunately no two of the pictures were alike." (Becker 2000, p. 233). The British lawyer and talented amateur astronomer Arthur Ranyard (1845–1894) also tested the method using various sensitive plates with different exposures and obtained unsatisfactory results (Becker 2000). Professor George Ellery Hale (1868–1838; Fig. 3.54), then of the Yerkes Observatory, would, at Pike's Peak (locality 5 in Fig. 3.40), also make an attempt, but atmospheric conditions proved uncooperative (Campbell 1895).

It was only during the short time of a total solar eclipse that the solar corona could be studied until 21 June 1931, when French Meudon Observatory astronomer Bernard Ferdinand Lyot (1897–1952) invented the coronagraph, for photographing

Fig. 3.54 George Ellery Hale in about 1912 (courtesy: en.wikipedia.org)

the corona outside of an eclipse, and this innovative instrument was used successfully in the Pyrenées. It was limited, however, in that it could only photograph the inner corona (Lyot 1939).

In more recent times, space observatories, including SOHO, the Japanese Yohkoh X-Ray Satellite and the Transition Region and Coronal Explorer (TRACE) Satellite, have enabled astronomers to conduct solar studies outside of eclipses, particularly in parts of the spectra not visible from the Earth (Pasachoff 2001, pp. 40–47). Now these solar telescopes have been succeeded by the American XRT (X-Ray Telescope) on the Japanese Hinode Satellite and NASA's SDO (Solar Dynamics Observatory) (xrt.cfa.harvard.edu and sdo.gsfc.nasa.gov).

3.2 Transits of Venus and Refinement of the Value of the Astronomical Unit

3.2.1 The Astronomical Unit: Its Significance and Early Determinations

Agnes M. Clerke (1902, p. 227) states in her classic nineteenth-century *A Popular History of Astronomy* that the astronomical unit, the mean semi-diameter of the Earth's orbit, is "… our standard measure for the universe." It is upon this measurement that all

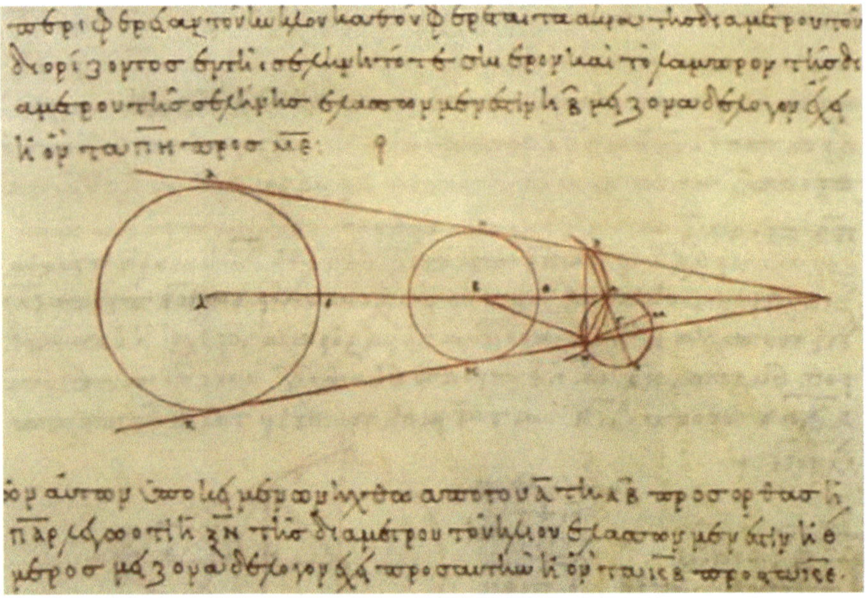

Fig. 3.55 A tenth-century Greek copy of Aristarchus of Samos' second century BC calculations of the relative sizes of the Sun, Earth and Moon (courtesy: en.wikipedia.org)

other measurements in our Solar System and beyond depend. Indeed, in the mid-1800s Sir George Biddell Airy (1857, p. 208), Great Britain's Astronomer Royal, described the determination of the astronomical unit as "… the noblest problem in astronomy." Up until the 1600s, attempts at calculating this distance led to figures that were much smaller than is now known to be the case. The Greek Aristarchus of Samos (c. 310–230 BC) using clever geometry, with inaccurate data to implement it (see Fig. 3.55), concluded that the Earth-Sun distance was at least eighteen times, but not more than 20 times the Earth-Moon distance. Today, knowing the Earth-Moon distance to be approximately 384,000 km, we would determine Aristarchus's range for the Earth-Sun distance to be about 6,900,000 to 7,700,000 km, considerably less than the actual value of 149,600,000 km. Another Greek, Hipparchus of Nicaea (ca. 190–120 BC), taking advantage of a solar eclipse in different degrees of totality at two different sites, applied trigonometry to the parallactic shift to calculate that the Earth-Moon distance was between 62 and 74 times the radius of the Earth. Using the radius of the Earth now known to be about 6,378 km, Hipparchus' range for an Earth-Moon distance would be from 395,000 to 472,000 km, a fair approximation for the time. The value of the radius of the Earth was well determined by the 1600s. Combining Hipparchus' determination with the premise of Aristarchus, the value of the astronomical unit could be calculated to be as low as (18 × 395,000) kilometers = 7,110,000 km = 4,400,000 miles. Up until the first part of the seventeenth century, this value for the astronomical unit was commonly accepted.

Fig. 3.56 A painting of Johannes Kepler in 1610 (courtesy: en.wikipedia.org)

It was in the seventeenth century that Johannes Kepler (1571–1630; Fig. 3.56) stated his three truisms, later to be called 'laws', that provided a basis for a more accurate determination of the astronomical unit. According to his third law, for all the planets the squares of the periods of revolution are proportional to the cubes of the semi-major axes of their orbits. Therefore, if one could determine the absolute distance between any two members of the Solar System, one could further derive the distance between any other two, including between the Earth and the Sun.

In 1627 Kepler published his *Rudolphine Tables* of planetary motion, named in honor of his patron, the Holy Roman Emperor Rudolph II of Prague. In these tables he predicted a transit of Mercury would occur on 7 November 1631 and a transit of Venus on 6 December 1631. Interestingly, and incorrectly, Kepler predicted that there would not be another Venus transit for 130 years. Transits of Venus are now known generally to occur in patterns of pairs about 8 years apart, separated alternately by about 105.5 and 121.5 years. Due to the 3.4° tilt of Venus' orbit with respect to that of the Earth's, a transit can only occur when both planets are near the nodes of their orbits (Mason 2004). Somehow Kepler missed the transit of 1639 in his calculations.

The transit of Mercury of 1631 was the first planetary transit ever known to have been observed. Only one published account of the observation of this event was made, that of Pierre Gassendi (1592–1655; Fig. 3.57) in Paris. His goal was to establish the diameter of the planet and he was to be surprised by its small size.

Fig. 3.57 A painting of Pierre Gassendi (courtesy: en.wikipedia.org)

Kepler had predicted an apparent diameter for Mercury of about one-tenth that of the Sun, but it was much much smaller and Gassendi at first mistook Mercury for a sunspot. He realized otherwise, however, when he saw its speed in crossing the solar disk (Van Helden 1976).

It was another astronomer, born later that century, Edmund Halley (1656–1742; Fig. 3.58), who promoted the use of the parallax observed during a transit of Venus as a tool for the calculation of the Earth-Sun distance. In 1716 he wrote a proposal which he presented to the Royal Society:

> … Scarce any problem will appear more hard or difficult than that of determining the distance of the sun from the earth, very near the truth; but even this, when we are made acquainted with some exact observations, taken at places fixed upon and chosen beforehand, will, without much labor be effected. And this is what I am now desirous to lay before this illustrious Society (which I foretell will continue for ages), that I may explain beforehand to young astronomers, who may perhaps live to observe these things, a method by which the immense distance of the sun may be truly obtained within a five-hundred part of what it really is. (quoted in Proctor 1882b, pp. 31–32).

Halley's method required observations of the entire transit. Then, if two observers located at about the same longitude but widely separated in latitude were to simultaneously observe the transit, they would see Venus cross the face of the Sun along slightly different transects. Since the speed that Venus crossed the Sun's disk was nearly identical at all sites, the length of each transect would be proportional to the time between ingress (at second contact) and egress (at third contact)—in practice, second and third contacts were preferred as they were more easily observed than the

Fig. 3.58 A portrait of Edmund Halley painted by Thomas Murray in about 1687 (courtesy: en.wikipedia.org)

external contacts (first and fourth). Having plotted the two transects, trigonometry then would reveal a figure for the solar parallax, and this in turn would provide a value for the 'astronomical unit'—the all-critical Earth-Sun distance. Halley's method required minimal special equipment, "... some common telescopes and clocks, only good of their kind ..." (Proctor 1882b, p. 35), as observers only had to time the transit. Its main disadvantage though was that it could only be used at longitudinally-similar locations, and under environmental circumstances where the entire transit would be visible. Nonetheless, Halley recommended that his method be used during the upcoming transits of 1761 and 1769 even though he knew he would not live to see the events himself.

The French astronomer Joseph-Nicolas Delisle (1688–1768; Fig. 3.59) reviewed Halley's method before the 1761 transit. He re-computed Halley's predicted path and concluded that Venus would actually cross the Sun about 6 arcminutes lower—meaning the transit would be shorter, therefore decreasing the number of suitable observing stations. Actually, Richard Proctor (1882b, p. 50) would later state that Halley's method: "... commonly fails at the earlier transit of a pair separated by eight years. Delisle proposed an alternate method that allowed observation of either the ingress or the egress contacts from sites that may be widely separated by longitude. The difference in the contact times would lead to an indirect measurement of Venus' rate of travel in miles per minute. This difference in absolute time, with the knowledge of

Fig. 3.59 Joseph-Nicolas Delisle (courtesy: en.wikipedia.org)

the longitude of each observing site, would lead to an indirect measurement of Venus' distance. The disadvantage of this method was that it required an accurate longitude for each observing site and the contact times had to be absolute (Airy 1857). Nonetheless, 'Delisle's Method' as it was known, would become the norm for future transits.

3.2.2 A Recap of Earlier Recorded Transits of Venus

3.2.2.1 The Transit of 6 December 1631

The first anticipated transit of Venus was that of 6 December 1631, predicted by Kepler in his *Rudolphine Tables*. He did not expect it to be visible in Europe although he encouraged others to look for it just in case there were some minor errors in his calculations. Pierre Gassendi, who had been successful in his observation of the transit of Mercury that same year, did look for this transit of Venus over a period of several days. As the event occurred during his nighttime, he was unsuccessful (Proctor 1882b).

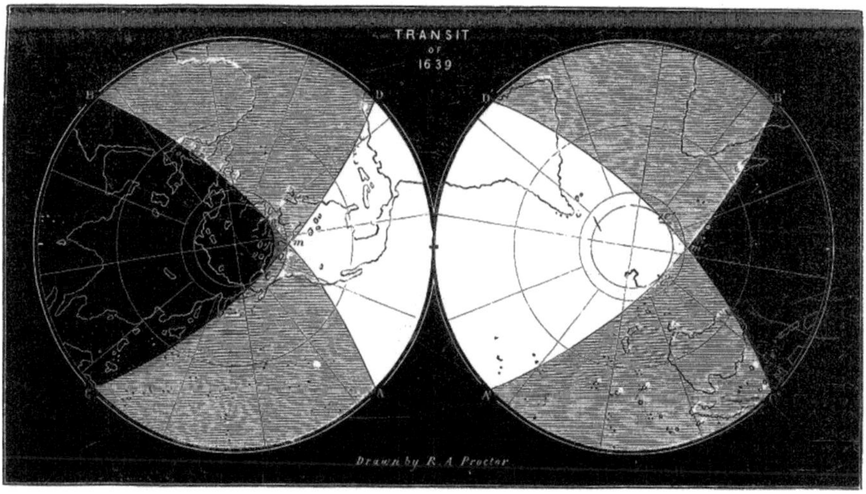

Fig. 3.60 Visibility of the transit of Venus in 1639 (after Proctor 1882b: Plate III)

3.2.2.2 The Transit of 4 December 1639

The next transit of Venus, missing from Kepler's *Rudolphine Tables*, was the first to be observed and documented, due to the efforts of a young Jeremiah Horrocks (sometimes spelled 'Horrox') (1619–1641) of Toxteth, England. Horrocks discovered Kepler's error and calculated that there would indeed be a transit of Venus in 1639, visible from his home vantage point. He shared this information with a friendly contact, William Crabtree (1610–1645) of Manchester, England. These two young men were the only ones to witness the event (Applebaum 2012; Aughton 2004; Chapman 2004).

The British popularizer of astronomy, Richard Proctor, wrote extensively on the transits of Venus. He drew his own maps (see Figs. 3.60, 3.61, 3.63, 3.70 and 3.78) showing the areas of visibility for several of the events, including the 1639 transit (see Fig. 3.60). The light areas on these maps represent where all four contacts would be visible. In the dark regions the transit would not be visible at all. In the shaded areas one would only see the ingress or egress contacts. At Horrocks' location in England, in 1639, he could only see the ingress.

3.2.2.3 The Transit of 6 June 1761

Motivated by the recommendations of Edmund Halley, astronomers worldwide were very interested in this next transit of Venus. Its significance as a means of solving 'the noblest problem in astronomy' was recognized, and more than 100 observing stations were established (see Woolf 1959). Many countries were able to make these observations within their own borders, but the governments of Great Britain and

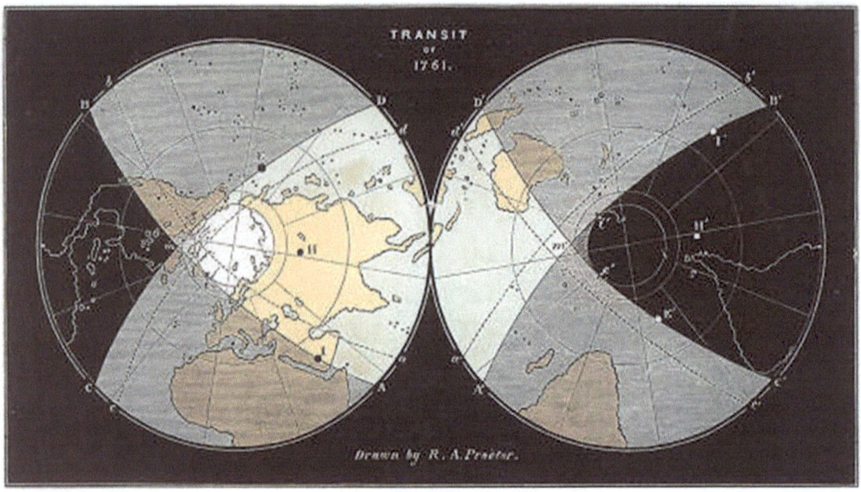

Fig. 3.61 Visibility of the transit of Venus in 1761 (after Proctor 1882b: Plate IV)

France subsidized multiple expeditions around the world to sites where different stages of the event might be visible. As Fig. 3.61 shows, the entire transit was predicted to be visible around the northern pole. In eastern Australia one might see the beginning (ingress) while in Western Europe one might see the end (egress). It was also intended that increasing the number of stations might better the odds of escaping failure due to instances of inclement weather.

It was the expeditions to remote venues that gave the astronomers the greatest hope of achieving their goal. The parallax methods suggested by Halley and Delisle were best applied where observing stations were far apart in latitude. Thus, sometimes expeditions would have to be sent to locations with primitive conditions. In addition, during this period of world history the Seven Years War raged between Britain and France and their respective allies, so there was always the threat of a ship meeting an enemy vessel whilst *en route* to an observing site. And although the astronomers might possess documents guaranteeing safe passage in enemy territory, these were not always honored.

Despite all these intricate preparations, success was ultimately limited. Some astronomers were frustrated in reaching their sites. Whilst succeeding in this, others were unable to make useful observations due to uncooperative weather. There was also the problem of determining the precise longitudes of the observing sites, which was necessary for the accurate calculation of the baseline needed to derive the solar parallax.

Ultimately, in the happy instances where all these obstacles were overcome, the scientists were frustrated by the unexpected phenomenon of the 'black drop' or 'black ligament' effect (see Fig. 3.62). Consequently, the *precise* determination of the times of the two internal contacts (i.e. the second and third contacts) often was not possible due to a distortion in the form of a blur or extension between the limbs of the Sun and Venus.

Final results were necessarily based on best estimates. Calculated values for the solar parallax ranged from 8″.28 to 10″.6, these extremes varying widely from the currently-accepted value of 8″.794148, which was determined by radar observations and adopted by the International Astronomical Union. There were hopes for better success at the next transit of Venus, which was only 8 years away, in 1769.

3.2.2.4 The Transit of 3 June 1769

There was greater activity surrounding the transit of 1769. The track for the event was more northerly, and the transit was predicted to be visible in its entirety over the Pacific Ocean, western America, and the South Pole region (see Fig. 3.63). The egress would be visible in eastern Asia and the ingress in eastern America and Western Europe. It was expected that lessons learned in 1761 would lead to more

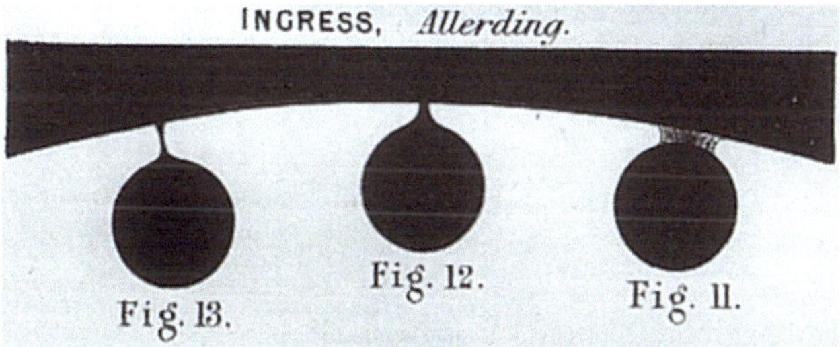

Fig. 3.62 Three drawings of the first egress contact (actually made during the 1874 transit), showing the 'black drop effect' and the problem of trying to determine exactly when the contact occurred (see Allerding 1883: Plate II)

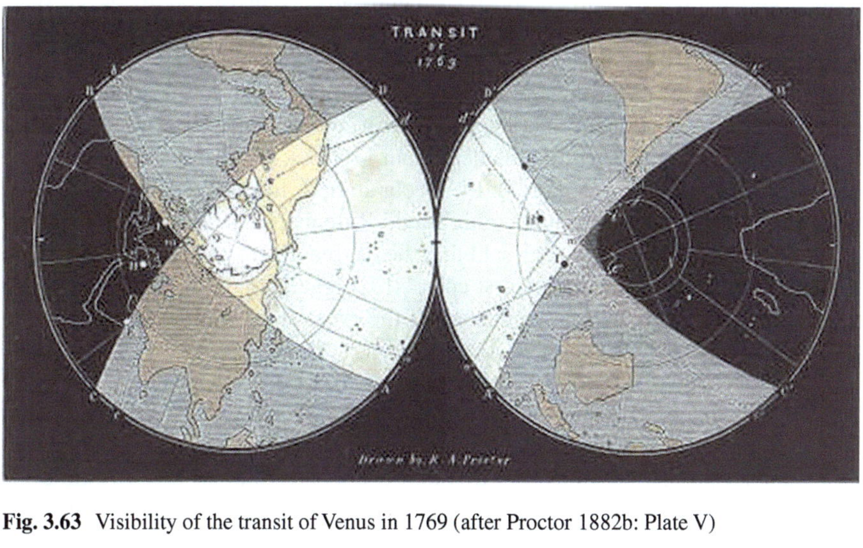

Fig. 3.63 Visibility of the transit of Venus in 1769 (after Proctor 1882b: Plate V)

Fig. 3.64 Abbé d'Auteroche (courtesy: transitofvenus.nl)

accurate results in 1769. The 'black drop' would no longer be a surprising phenomenon, and although its cause was still unclear, it was felt that greater care might be taken with its anticipation, yielding more consistent results. Furthermore, the Seven Years War was over, so access to far-flung observing sites hopefully would not pose so great a problem.

Again the British and the French led the international efforts. Some 'British' efforts of note took place in America, still British colonies, in this pre-Revolutionary year. A building was erected in Philadelphia in 1769 "… as an observatory to notice the transit of Venus." (Centennial of the Declaration of Independence of the United States 1873). Just north of Philadelphia David Rittenhouse, a clockmaker by trade, set up an observatory. It has been said that the degree of tension he experienced led to his fainting at the moment of first contact! Exaggeration or not, he was able to participate and document what remained of the event at his location (Hindle 1956, p. 157; Smith et al. 1769, pp. 289–326).

On the west coast of North America, there was a French expedition led by the Abbé Jean Chappe d'Auteroche (1722–1769; Fig. 3.64; Nunis 1982) to Saint Joseph, Baja California. Unfortunately the location at this time was in the midst of an epidemic, probably typhus, but d'Auteroche's party did successfully time the transit contacts. In order to establish their longitude accurately, they observed a lunar eclipse on 19 June and the eclipses of some Galilean moons of Jupiter from 20 June through 30 June. Chappe then became ill, and he died on 1 August 1769 (Proctor 1882b; Stone 1868).

An inspiration to the dedication to the science of astronomy would be the French astronomer from the Academy of Science in Paris, Guillaume Joseph Le Gentil de La Galasière (1725–1792; Hogg 1951). He was sent to the Southern Hemisphere for the earlier transit of 1761. He arrived in the Indian Ocean more than a year before the event to learn that his ultimate destination at Pondicherry in India was blockaded by the British, a consequence of the Seven Years War (Maor 2004, pp. 85–86; Sheehan and Westfall 2004, pp. 151–152). Anticipating assistance from a French fleet he waited in the area. Meanwhile he came down with dysentery and the fleet was destroyed in a hurricane. After his recovery he joined the French ship, *La Sylphide*, which was heading toward the blockaded Pondicherry. This time his ship was alternately frustrated by calm waters, delaying progress, and by monsoons, which threw it off course. Ultimately Le Gentil was obliged to observe the transit at sea, "... pour ne pas rester oisif à bord pendant que tous les astronomes étoient attentifs à cette observation, je la fis le moins mal qu'il me fut possible, & je vous l'envoie telle qu'elle est." (Le Gentil 1781, p. 188). In his correspondence with Monsieur de la Nux from the Academy of Sciences he sent his noted times for the four contacts though he knew they were of no value without a precise longitude. Undaunted, he suggested that he remain in the area for the next transit, and study the geography and natural history of the area. The Academy agreed. During this time the Seven Years War with England came to an end and the situation seemed more favorable (Fernie 2002, pp. 138–140). Le Gentil had planned to make his observations in Manila but the Academy ordered him back to Pondicherry, though only the egress of Venus would be visible there. Arriving at Pondicherry a good year before the transit, this time Le Gentil would be disappointed due to bad weather. When he finally returned to France after an absence of more than 12 years, it was to learn that he had been assumed dead, lost his position at the Academy and his estate was being divided among his heirs. However, things would turn out well for him. He would marry and have a family, regain his position at the Academy of Science and publish his story (Maor 2004, pp. 105–107).

In the early nineteenth century the German astronomer Johann Franz Encke (1791–1865; Fig. 3.65), Director of the Berlin Observatory, calculated the Sun's parallax to be about 8″.571 leading to a solar distance of 95,370,000 miles based on data from the transits of 1761 and 1769 (Stone 1868). In the mid-nineteenth century this figure came into question. Improved instrumentation used with other approaches, such as those involving the movements of the Moon and minor planets, the speed of light or observations of the placement of Mars among its neighboring stars, led astronomers to believe the Sun to be nearer than the 95,000,000+ value calculated by Encke (Transits of Venus... 1868). American astronomer Simon Newcomb (1867), would in the late nineteenth century, revisit data collected during these two transits, correct values for the longitudes where possible, and recalculate a value for the astronomical unit to be 93.0 ± 0.6 million miles, very close to the currently-accepted figure (see Dick et al. 1998).

Fig. 3.65 J.F. Encke (courtesy: en.wikipedia.org.)

3.2.3 Advances in Instrumentation

Subsequent observations of the transits of Venus would have the advantages of new instrumentation.

3.2.3.1 The Development of Astronomical Photography

In 1868 Walter De la Rue (1868) presented his recommendation to the Royal Astronomical Society for the use of photography in the upcoming transits of Venus. He felt that photography would provide permanent and accurate records of the observations made. It was also suggested that a series of photographs taken of Venus traversing the Sun would surely increase the accuracy by increasing the number of recorded observations (The application of photography to astronomy 1869).

The first practical photographic format, the daguerreotype, was invented in 1839. Frederick Scott Archer (1813–1857) made available a wet plate process, which was a much faster medium than the daguerreotype and could easily be printed after drying, but it could not be allowed to dry until it was completely developed (Frederick Scott Archer 1875). The American, Himes (1874), described a method for the preparation of dry-plates for photography that could more efficiently be prepared in light,

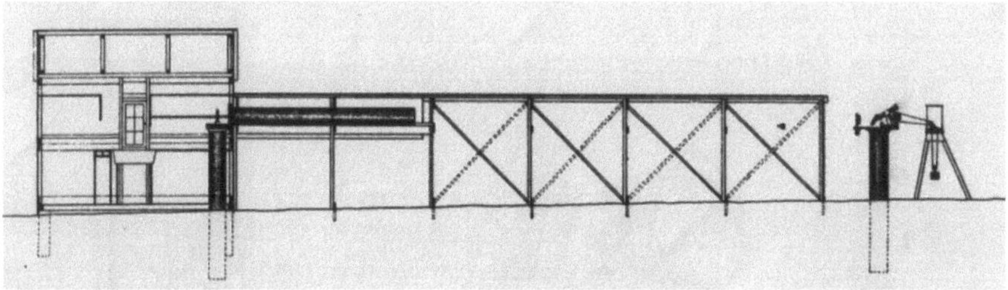

Fig. 3.66 A schematic view of the horizontal photoheliograph, showing (from left to right) the 'Photographic House' with the plate-holder and its supporting pier; framework shielding the incoming solar image; the heliostat and its supporting pier, and the associated clock drive (after Newcomb 1880)

sensitized and desensitized by the use of silver compounds, and he recommended their use for the upcoming Venus transit of 1874. In 1874 the British astronomer and photographer William (later Sir William) de Wiveleslie Abney (1843–1920) also described a dry plate process for solar photography that would be more convenient to use. Complicated recipes were provided requiring the use of eggs for albumen and modifications for different climates. The dry-plate format would be more widely used during the transit of 1882.

3.2.3.2 The Horizontal Photographic Telescope

The horizontal photographic telescope (see Fig. 3.66) was invented by Harvard College Observatory Director Joseph Winlock, and was favored by the Americans during the nineteenth century transits. It used a clock-driven mirror (Fig. 3.67) to bring the solar image to a long-focus objective lens which then directed the image to the photographic plate (Fig. 3.68) which was mounted in a firmly-supported plate-holder located in a prefabricated 'Photographic House' nearly 40 ft away. This type of photoheliograph produced relatively large and distortion-free images which could be photographed for later measurement (Dick et al. 1998; Janiczek 1983; Lankford 1987).

3.2.3.3 The Heliometer

The 'modern' heliometer, which functioned as a stand-alone telescope, was introduced by Joseph Fraunhofer of Germany in the nineteenth century. Here an object glass is divided diametrically into two halves, which can be manipulated by a screw in order to measure small angular distances between two objects with a built-in micrometer used to bring the two objects into coincidence (see Fig. 3.69). The Germans especially would find this device useful in their observations of the nineteenth-century transits of Venus (Radau 1874; Mauritius Expedition 1874).

Fig. 3.67 A close up of the heliostat and its clockwork drive (after Janiczek 1983, p. 58)

3.2.3.4 The Role of Spectroscopy

During the interim between the eighteenth century and the nineteenth century transits of Venus the science and technology of astronomical spectroscopy emerged. The remarkable American astronomer Henry M. Parkhurst (1825–1908) proposed that the near approaches of contact between celestial objects might be observed through a spectroscope thereby uncovering much of the mystery of what was being observed, but he did not believe this would be applicable to a Venus transit until the year 2490 (Scientific Intelligence 1874). However, an Italian expedition organized by Pietro Tacchini, that went to India to observe the 1874 transit, successfully used spectroscopy to confirm the existence of a Venusian atmosphere. They also demonstrated the value of spectroscopy in determining the times of the ingress and egress contacts (see Pigatto and Zanini 2001).

3.2.3.5 The Role of the Telegraph

Longitude determinations had been a major difficulty for some of the 1761 and 1769 transit expeditions. John Harrison had completed his most advanced clock in 1759. With an accurate clock one could establish the precise difference in longitude between two observing sites (Whitesell 2000).

In 1835 Samuel Morse demonstrated the electric telegraph at the University of the City of New York. This device would provide a means for observers at two different locations to compare their clock times by telegraphic communication. Such calibrated chronographs might then be used to determine the precise time of celestial phenomena (ibid.).

A correspondent on the American 1874 transit of Venus expedition to Nagasaki, Japan, described the extent of reach of the telegraph at that time. This enabled the

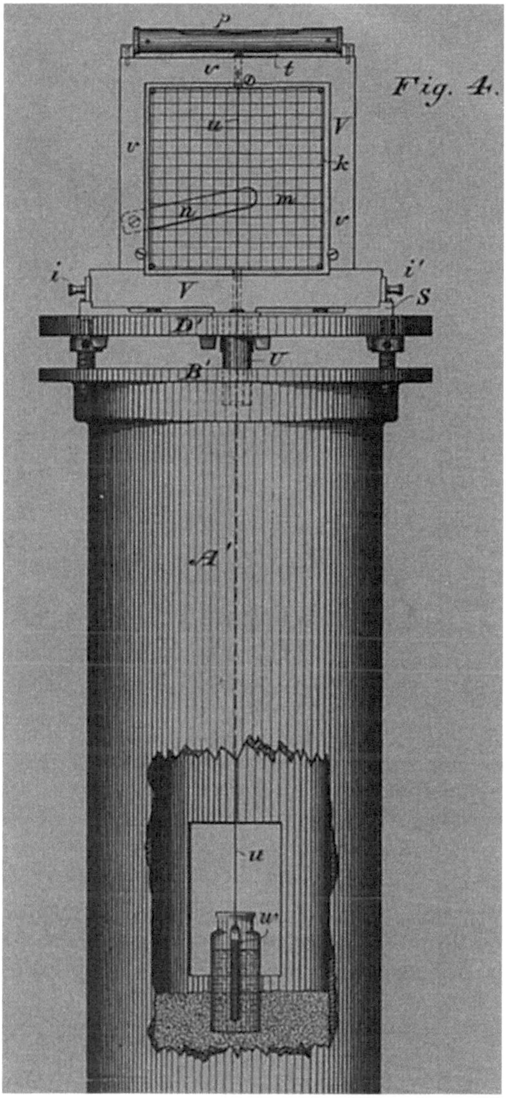

Fig. 3.68 A close up of the photographic plate-holder and its supporting pier (after Newcomb 1880)

determination of longitude, a necessary piece of data for their ultimate scientific calculations:

> For this determination of longitude, Nagasaki had been connected with Vladiwostock by the telegraphic method, by a series of observations in concert with Prof. Hall … Vladiwostock is connected by telegraph with St. Petersburg, and hence this station is thus connected with the European and American observatories … And at some subsequent time not far distant, when a cable is laid across from San Francisco to Japan, the last link will be put in, and one continuous chain longitude determination will encircle the world. (The American transit expedition 1875, pp. 245–246).

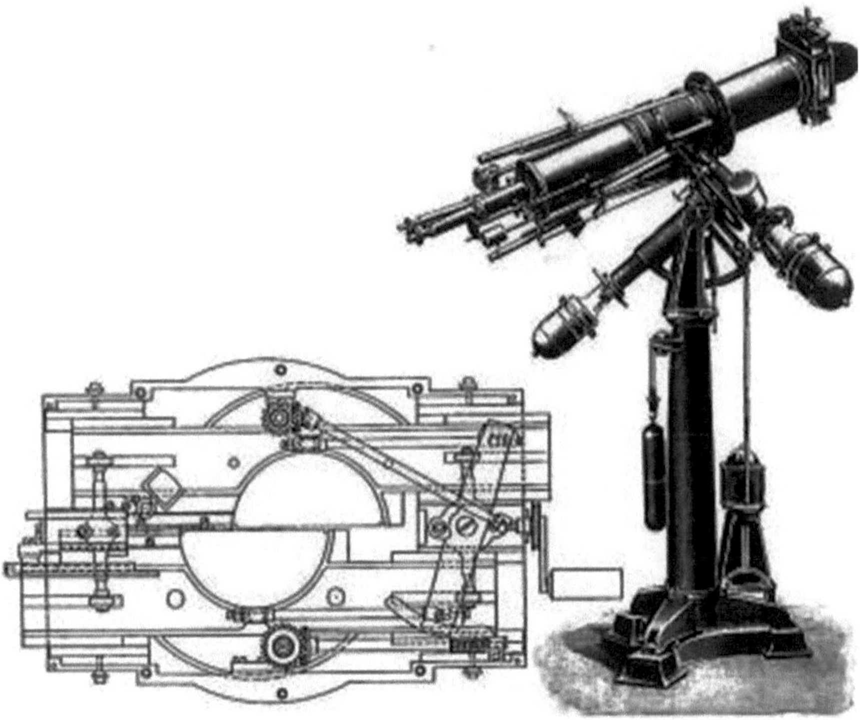

Fig. 3.69 The 107-mm Repsold heliometer built for the Dun Echt Observatory and used by David Gill to observe the 1874 transit of Venus from Mauritius. The *inset* shows details of the two objective semi-lenses (after Willach 2004, p. 13)

Astronomer Royal George Biddell Airy (1857, p. 215) would praise the immense advantages of 1882 transit observations in the United States due to the extensive presence of "… the galvanic telegraph".

3.2.4 The Transit of Venus of 9 December 1874

By 1874, the value of the solar parallax had not been established to the satisfaction of the scientific community. Although other methods existed for its determination, the promise of the transit of Venus could not be ignored. The 1874 event had been anticipated since the last transit, in 1769. Its significance was also much promoted by Astronomer Royal Airy. It was believed that expeditions would be better prepared and equipped than they had been previously. One also might expect better cooperation among many nations participating in such expeditions in this time of relative peace.

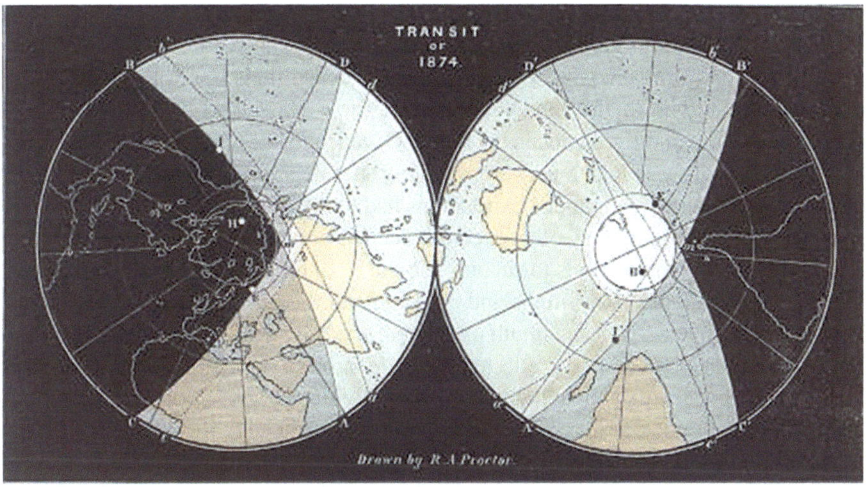

Fig. 3.70 Visibility of the transit of Venus in 1874 (after Proctor 1882b: Plate VI)

Airy would delineate what he felt would be the best sites for observations based on the method to be used. For the method of duration, i.e. Halley's method:

> The most northerly stations are to be found in Siberia, Tartary and Thibet (which will scarcely be visited by astronomers in December) and in North British India. The most southerly stations will be Kerguelen's Island, Van Dieman's Land and New Zealand.

For the method of difference of absolute time, i.e. Delisle's method:

> For the ingress, favourable positions will be found in Owhyhee (where the displacement tends to throw Venus upon the Sun's disk, or to accelerate the ingress), and at Bourbon, Mauritius, and Kerguelen's Island (where the displacement tends to throw Venus from the Sun's limb, or to retard the ingress). For the egress, Sicily, Italy and portions of Europe west of the Black Sea, are so situate as to throw Venus upon the Sun's disk, or to retard the egress; and New Zealand, New Caledonia, Van Dieman's Land, and Eastern Australia, are all well situated for accelerating the egress. But it is doubtful whether the longitude of any of the stations named, except those in Europe, are yet known with sufficient accuracy. (Airy 1857, pp. 214–215).

The transit was expected to last more than 4 h (Dick et al. 1998).

3.2.4.1 Foreign Expeditions

Since the entire transit was best observed in the area of Australia, New Zealand and points west (see Fig. 3.70), many astronomers made the long voyage to these parts of the world, which often were unexplored and quite primitive in nature. The Russians stayed in their own 'backyard', establishing 27 stations across their homeland (Struve 1873; Werrett 2006). The Australians attracted foreign expeditions, but their colonial Governments also provided funding so that local observatories in Adelaide, Sydney and Melbourne could set up their own country observing stations. All of these venues met with at least partial success in their endeavors (see Lomb 2011; Orchiston 2004b).

Great Britain organized the most foreign expeditions (see Ratcliff 2008), establishing stations in the Sandwich Islands (now Hawaii—see Chauvin 2004), Egypt, Rodriguez and Kerguelen in the Indian Ocean, and New Zealand (see Orchiston 2004b). Some of these locations were remote from telegraph lines which might have been used to establish longitude. Expeditions had to arrive at these sites early enough to determine the longitude by other methods such as lunar transits. Astronomer Royal Airy gave specific instructions to accomplish this: "… requiring 100 double observations of lunar altitudes or azimuths and 30 transits of the Moon across the meridian." Parties were to use: "… an altazimuth instrument specially designed for this expedition by Airy." Father Stephen Perry S.J., led the expedition to the inhospitable Kerguelen Island. He recorded that the reduction of a single lunar observation for determination of longitude took at least six man-hours. Perry used a six-inch refractor for his transit observations. His efforts were partially frustrated due to overcast skies. When the sky cleared he did obtain data for the third and fourth contacts (Hingley 2005). From Great Britain also came the most successful privately-funded expedition of the 1874 transit. Lord Lindsay (James Ludovic Lindsay) mounted an expedition to Mauritius, using his own yacht, the *Venus*, to carry out a number of astronomical observations, including work on the solar parallax (Brück 2004).

The French had six expeditions (Débarbat 2004; French preparations for the transit of Venus 1874; Lauga 2004). Jules Janssen, leader of the French expedition to Japan in 1874, designed for this event a revolving photographic device which could automatically take a "… series of 48 or more exposures controlled by a spring-driven clockwork …" (Launay and Hingley 2005; Sicard 1998). Americans and British expeditions, and two of the Australian observatories (see Lomb 2011), also used this device or a British derivative of it. It was used on Kerguelen Island to capture the transit egress, but the results were deemed disappointing (Hingley 2005).

The Germans had six expeditions (Duerbeck 2004). Their heliometer studies of the transit were conducted at Tschifu (China), Kerguelen, Auckland, the Mauritius Islands, Isfahan (Persia) and Luxor (Egypt). Photographic observations were also made but were disappointing and this method would be abandoned for their expeditions in 1882 due to the low accuracy achieved (see Duerbeck 2007).

Many other countries participated in these international efforts on a smaller scale (e.g. see van Gent 1993; Kopper 2004; De Mourão 2004; Pigatto and Zanini 2001).

3.2.4.2 The American Expeditions

This was the first time that the United States would participate in transit of Venus expeditions as a sovereign nation. Jessica Ratcliff (2008, p. 46) notes:

> The major new role played by the United States, in comparison to its small presence in the transit activity of the 1760s, is perhaps one of the most dramatic shifts of the political-scientific landscape of transit participation over the century since the last transit of Venus.

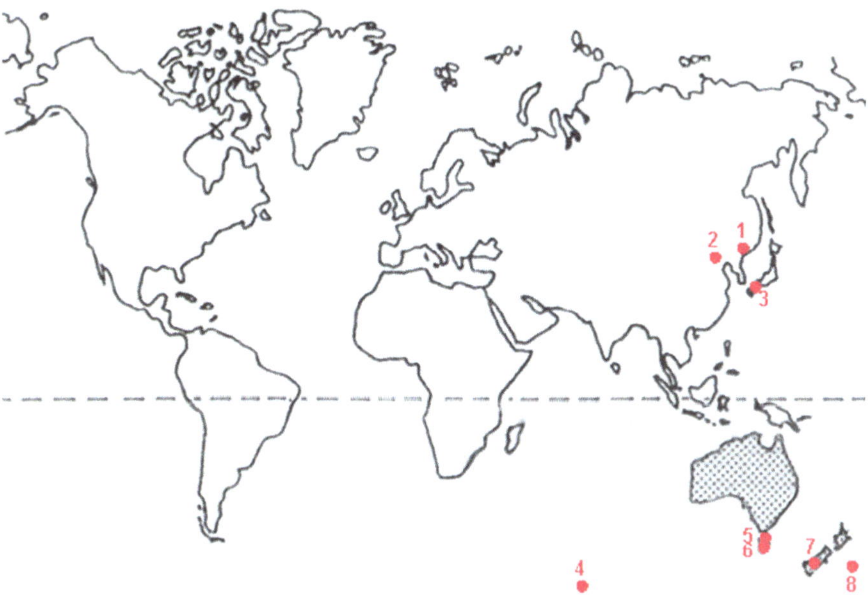

Fig. 3.71 The distribution of U.S. transit of Venus stations in 1874 (after Cottam, Orchiston et al. 2011). Key: 1 = Vladivostok, 2 = Peking, 3 = Nagasaki, 4 = Kerguelen Island, 5 = Campbell Town, 6 = Hobart, 7 = Queenstown, and 8 = Chatham Island

The United States equipped eight expeditions (Dick et al. 1998), three in the Northern Hemisphere (Beijing = Pekin, China; Nagasaki, Japan; and Vladivostok − Wladiwostok, Russia) and five in the Southern Hemisphere (Chatham Island, off New Zealand; Malloy Point (Kerguelen), French Southern and Antarctic Lands; Hobart, Tasmania; Campbell Town, Tasmania; and Queenstown, New Zealand) (see Fig. 3.71).

The United States created a Government Commission to organize these expeditions. It was headed by the Superintendent of the U.S. Naval Observatory, Rear Admiral Benjamin F. Sands (1811–1883; Fig. 3.72), and included the President of the National Academy of Sciences, Joseph Henry (1797–1878; Fig. 3.73), Harvard University Professor of Mathematics and Astronomy, Benjamin Peirce (1809–1880; Fig. 3.74), and the two U.S. Naval Observatory astronomers, Simon Newcomb (1835–1909; Fig. 3.75) and William Harkness. This Commission made decisions regarding methods and equipment, and it was decided there would be consistency in both. Photography was chosen as the selected method, and Newcomb proposed a "… fixed horizontal telescope of nearly 40 ft focal length, through which sunlight was directed by a heliostat." (Dick et al. 1998, p. 229). American Alvan Clark constructed the photoheliograph lenses and the heliostat mirrors. In addition, visual observations would be made using 5-in. refractors (Fig. 3.76), also constructed by Alvan Clark. All sites would include a transit instrument (Fig. 3.77), which would be used for system alignment as well as for time-keeping (ibid.).

Fig. 3.72 Rear Admiral Benjamin F. Sands (courtesy: en.wikipedia.org)

Fig. 3.73 A portrait of Joseph Henry painted by Henry Ulke in 1879 (courtesy: en.wikipedia.org)

Fig. 3.74 Benjamin Peirce (courtesy: en.wikipedia.org)

Fig. 3.75 Simon Newcomb (courtesy: en.wikipedia.org)

Fig. 3.76 One of the 5-in. equatorially-mounted Clark refractors used by the Americans (U.S. Naval Observatory; photograph by the second author)

The composition of the individual parties also was standardized: there would be one chief, one astronomer, one head photographer and two assistant photographers as minimum. Occasionally there would be additional help, sometimes recruited locally (ibid.).

The transit was not to occur until December 1874, but starting in May practice observing sessions were arranged to assure efficient use of apparatus and time, as well as to maximize accuracy in the determination of the contact times. To this end, an artificial Venus would move across an artificial Sun on the wall of a building somewhat less than a mile from the Naval Observatory (ibid.).

All southern station parties boarded the *S.S. Swatara* and departed the United States on 7 June 1874. On 7 September the Kerguelen Island party disembarked. On 1 October the next two parties disembarked at Hobart, Tasmania (Orchiston and Buchanan 1993), and one of these would continue on to Campbell Town. On October 16, the next party disembarked at Bluff Harbor, New Zealand (Orchiston et al. 2000). The final group was dropped off at Chatham Island to the east of New Zealand on 29 October (Dick et al. 1998).

Fig. 3.77 One of the Stackpoole broken-tube transit telescopes used by the Americans (U.S. Naval Observatory; photograph by the second author)

The three northern expedition parties left in two groups from San Francisco to arrive at Nagasaki, Japan. One party remained there and the other two continued on to their respective sites at Vladivostok and Beijing, both arriving at their destinations on 9 September (ibid.).

There was much preparation required for the transit once the destinations were reached. First the specific sites had to be selected. In particular, the Bluff Harbor party opted to move inland to Queenstown in hopes of finding better environmental conditions there (Orchiston et al. 2000). Prefabricated buildings were set up, and piers erected for specific instruments. Once established, longitude determinations and practice in the use of all equipment was necessary. Thanks to all of the meticulous planning and preparations, by 9 December all parties were ready (Dick et al. 1998).

All sites, northern and southern, had some degree of success in that all obtained some usable photographs of the transit. It was later generally agreed that the quality of the American photographs was superior to those obtained by expeditions organized by other nations. Even the Americans, however, were not able to avoid the 'black drop' problem, and it was felt that those photographs of Venus fully on

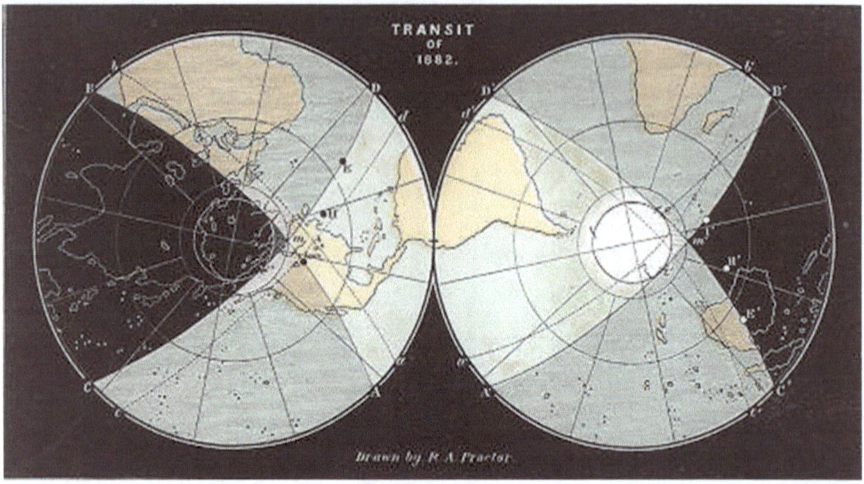

Fig. 3.78 Visibility of the transit of Venus in 1882 (after Proctor 1882b: Plate VII)

the face of the Sun might prove to be more usable. It would be years before all the data was reduced (ibid.). In fact, as late as 1880, Professor Young (1880, p. 88) admitted: "The results of the transit of Venus observations have not yet been so fully published as might have been expected." Indeed the Americans did not publish any official figure for the solar parallax from these efforts. However, David Todd, then at the Nautical Almanac Office, published a 'provisional' value of $8.883 \pm 0.034''$, translating to a value for the astronomical unit of 92,028,000 miles. This was based on data taken from *Observations, Part One, "General Discussion of Results"* of 1880, the only relevant official publication completed (Todd 1881b).

3.2.5 The Transit of Venus of 6 December 1882

Disappointment in the results of the observations of the 1874 transit of Venus might have dampened some of the enthusiasm for the upcoming transit of 1882 but there were reasons for renewed resolve. This transit would be visible in much of Europe and the Americas (see Fig. 3.78). It would last longer at about 6.3 h, as opposed to the approximately 4.6 h in 1874. This meant the area where some part of the transit would be seen would be greater. Maybe most significant was the recognition that this would be the last transit for more than a century. If there was any doubt, this transit could not be ignored (Airy 1880). The U.S. Congress appropriated $177,000 dollars for American efforts. Instruments would be improved and there would be expeditions this time both within and outside of American borders (Dick 1995).

3.2.5.1 Foreign Expeditions

In anticipation of the 1882 transit an international conference was held in Paris, in October 1881, to co-ordinate efforts. Fourteen nations participated (Orchiston and Buchanan 1993). Discussion on methodology led to a general acknowledgement that photography had not led to satisfactory results in 1874 and its use should be less significant in 1882 (Sheehan and Westfall 2004).

Some countries, such as Portugal and Spain, that had not participated in previous transit parties, did attend the conference and would have their own parties in 1882. Some others, such as Norway and Chile, sent representatives to the conference but ultimately did not mount their own expeditions. Great Britain had a Transit Committee that decided to send numerous expeditions around the world (Ratcliff 2008), including to Canada in North America, and to some sites that would not have access to all four contacts, such as South Africa (Koorts 2004), and Burnham—yet again—in New Zealand (Orchiston 2004b). Russia and the United States declined to participate in the Paris meeting. American Simon Newcomb did not have much faith in the established procedures, having been frustrated in his efforts in 1874 (Tebbutt 1883), while Russia had decided that observations of minor planets at opposition would be less costly than the Venus transit efforts.

3.2.5.2 Foreign Expeditions to the United States

Although the United States was not represented at the conference at Paris, it would host transit parties from Germany, Belgium and France (Sheehan and Westfall 2004, pp. 270–272).

Germany had two expeditions to the United States, one to Hartford, Connecticut, and one to Aiken, South Carolina (Duerbeck 2004). Due to the disappointing results provided by the photographic method during the transit of 1874 the Germans decided to depend on measurements of the planet's position on the solar surface as determined with the heliometer.

Expedition I, based on the grounds of Trinity College in Hartford, Connecticut (see Fig. 3.79), was led by the astronomers Gustav Müller (1851–1925) from the Potsdam Astrophysical Observatory and Friedrich Deichmüller (1855–1903) from Bonn Observatory (ibid.). The morning of the transit the sky was overcast. Having lost the ingress contacts Müller reported:

> The ingress could not be observed, and only for one moment Venus was seen between first and second contact halfway in the Sun. Only after ingress the clouds started to disperse with rapidity, and our mood started to rise. About one hour after external contact the clouds were so thin that we could start the heliometer measurements ... Soon the sky improved, and remained quite good until the end ... (ibid.).

They obtained eight full sets of heliometer readings (Knapp 2004, p. 14).

Fig. 3.79 German astronomers at Hartford, Connecticut (after *Frank Leslie's Popular Monthly* 1883)

Dr Julius Heinrich Franz (1847–1913), Principal Astronomer at the Royal Observatory in Königsberg, headed Expedition II to Aiken, South Carolina. There the property of Henry Smith was selected for the transit station (see Fig. 3.80) as it was far enough away from the railroad tracks to avoid the occasional jarring of the earth due to passing trains.

Members of the public were very interested in the goings-on at the Smith estate but the Germans stationed guards to keep curious citizens away. It was said that even the Mayor of Charleston could not visit the estate while the scientific work was taking place (Aiken and the transit of Venus 1935). Aiken had been selected as a suitable site due to its usually fair climate, however unexpected rain prevented observation of the first two contacts. It did clear thereafter, allowing the Germans to make some satisfactory heliometric measurements during the remainder of the transit. A total of 48 observations, 3 sets of 16 each were made.

A marker, donated by the Germans, was placed at the site, the residence of Henry Smith, to commemorate the event. This marker was later donated by John Weems, then owner of the property, to the Aiken County Historical Museum. There it now stands surrounded by the associated observatory-inspired structure, with a descriptive plaque, as shown in Figs. 3.81 and 3.82 (The transit of Venus, *Journal*... 1995).

Fig. 3.80 The German transit station at Aiken, South Carolina. According to Duerbeck (2004), Franz is second from left (courtesy: Aiken County Historical Society)

Fig. 3.81 The 1882 transit of Venus observatory structure at Aiken, South Carolina surrounding cracked commemorative marker (photograph by the first author)

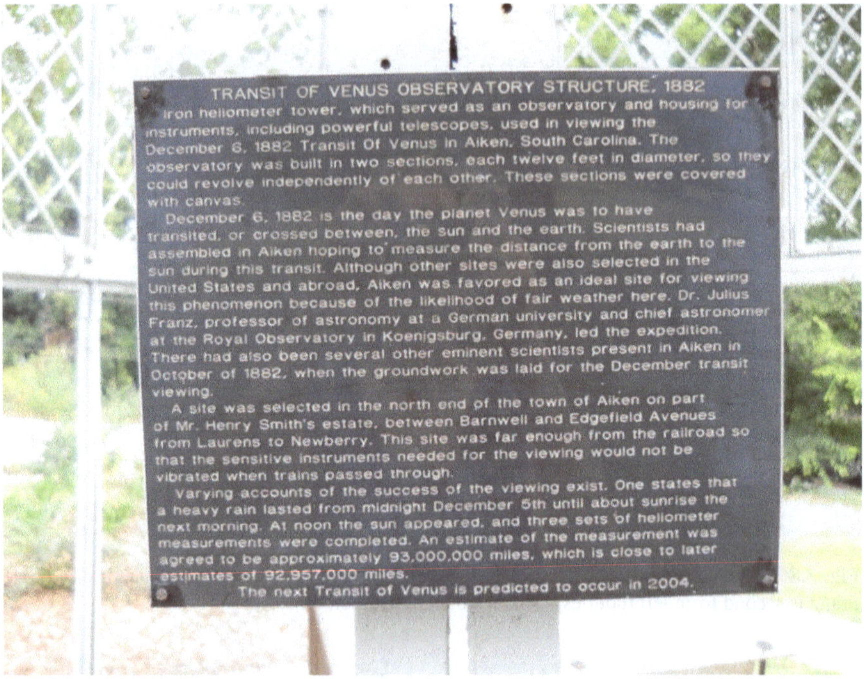

Fig. 3.82 The plaque at the German transit of Venus expedition site at Aiken, South Carolina (photograph by the first author)

The limestone slab, which is 27×31 in. and 4 in. thick and is now cracked, contains the following inscription:

Venus – Durchgang 1882
Deutsch Station II
5 h 26 m 52 s 6 W 33° 31′ 51″ N

The English translation is:

The Transit of Venus 1882
German Station No. II
5 h 26 m 52 s6 W 33° 31′ 51″ N)

San Antonio, Texas, would host two 1882 transit of Venus expeditions. One of the four official American parties was located on the grounds of what is now known as Fort Sam Houston (Dick et al. 1998). The Belgian nation was participating in a major scientific expedition for the first time, with transit parties located both here in San Antonio and in Santiago, Chile (see Sterken et al. 2004). Both Belgian parties were organized by the Director of the Royal Observatory in Brussels, Jean-Charles Houzeau (1820–1888; Fig. 3.83), who would himself lead the San Antonio party (Sterken 2009). The Belgians were located about 500 m to the west of the Americans,

Fig. 3.83 Jean-Charles Houzeau (after Sterken et al. 2004)

on private property (Sterken et al. 2004). The methodologies of the two countries were quite different: following the published instructions for all of their official expeditions the Americans would rely on the photographic method, while the Belgians at both sites would use Houzeau's invention, a twin-objective heliometer with unequal focal lengths (see Fig. 3.84).

The heliometer had two objectives with different focal lengths whereby large and small images of both the Sun and Venus were produced. A large solar image was projected on a screen (seen below the heliometer tube in Fig. 3.84). A smaller solar image produced by the short-focus objective was made to coincide with that of Venus by micrometer adjustment. The difference in micrometer readings between the "… small Sun centered on the crosshairs, being the center of the large Sun … [and the] small Sun centered on large Venus … " enabled determination of the distance between the centers of both objects (Sterken et al. 2004, p. 312). Houzeau's assistant, Albert Benoit Marie Lancaster (1849–1908) reported on the progress of the day. At 6:15 a.m. Houzeau went to the American site to compare chronometer times, but upon returning to the Belgian site there was early frustration as the first two contacts were lost due to clouds. At about 9:30 a.m., 12 min before the minimum distance of centers, the sky cleared and 124 μm readings were made (ibid., p. 319). When these were combined with the results obtained by the Belgian party in Chile (which enjoyed perfect weather), Houzeau calculated a final solar parallax of $8''.911 \pm 0''.084$ (Sterken and Duerbeck 2004).

In October 2005 an historical marker was inaugurated and placed at the Belgian transit of Venus observation site in San Antonio (see Fig. 3.85). The original structure, a wooden house that was occupied by the party, is now gone and has been

Fig. 3.84 The heliometer with unequal focal lengths (after Sterken et al. 2004)

replaced by the Bullis House Inn (see Fig. 3.86), a bed and breakfast that was built between 1906 and 1909, which is now in itself a Texas state historic landmark. Note the unfortunate error on the marker, which states that 124 photographic plates were taken. The Belgians actually used micrometric rather than photographic data, based on the superposition of the projected images of the Sun and Venus (Sterken 2009, p. 584). Also, reference is incorrectly made to "Sir" Edmund Halley.

The French also sent an expedition to the United States (Débarbat 2004; *Passage de Vénus du décembre 1882: rapports préliminaires*, 1883). The report on their efforts at Fort Marion in Saint Augustine, Florida, was made by the three members, Colonel François Perrier (1833–1888; Fig. 3.87), Commandant Bassat and Captain Gilbert Etienne Defforges (1852–1915). These three officers in the French Army took separate readings using three different telescopes, an 8-in., a 6-in. and a 3-in., respectively. They achieved fair agreement, especially on the fourth contact.

Captain Defforges reported that 200 photographs were taken of the planet while it was crossing the disk of the Sun. He also was responsible for establishing the longitude at the site, working in conjunction with E.D. Preston of the United States

Fig. 3.85 The historical marker at the 1882 Belgian transit of Venus expedition site to San Antonio, Texas (photograph by the first author)

Coast and Geodetic Survey, who communicated with him telegraphically from Savannah before the transit. They made another series of tests for confirmation, exchanging their locations after the event. Commandant Bassat had already established latitude by means of the observation of a number of familiar stars (ibid.).

The French enjoyed good weather for the entire transit and Colonel Perrier noted, with satisfaction, the encroaching clouds soon afterwards: "Le temps est à la pluie et à la tempête!!!" (*Passage de Vénus du 6 decembre 1882*, 1883, p. 56).

3.2.5.3 The American Expeditions

The Americans organised several northern parties for the 1882 transit of Venus and these were all in their home country: at San Antonio, Texas; Cedar Keys, Florida; Washington D.C.; and Fort Selden at Cerro Roblero, in the New Mexico Territory in the west (Dick et al. 1998).

Fig. 3.86 Bullis House Inn, site of the 1882 Belgian transit of Venus expedition to San Antonio, Texas (photograph by the first author)

Fig. 3.87 Colonel François Perrier (courtesy: en.wikipedia.org)

Fig. 3.88 A view of part of the grounds of Fort Sam Houston, San Antonio (photograph by the first author)

The Americans at the San Antonio site, on the grounds of the current military base of Fort Sam Houston (see Fig. 3.88), were under the leadership of Asaph Hall from the U.S. Naval Observatory. First contact should have occurred at about 7:20 a.m. but was lost to them due to clouds, just as it was to the Belgian astronomers 500 m away. The Americans captured their first photograph of Venus as the sky began to clear at about 10:17 a.m. By the time the transit terminated at about 1:30 p.m. they had obtained 204 photographs (Viewing Venus 1882). Having sent a telegram shortly after the event, Professor Hall subsequently reported in more detail on his successes and frustrations in a letter to Vice-Admiral Stephen Clegg Rowan (1808–1890, see Fig. 3.89), President of the Transit of Venus Commission, written on 8 December (Hall 1882).

Besides the standard make-up members of all American expedition sites, Hall was able to take advantage of some on-site military personnel, not astronomers, cited in his letter to Vice-Admiral Rowan:

> Major Clous and Capt. Livermore made observations of the diameter of Venus with our double-image micrometer. Lt. Shunk assisted Mr. Woodward [assistant astronomer] in managing the heliostat and chronograph and was of very good service. (Hall 1882).

John Walter Clous (1837–1908) was acting Judge Advocate there in San Antonio at the time of the transit. Captain William Roscoe Livermore (1843–1919) was the base's Chief Engineer Officer while William Alexander Shunk (1857–1936) was career military on a temporary assignment in San Antonio (Jacqueline Davis, 2011,

Fig. 3.89 Vice-Admiral Stephen Clegg Rowan (courtesy: en.wikipedia.org)

personal communication). Hall and his group would remain there for several more days to confer with the Belgians and to make other observations to ensure the accuracy of their position and chronometers (Hall 1882). On 3 December 2004 an historical marker (see Figs. 3.90 and 3.91) was dedicated on the grounds of Fort Sam Houston, near the site of the American expedition (Maley 2005). The field where they made their observations is now an area of base officer housing. The marker is placed off a driveway a short distance from the precise location of their work, which is now occupied by a private residence (Jacqueline Davis, 2011, personal communication).

Professor Eastman from the U.S. Naval Observatory was the leader of the observation site at Cedar Keys, Florida (Professor J.R. Eastman dies 1913). As reported in his telegram (see Fig. 3.92a, b) the expedition members at that site succeeded in timing the last three contacts. The circumstances were described in more detail in a letter of the same date to Vice-Admiral Rowan. (Eastman 1882). After the first contact the sky became clear so many photographs were taken. The dry plates would soon be used up so it was decided to take some photographs using the wet plate process as well. Eastman related: "We then alternated groups of dry and wet plates until about five minutes before third contact we had exposed 150 dry plates and 30 wet plates."

Fig. 3.90 Base officer housing at Fort Sam Houston, showing the 1882 historical marker in the center of the scene (photograph by the first author)

Fig. 3.91 A close up of the historical marker for the American transit of Venus expedition at Fort Sam Houston (photograph by the first author)

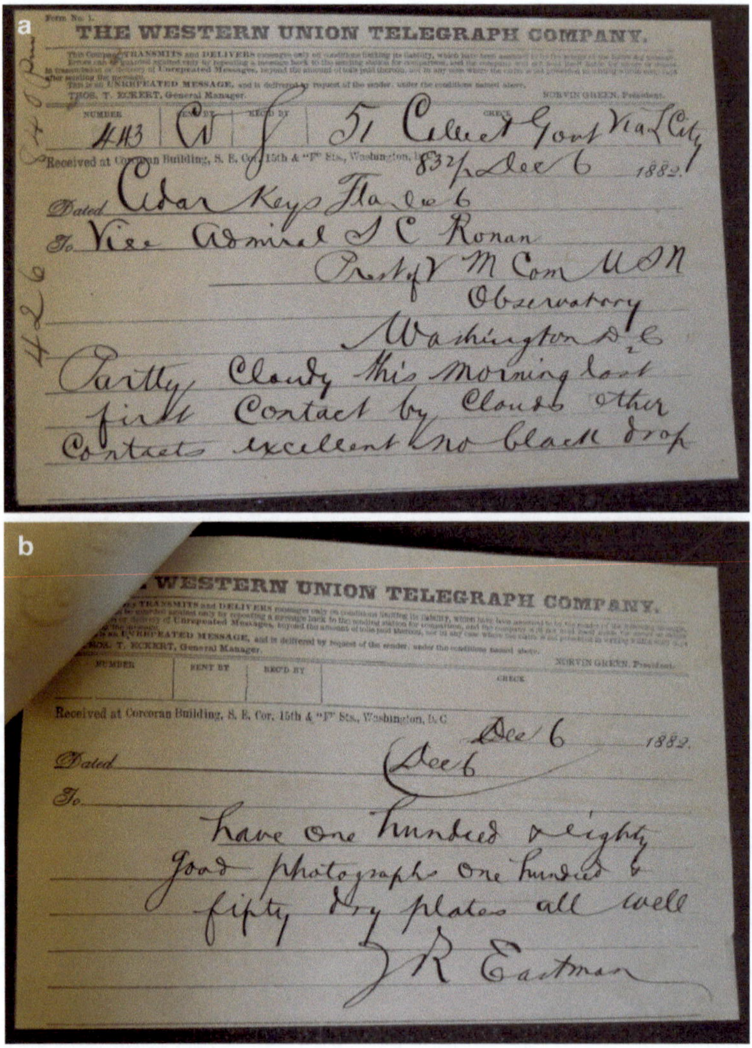

a

THE WESTERN UNION TELEGRAPH COMPANY.

THOS. T. ECKERT, General Manager. NORVIN GREEN, President.

413 | CW | 51 Collect Govt Via L City

Received at Corcoran Building, S. E. Cor. 15th & "F" Sts., Washington, D.C. 832p Dec 6 1882

Dated Cedar Keys Fla Dec 6

To Vice Admiral S C Rowan

Prest of V M Com USN

Observatory

Washington D C

Partly Cloudy this morning lost first contact by clouds other contacts excellent no black drop

b

WESTERN UNION TELEGRAPH COMPANY.

NORVIN GREEN, President.

Received at Corcoran Building, S. E. Cor. 15th & "F" Sts., Washington, D.C. Dec 6 1882

Dated Dec 6

To

have one hundred & eighty good photographs one hundred & fifty dry plates all well

J R Eastman

Fig. 3.92 (**a** and **b**): Eastman's telegram to Vice Admiral S.C. Rowan (courtesy: National Archives, Washington D.C.; photograph by the first author)

Evidence of the degree of cooperation expected from all in this scientific endeavor was in the stated expectation that Eastman would communicate with both the Coast and Geodetic Survey party in Savannah and the French party at St. Augustine to help the French establish their longitude. However, as Eastman had not as yet heard from either party he expressed his willingness to make this determination after the fact. Following the intense labors of the previous days he decided to take 10 days vacation (Eastman 1882).

Fig. 3.93 George Davidson (courtesy: en.wikipedia.org)

The United States Coast and Geodetic Survey's Professor George Davidson (1825–1911; Fig. 3.93) had charge of the American observation site at Fort Selden, New Mexico. On the day of the transit Davidson sent a telegram (Fig. 3.94a, b) to Rowan of the Commission reporting complete success. All four contacts were seen, measurements were taken of the diameter of Venus and 216 "... splendid ..." photographs were taken. On the same date he also sent a short note to Julius Hilgard (1825–1890), Superintendant of the Coast and Geodetic Survey, conveying the same happy information (see Fig. 3.95).

William Harkness was in charge of the efforts in Washington D.C. He was one of only two remaining members of the American Transit of Venus Commission that had begun in 1871 and ended in 1891, anticipating the two transits of the century, the other being Simon Newcomb (Dick 2004). Harkness' party observed all four contacts at their site, the U.S. Naval Observatory.

The Americans never published a determination of the solar parallax based on their 1874 results but this time Harkness would use the 1882 observations to do so:

> Professor Harkness, U.S.N., reports that the photographs of the last transit of Venus (more than 1400 photographs being available) lead to the following value of the solar parallax; $\pi = 8''.842 \pm 0.''0188$. With 3963.296 miles as the equatorial radius of the earth, the resulting mean distance of the sun is 92,455,000 miles, with a probable error of 123,400 miles. (From the *Report of the Supt. U.S. Naval Observatory*, June 30, 1889). (Harkness 1891, p. 46).

In 1894 Harkness would publish an updated figure of $8.''809 \pm 0.''0059$, translating into a value for the astronomical unit of 92,797,000 miles (Dick 2004).

The four Southern Hemisphere sites selected by the Americans were in South Africa, Patagonia, Chile and New Zealand. Simon Newcomb led the expedition to South Africa and established an observing station alongside the Huguenot Seminary

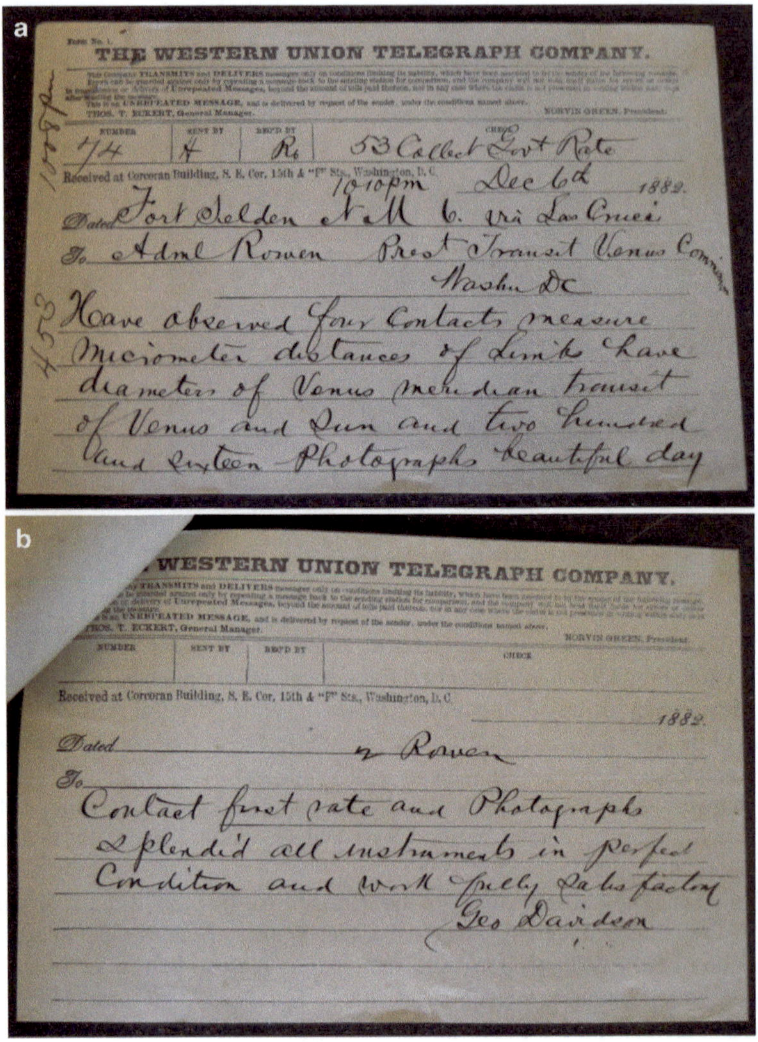

a

THE WESTERN UNION TELEGRAPH COMPANY.

NUMBER	SENT BY	REC'D BY	CHECK
74	H	Ro	53 Collect Govt Rate

Received at Corcoran Building, S. E. Cor. 15th & "F" Sts., Washington, D.C. 10 10 pm Dec 6th 1882.

Dated Fort Selden N M 6. via Las Cruces

To Adml Rowan Prest Transit Venus Commission Washn DC

Have observed four contacts measure micrometer distances of limb have diameters of Venus meridian transit of Venus and Sun and two hundred and sixteen Photographs beautiful day

b

WESTERN UNION TELEGRAPH COMPANY.

Received at Corcoran Building, S. E. Cor. 15th & "F" Sts., Washington, D.C. 1882.

Dated Rowan

To

Contact first rate and Photographs splendid all instruments in perfect condition and work fully satisfactory

Geo Davidson

Fig. 3.94 (**a** and **b**): Davidson's telegram to Vice Admiral S.C. Rowan (courtesy: National Archives, Washington D.C.; photograph by the first author)

for Girls, encouraging local participation. Here only the first and second contacts would be visible. Left behind were their instrument-mounting piers that Newcomb hoped would still be there for the next transit in 2004. Regrettably, they were not (Koorts 2003). Lieutenant Samuel W. Very, U.S.N. was chief astronomer for the site at Santa Cruz, Patagonia, where all four contacts were observed. Professor Lewis Boss (1846–1912; Fig. 3.96; Boss 1918), Director of the Dudley Observatory, led an observing party to a site about four miles from Santiago, Chile, where again all four contacts were observed (Dick 2003). Edwin Smith (1851–1943; Fig. 3.97) from the United States Coast and Geodetic Survey was the chief astronomer of the

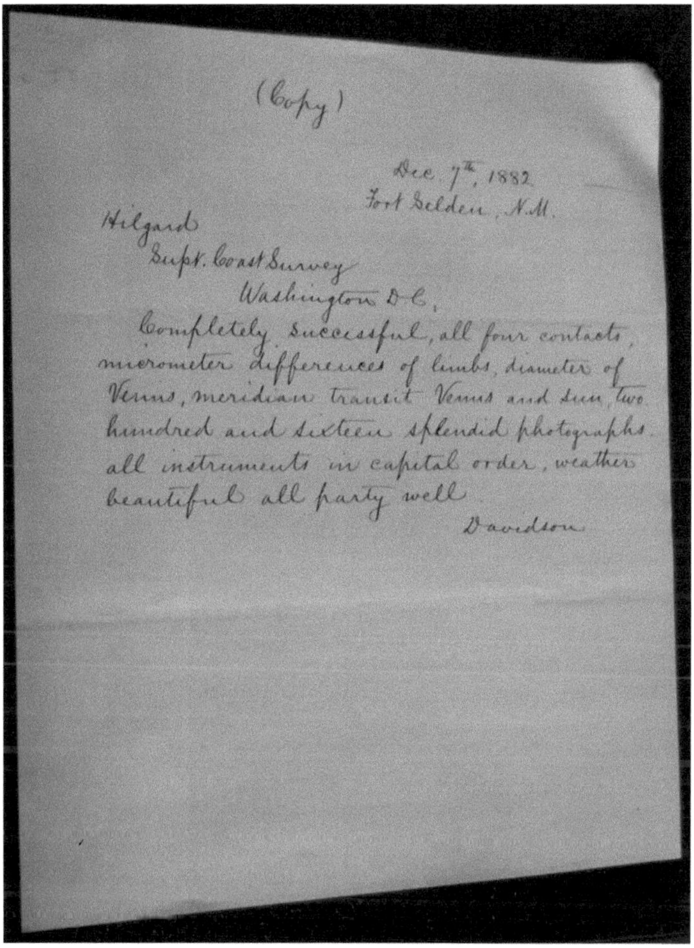
(Copy)

Dec. 7th, 1882
Fort Selden, N.M.

Hilgard
Supt. Coast Survey
Washington D.C.

Completely successful, all four contacts, micrometer differences of limbs, diameter of Venus, meridian transit Venus and sun, two hundred and sixteen splendid photographs. all instruments in capital order, weather beautiful all party well

Davidson

Fig. 3.95 Davidson's letter to Julius Hilgard (courtesy: National Archives, Washington, D.C.; photograph by the first author)

final group at Auckland, New Zealand, where only the two internal contacts were observed (Orchiston 2004b).

Not included among the eight 'official' expeditions sites were those under Charles A. Young at Princeton University and David Todd at the Lick Observatory on Mt. Hamilton, California. Todd, a Professor of Astronomy at Amherst College, was invited to observe the transit from the Lick Observatory by Captain Richard S. Floyd, the President of the Lick Trustees. Todd accepted, and the clear skies enabled him to obtain 147 photographs, 125 of which were deemed measurable. Princeton astronomer Charles A. Young stated that Todd's photos may have been the best obtained (Sheehan and Misch 2004, pp. 125–130). In 2004 Misch and Sheehan (2004b) found 142 of the original negatives in Lick's plate archive and reconstructed

Fig. 3.96 Lewis Boss (after *Popular Science Monthly* 1912)

Fig. 3.97 Edwin Smith (courtesy: www.gaithersburgmd.gov/observatory/)

Fig. 3.98 Edward Charles Pickering (courtesy: en.wikipedia.org)

a movie of the event (Misch and Sheehan 2004a). Young and Todd followed the instructions of the U.S. Transit of Venus Commission and their data were included in the official report. Ultimately the southern stations contributed 587 measurable plates and the northern stations (including Princeton and Lick) 793 (Dick 2003).

Unlike most of the European transit teams, the Americans had decided to stay with the photographic method. Most parties used the improved dry colloidal emulsion plates. The Americans were generally fortunate with regard to weather conditions. Several parties from both hemispheres saw all four contacts. Seventeen hundred photographs were taken, and the majority could be measured (Dick et al. 1998).

In America there was also cooperation from many established observatories across the country, as well as from private individuals. Instructions and time signals were available to anyone who would contribute to the effort (ibid.).

Due to the high possibility of inclement weather, the Harvard College Observatory had not been selected as a primary site by the Transit Commission. However, the Observatory's Director, Edward Charles Pickering (1846–1919; Fig. 3.98), had some success there and reported his results to the American Academy of Arts and Sciences. Several of his observers noted all four contacts (Pickering 1882–1883). Maria Mitchell and her students observed from the grounds of Vassar College as she had been denied permission to participate in any of the Government expeditions. Her group used a small version of the official photoheliostat, as well as an equatorial similar to those used by the U.S. expeditions, and succeeded in photographing the event (Sheehan and Westfall 2004).

Ninety-three reports from individuals, amateurs and professionals, were received by the Transit of Venus Commission (Dick et al. 1998), and these are now in the National Archives in Washington, D.C.

Again reduction of data would be a time-consuming undertaking. The ‘black drop’ effect often was reported (Howlett 1883) but not always (Horner 1883). Frequently reported was the presence of a Venusian atmosphere (Prince 1883), but again not always (Howlett 1883). These would continue to complicate accurate measurements and interpretation of the results. Many explanations have been proposed for the black drop effect. Schaefer (2001) would conclude that the primary cause is a combination of smearing due to atmospheric seeing and the usual Airy diffraction pattern of a telescope, factors dooming the accurate results desired from these expeditions. More recent studies have eliminated any potential contribution from terrestrial atmospheric turbulence thanks to observations of the transits of Mercury in 1999 and Venus in 2004 made with NASA’s Transition Region and Coronal Explorer (TRACE). It has been established that the effect can be explained as a combination of a point spread function due to the telescope and solar-limb darkening (Pasachoff et al. 2005, pp. 242–243). Simon Newcomb did not have much faith in the endeavors of the transit of Venus expeditions. In his 1895 monograph, *The Elements of the Four Inner Planets and the Fundamental Constants of Astronomy*, he ranked the value of results obtained by numerous other methods above those obtained using the transits of Venus (Newcomb 1895, p. 166). However, in the previous year William Harkness from the Naval Observatory addressed the American Association for the Advancement of Science. He stated his final best estimate for the solar parallax to be $8''.809 \pm 0''.0059$ corresponding to a value for the astronomical unit of $92{,}797{,}000 \pm 59{,}700$ miles (Dick et al. 1998). For comparison, the currently-accepted figure is $8''.794148$.

Chapter 4
Popular Astronomy and the Solar Eclipses of 1868, 1869 and 1878

4.1 The Solar Eclipses of 1868, 1869 and 1878

Though numerous solar eclipses had been studied and documented in previous centuries, it was not until the middle of the nineteenth century that the tools of photography and spectroscopy became available, enabling scientists to make more of these rare and brief events. In 1860 documentation of an eclipse was made by photography. During the eclipse of 18 August 1868 spectroscopy was first used to determine more about the nature of the solar corona, during an eclipse that attained totality of almost 7 min duration at some sites.

Astronomers from the United States would have the opportunity to apply these new scientific advances to the study of eclipses on their 'own turf' during the total solar eclipse of 7 August 1869. In this case the line of totality would pass right through the heartland of their country. This was very much an American event as the Europeans waited to see what these relative newcomers to such scientific endeavors would contribute to the understanding of the Sun and its corona.

The next solar eclipse to occur on American soil would take place on 29 July 1878, and the line of totality would cross the western regions of the country and its territories.

A significant amount of information on the observation, science and significance of these events was published in the popular periodicals and newspapers of this period. This is apparent even in the cases of periodicals with a non-scientific emphasis. Various examples follow.

4.2 A Sampling of Popular Periodicals

There were a number of publications reflecting a variety of interests during this period that were designed to appeal to the non-scientific public. Some were intended to appeal to particular segments of the public, on the basis of such factors as

S. Cottam, W. Orchiston, *Eclipses, Transits, and Comets of the Nineteenth Century*, Astrophysics and Space Science Library 406,
DOI 10.1007/978-3-319-08341-4_4

geographical region, profession, sex and age. On occasion these periodicals would have articles for those interested in scientific topics, including articles on eclipses which provided scientific information, periodic updates and relevant human-interest stories. The brief descriptions of these journals that are provided hereafter will not be repeated in the next chapter in cases where these same journals provide articles on the transits of Venus.

Solar eclipses are events that occurred frequently enough that articles in the popular literature would often provide information in an overlapping manner. Thus, the eclipses of 1868, 1869 and 1878 would often be mentioned within articles on other eclipses. The articles discussed here for the most part cover the date range 1868–1878 (inclusive), and are in periodicals that have been grouped into three broad thematic categories: general, religious and scientific. Within each category the periodicals are presented in alphabetical order by title, and within each title chronologically by publication date.

4.2.1 General Periodicals

The *Albion* was a weekly magazine published in New York from 1822 to 1875. It drew its contents heavily from British periodicals (Mott 1930, pp. 130–131). On 18 April 1868 a short item appeared titled "The Eclipse of the Sun". This article dealt with the eclipse of 1868 during which totality would last almost 7 min at some venues. The Royal Society would have an expedition present at their site in India and hoped ultimately to be able to answer some "… highly important questions …" (p. 192). On 8 August 1868 a lengthier article, "The Solar Eclipse of Aug. 17, 1868", was printed. This time the interested reader could learn the reason for the long duration of the eclipse. He or she would learn that photography and the newer science of spectroscopy would be used, and that there was also hope that a planet between the orbit of Mercury and the Sun might be found to account for perturbations apparent in Mercury's orbit. In October of that year some of the early results obtained in India were published. It was determined by using spectroscopy that the 'flames,' i.e., prominences, were self-luminous and of a gaseous nature, and by polariscopy that the light of the corona was, at least in part, reflected light from the Sun (Late Observations in India 1868).

On 10 July 1869 the *Albion* mentioned the American eclipse to take place that same year. Several sites and scientific expedition party constituents were enumerated (Home news 1869). On 14 August 1869 some of the results received from these sites were printed. Satisfactory photographs were obtained by Joseph Winlock's group at Shelbyville, Kentucky. No reports up to that point in time indicated the presence of an intra-Mercurial planet (The solar eclipse 1869). On 9 October 1869 a brief summation of the results of Professor Hall, commander of the expedition at the far west site of the Behring Strait, was printed. He had somewhat limited success due to the presence of clouds (Home news 1869).

On 17 December 1870 the *Albion* printed an article anticipating the next solar eclipse, which would take place on 22 December of that year in the Mediterranean. It was a reprint of a letter by the British amateur astronomer John Russell Hind (1823–1895; Fig. 4.1) to the Editor of the *London Times* with specifics of location and

Fig. 4.1 J.R. Hind in the 1860s (courtesy: en.wikipedia.org)

contact times. On 23 March 1872, under the feature "Current Notes", there was a paragraph on an expedition that observed the eclipse of 12 December 1871 from Cape Sidmouth, Australia. Due to heavy cloud cover the corona was only faintly seen.

The *American Literary Gazette and Publishers' Circular*, named by George Childs of Philadelphia in 1855, was intended as a means to advertise new publications to booksellers (Mott 1938a, pp. 158–159). On 15 May 1871 it published notice of two new works by the British astronomer and popularizer Richard A. Proctor, *The Sun* and *Light Science for Leisure Hours*. The former was described as including results of spectrum analysis applied to the solar corona and prominences during a solar eclipse (Advertisement 9 – No title 1971).

Appletons' Journal of Literature, Science and Art was a periodical published by D. Appleton & Company from 1869 to 1876, containing short articles reflecting current interests (Mott 1938b, pp. 417–421). On 4 September 1869 the article "The Lessons of the Eclipse" was published. This piece waxed philosophical about the recent understanding that man had about the world around him. In particular it had only been within the last couple of centuries that the phenomenon of an eclipse was understood. On 16 October 1869 the entire Presidential Address of Cambridge University's Professor George Stokes (1819–1903; Fig. 4.2) read before the British Scientific Association was reprinted. Most of his oration focused on astronomical topics and most of these dealt with eclipses and the new science of astrophysics. Stokes recounted the proof of the solar nature of protuberances (i.e., prominences) from the photographs of Warren De la Rue and the discovery by both Jules Janssen and J. Norman Lockyer that these prominences, with an appropriately-modified

Fig. 4.2 George Stokes in the 1860s (courtesy: en.wikipedia.org)

spectroscope, could be studied at times other than during an eclipse (Address of President Stokes … 1869).

"Scientific Notes" was a regular feature of *Appleton's Journal*. On 5 March 1870 this feature included a paragraph quoting American educator Professor Joel Dorman Steele (1836–1886) in response to a disquieting newspaper story about the relationship of solar prominences to the earthly phenomena of earthquakes and volcanic eruptions. On 8 July 1871 there were two items regarding solar science within "Scientific Notes". One discussed speculation regarding the source of solar energy, in particular Richard Proctor's theory that it is in part due to the downfall of meteors into the Sun. The other item discussed Janssen's investigations of the great eclipse of 1868 in India leading to his announcements of a hydrogen atmosphere around the Sun and the discovery of a means of observing solar prominences at times other than during total solar eclipses.

On 27 April 1872 *Appleton's Journal* printed the illustrated four-page article "The Language of Light". It dealt initially with the discovery by Gustav Kirchhoff in 1859 that the dark lines of the solar spectrum, observed by the British scientist William Hyde Wollaston (1766–1828) and then carefully mapped by Joseph Fraunhofer, could be associated with specific elements. The article ended with a discussion of the work of Janssen and Lockyer subsequent to the 1868 eclipse and their continuing study of solar prominences.

The *Appleton Journal's* feature "Scientific Notes" on 13 July 1872 included a paragraph on the efforts of Janssen and others during their observations of the eclipse of 1870 in the Mediterranean, 1 year after the American eclipse of 1869.

Fig. 4.3 Cyrus Field circa 1870 (courtesy: en.wikipedia.org)

Janssen was on one of the 64 balloons that left from Paris carrying passengers to participate in this event. These balloonists were escaping the siege of Paris by the German army during the Franco-Prussian War (MacPherson 1908, p. 73). Five of the balloons were captured by Prussians, and two were lost at sea, but Janssen's escape from Paris was uneventful. Another paragraph dealt with some recent advances in the understanding of sunspots. A periodicity had been recognized and documented by numerous astronomers and physicists. A consequential periodicity in light and heat on the Earth, inversely related to sunspots, was claimed to account for the success of vineyards.

In a paragraph in "Scientific Notes" of 8 March 1873 *Appleton's Journal* relates the content of a tongue-in-cheek paper by Mr J.R. Hind to the Royal Astronomical Society stating that in the year 2151 parts of England would experience a total solar eclipse and it would not be necessary for scientists to incur the expenses of a trip to the South Seas. It is interesting that Hind chose to mention the 2151 total eclipse, and skipped over those of 1927 and 1999 that would also be visible in parts of England. Another paragraph in the same feature, not specifically relating to eclipses but significant in all matters regarding international cooperation in astronomy, relays the communication between Professor Joseph Henry from the Smithsonian Institute and the American businessman Cyrus West Field (1819–1892; Fig. 4.3) that announcements of important astronomical observations can now be made available quickly and free of charge via the Atlantic Cable.

Included in another regular feature "Science and Invention" on 18 July 1874 *Appleton's Journal* printed several paragraphs on the experiences of Mr Edward James Stone (1831–1897) observing an eclipse at the Cape of Good Hope. His main

Fig. 4.4 John Humphrey Noyes in about 1850 (courtesy: en.wikipedia.org)

instrument was, as would be expected, the spectroscope with which he confirmed Charles Young's observation that just as the Moon covered the Sun the dark Fraunhofer lines flashed brightly. This 'reversing layer' had been described by Mabel Todd (1856–1932) as one of the most notable outcomes of the eclipse stations during the 1870 eclipse. Young described what he saw at this eclipse in Spain: "… the moment the Sun is hidden, through the whole length of the spectrum, in the red, the green, the violet, the bright lights flash out by hundreds and thousands, almost startlingly …" (Todd 1894, p. 135). Stone also confirmed that the solar corona "… is composed of incandescent gas, shining partly by its own glow and partly by reflected sunlight …" (Todd 1894, pp. 94–95). A drawing was made of this corona by a lady several hundred miles away demonstrating agreement with Stone's impressions and confirming again that this phenomenon was not of terrestrial origin.

Beadle's Monthly, a Magazine of Today was a short-lived journal published for 18 months during the years 1866 and 1867. It attempted to imitate the general content of the popular *Harper's Magazine* (Mott 1938a, pp. 466–468). In May 1867 it printed an article by David Trowbridge titled "Words from the People" that related, in a non-technical manner, much about the contemporary knowledge regarding the constitution and nature of sunspots, alluding to the eclipses of 1842, 1851 and 1868.

The Oneida *Circular*, founded by the charismatic utopian socialist John Humphrey Noyes (1811–1886; Fig. 4.4), was first printed on 6 November 1851. Originally a weekly, it would continue after a year as a bi-weekly publication until 1876. Its chief topics were doctrines of liberalism, communism and free love

(Mott 1938a, pp. 78 and 207). On 4 May 1868 there was a paragraph under the feature "Scientific" informing the reader of the upcoming eclipse of 18 August of that year when spectroscopic studies would be performed on the red flames (i.e. prominences) ejected from the Sun. On 2 August 1869 there was a short item stating the times, provided by Professor C.H.F. Peters of Hamilton College, for the start and end of the upcoming solar eclipse on the seventh at the Observatory in Clinton. The eclipse would be partial at that site (Article 1 – No title 1869).

The Eclectic Magazine of Foreign Literature was a monthly periodical published from 1844 to 1898 to foster "… a taste in foreign literature …" (Mott 1938b, p. 259). During its existence it included many articles on solar eclipses. A frequent feature of the publication was "Science". In November 1866 this feature described the unsuccessful attempts to photograph solar prominences during the total eclipse that took place in the Southern Pacific on 15 April of that year. It is interesting that in Fred Espenak's NASA web-site this eclipse is described as 'partial'. Although the photography was unsuccessful a visual description was shared. This information was obtained from a communication by Father Angelo Secchi to *The Bulletin International.*

In April 1868 the readers of the *Eclectic Magazine's* "Science" were informed of the upcoming eclipse of that year, its path and duration, as well as some possible plans for expeditions by the international community. This article was taken from *Galignani's Messenger* of 15 February. Again, in May 1868 "Science" included an item of the highly-anticipated eclipse of that year. Two sites had to date been established along the path of totality in India, one to be overseen by Major Tennant and the other by Lieutenant Herschel. It was hoped there would be other scientific parties present to cover the entire length of the path, and ultimately to settle the question of the "… real nature of the red protuberances seen around the sun …" (p. 647).

In September 1868 the *Eclectic Magazine* … reprinted an article from *Popular Science Review*, "The Great Eclipse of August 17, 1868". Here the track and duration of the eclipse in specific regions was described in detail. Tennant and Herschel led the two English expeditions to two sites in India. Particulars of the equipment provided were enumerated. Both parties would be involved in spectroscopic research. Herschel had received specific instructions "… to confine his attention to observations of the spectra of the corona and red prominences …" (p. 1154). The French had sent Jansen (*sic*) to head an expedition, and the Pope had sent Father Secchi. Pogson, the Superintendent of the Madras Observatory, was already in place to head a third group for the English (ibid., pp. 1152–1155).

In the October 1868 issue of the *Eclectic Magazine* … was a 10-plus page article "Great Solar Eclipses" taken from *The Cornhill Magazine.* Subsequent to the interest stirred up by the great eclipse of that year, this article provided information on historical eclipses, beginning with that of 585 BC which was said to have brought about a cessation of hostilities between the Lydians and the Medes, who had been at war for years. At the eclipse of 2 May 1733 the 'red prominences' were described for the first time, which in the year of this article would be studied with a spectroscope. The first eclipse to become the object of formal expeditions was visible in parts of Europe and occurred on 8 July 1842. Participants included observers from

England, Russia, France and Germany, all achieving great success. The next eclipse to garner such activity occurred on 28 July 1851 where "… the red prominences were seen with remarkable distinctness …" (p. 1187). Drawings were made at multiple sites and compared. On 18 July 1860 an eclipse occurred, best observed by the Europeans in Spain. The prominences again were the focus of many observations. Results of the spectroscopic studies at the most recent eclipse of 1868 were greatly anticipated (ibid., pp. 1181–1191).

In November 1868 the *Eclectic Magazine* … published the article "Lunar Assistance", from the journal *All the Year Round,* which dealt with generalities of lunar and solar eclipses. In the same issue was the feature "Science" with a summary report of "The Great Solar Eclipse". Telegrams had been received from Major Tennant of the Royal Astronomical Society and Jules Janssen of the French expedition. Tennant had successfully photographed the event, while Janssen had studied the prominences spectroscopically and determined these were gaseous in nature.

In January 1869 *Eclectic Magazine* … published "The Hindu View of the Late Eclipse", an article of several pages describing some of the superstitions and ceremonies among the Hindu locals relating to their experiences during the eclipse in India in 1868. According to the author there were millions who believed the Sun was being swallowed up by the dragon *Rahu*.

In March 1869 the *Eclectic Magazine* … reprinted an article by Richard Proctor taken from the *Temple Bar*, "Strange Solar Discoveries", where some basics of solar physics were described for the public. Proctor discussed Kirchhoff's discovery regarding the solar spectrum, the theories of sunspots and Lockyer's and Janssen's almost simultaneous discoveries of the possibility of studying solar prominences at times other than during total solar eclipses.

In April 1869, in its regular feature "Science", the *Eclectic Magazine* … included a short piece, "The Sun", taken from "The Great Solar Eclipse of August 18, 1868", in the *Quarterly Journal of Science*. Here some suggestions were printed regarding the photography to be employed at the upcoming eclipse of 1869 to take place in America. In May of 1869 an article by the Savilian Professor of Astronomy at Oxford University Charles Pritchard (1808–1893; Fig. 4.5) taken from *Good Words* and entitled "Perceiving Without Seeing – A Romance in Astronomy" was printed. This provided an historical perspective on some of the methods used and results obtained in astronomical studies. Pritchard described the use of photography and spectroscopy in the study of solar prominences, as well as how spectroscopy might be used to determine the movement of binary stars.

In September 1869 there was another short article in the *Eclectic Magazine* … on the 1868 eclipse, "Borneo Observations of the Eclipse of August, 1868". From Borneo, the most eastern expedition site, J. Pope Hennessy observed the event and the article quoted his description of the three prominences that were seen. In February 1870 "The Sun's Crown", from *Spectator,* was reprinted. Whereas the focus of expeditions in 1868 was mainly the solar prominences, during the eclipse of 1869, in America, this had shifted to the corona itself.

In November 1870 the *Eclectic Magazine* … reprinted "What We Hope to Learn from the Approaching Solar Eclipse", taken from *St. Paul's*. Background information

Fig. 4.5 Charles Pritchard (after Pritchard 1897: Frontispiece)

was provided on the application of spectroscopy during the eclipse of 1868. Successes had been achieved in answering questions about the prominences. This time the corona would "… undoubtedly occupy the chief attention of observers during the approaching total eclipse of the sun …", that of December 1870 (p. 554). The source of this phenomenon was still in question. Was it solar, lunar or terrestrial? Was its appearance due to meteoric systems? It was hoped that the effective use of photography and spectroscopy would lead to some answers (ibid., pp. 551–558).

In January 1871 the *Eclectic Magazine* … reprinted "The Sun's Corona", from the *Cornhill Magazine*. This gave an historical perspective on the understanding of the corona beginning with the description by the Irish physician Dr Thomas Wyberd of the total eclipse of 29 March 1652: "… the sun was totally eclipsed, and the appearance was due to a corona of light round the moon, arising from some unknown cause." (p. 53). The observations of French astronomers in 1724 and Swedish in 1733 led them to believe the corona to be a solar appendage. Herve Fayé (1814–1902; Fig. 4.6; from Paris Observatory) and Britain's Norman Lockyer would later support the theory that the corona was actually a terrestrial phenomenon. It was to be hoped that the eclipse of December 1871 would provide an opportunity for some answers (ibid., pp. 52–59).

In April 1871 the *Eclectic Magazine* … reprinted another article from the *Cornhill Magazine*, "The Late Solar Eclipse", referring to the eclipse of December 1870. Here, again, the reader would get some historical background, beginning with

Fig. 4.6 Hervé Faye (courtesy: en.wikipedia.org)

the eclipse of 1842, which was the first to inspire the initiation of formal expeditions. At this time the presence of red-colored prominences was just suspected, but these were to be photographed by De la Rue and Secchi in 1860, and in 1868 their nature would be ascertained by spectroscopy. In 1868 Janssen and Lockyer also discovered that these phenomena could be studied outside of total solar eclipses. After that, the focus of spectroscopic study shifted to the corona. In 1868 Tennant had determined the solar spectrum to be continuous, but in 1869 there were numerous observations of lines, though there was no agreement on the number seen. Professor Young's observation of coronal spectral lines, corresponding to those seen in an aurora, led him to conclude that the corona was a 'perpetual solar aurora'. Observations of the neophyte American astronomers were not readily accepted and confirmation would be an interest at the Mediterranean eclipse of 1870. Lockyer, who had been dubious of the American results, concluded in Sicily that the observations in 1869 had indeed been confirmed. The corona demonstrated a mixed spectrum of bright lines on a continuous background. There was no agreement on the source of these bright lines.

In July 1871 the *Eclectic Magazine* ... reprinted Richard Proctor's article from *St. Paul's*, "The Sun's Atmosphere at Length Discovered". After some introductory background, Proctor focused on the spectroscopic work of the American Professor Young. The latter's observations of a momentary spectrum of bright lines at the edge of the Sun's disk just before totality indicated the presence of a shallow atmosphere. The *Eclectic* printed several articles over the next few years by the popularizer Richard A. Proctor, which helped to sustain an interest among the American public in the subject of eclipses until the next one to take place on their own soil, in 1878 (Proctor 1871h). In August 1872 there was his article "The Late Solar Eclipse", dealing with the historical descriptions of layers of the corona up to the time of the eclipse of 1871. The bright-line spectrum of a very shallow atmosphere seen by Professor Young in 1870 was confirmed by several observers in 1871. Proctor remarked that above that atmosphere is a red envelope, the chromosphere at about 5,000 miles in depth, mainly composed of hydrogen. Above that layer is the first of two layers of the corona, the inner corona, a fairly uniform ring in depth around the Moon, felt to be an actual appendage of the Sun. The light at this level seemed to resemble somewhat that of the *aurora borealis*. Beyond that is the outer radiated corona, its nature as yet controversial. Janssen, from an excellent site in India, successfully detected in the corona the chief solar dark lines as well as several bright lines. Janssen felt the dark line spectrum was due to an absorptive effect within this layer of the corona. Proctor thought it was more likely an effect of reflected sunlight. There were still questions to be answered (Proctor 1872a).

In March 1874 the editor of the *Eclectic Magazine* ... published an item "Richard Anthony Proctor" about the popularizer himself. Following a short biography there was an enumeration of his printed works, starting with "Double Stars", an article in the December 1863 issue of the *Cornhill Magazine*. Proctor wrote books and frequent items for periodicals and newspapers on scientific subjects for the non-scientific public. He had a talent for constructing astronomical maps. Though he was a popularizer of science he also did original research and between 1868 and 1873 contributed many papers to the Royal Astronomical Society that dealt with the solar eclipses of this period and the study of the corona.

In December 1874 the *Eclectic Magazine* ... printed Proctor's article "The Sun a Bubble", taken from the *Cornhill Magazine*. Here he discussed theories, past and present, of the nature of the Sun, beginning with that of Sir William Herschel (1738–1822; Fig. 4.7), who described the Sun as a solid, possibly habitable, globe. Sir William and his son, Sir John Herschel (1792–1871; Fig. 4.8) went so far as to suggest the possibility of life. Evidence during solar eclipses of dynamic solar prominences would seem to negate this theory. Professor Young of America, studying the solar limb with a spectroscope, concluded the Sun to be bubble-like, with the difference that it tolerated blasts of heated gas from within, which accounted for the appearance of solar spots.

In December 1876 a Proctor article from the *Popular Science Review* titled "Astronomy in America" was reprinted by the *Eclectic Magazine* ... Here the author would describe, in most favorable terms, the progress of this young country in the science of astronomy. He described the building of the early observatories and the support of important people, such as President John Quincy Adams. He told of the success of O.M. Mitchel in obtaining sufficient donations from the non-scientific public for the establishment of an observatory in the city of Cincinnati. He described some successes in astronomical research at the American total solar eclipse of 1869. The last paragraph of this article is an interesting shift to his thoughts on the negative influence of formal religion against scientific advancement,

Fig. 4.7 Sir William Herschel (courtesy: en.wikipedia.org)

Fig. 4.8 Sir John Herschel (courtesy: en.wikipedia.org)

citing Puritans and Roman Catholics, who would denounce any discovery not in line with their orthodox views.

Yet another article by Proctor would be reprinted by the *Eclectic Magazine* … in May 1877, this time from the *Belgravia Magazine*, entitled "Suns in Flames". Spectroscopy was described as it was used in the study of novae, the search for the planet Vulcan and the study of the solar prominences and the corona.

In February 1879 *The Eclectic Magazine* … printed the anonymous article "The Sun's Long Streamers", from the *Cornhill Magazine*, dealing with the latest theories on the solar corona based on the results noted during the American solar eclipse of 1878. The theory, suggested as likely here, was that many comet and meteoric streams travel close to the Sun, and the outward streamers may be the result of the Sun's repellant action.

Every Saturday was a weekly publication issued from 1866 to 1874. Printed in Boston, it drew early in its history from English periodicals, and occasionally from French and German ones. In 1870 it shifted from this eclectic nature and began to write about politics and arts closer to home. At this time it also began to include some excellent wood-engraving illustrations although these were abandoned the next year (Mott 1938b, pp. 357–360). On 1 August 1868 there was an article titled "The Coming Eclipse". Here the significant historical eclipses of 1706, 1715, 1842 and 1850 were briefly described leading up to the great eclipse of 1868 where spectroscopy would be used to answer questions regarding the nature of the prominences and the corona, and polariscopy would be used to determine whether these instances of solar light are original or reflected. Several expeditions to accomplish these tasks were mentioned; all were European.

On 4 February 1871 and the following week on 11 February 1871, *Every Saturday* included some beautifully-illustrated articles about the eclipse of 1870. The first of these is "The Eclipse of the Sun", which describes a partial solar eclipse in London. The author reflects on his personal experiences of the total solar eclipse of 1860, which he observed in Northern Spain, for comparison. Included are illustrations made from photographs at several stages of the partial eclipse seen on 22 December 1870. The second article, "Illustrations", is a collection of nine articles with accompanying engravings. Included is an illustration relevant to one of the articles depicting the same partial solar eclipse described in the article of 4 February but seen as total from Gibraltar (see Fig. 4.9).

On 2 March 1872 *Every Saturday* published the article "Important Discovery during the Late Eclipse", referring to the event of December 1871. By a method not described, the Italian spectroscopist and Director of the Observatorio del Campidoglio Professor Lorenzo Respighi (1824–1889; Fig. 4.10) obtained, not just Professor Young's bright green line of the inner coronal ring, but an actual image of the source of this light, resembling the inner ring.

Finally on 24 October 1874 *Every Saturday* reprinted the previously-mentioned article from the *Eclectic Magazine of Foreign Literature* and *Cornhill Magazine*, written by Richard A. Proctor, "The Sun a Bubble".

Flag of Our Union was a general interest publication out of Boston published from 1846 to 1870. Among its contributors were Edgar Allan Poe and Horatio Alger

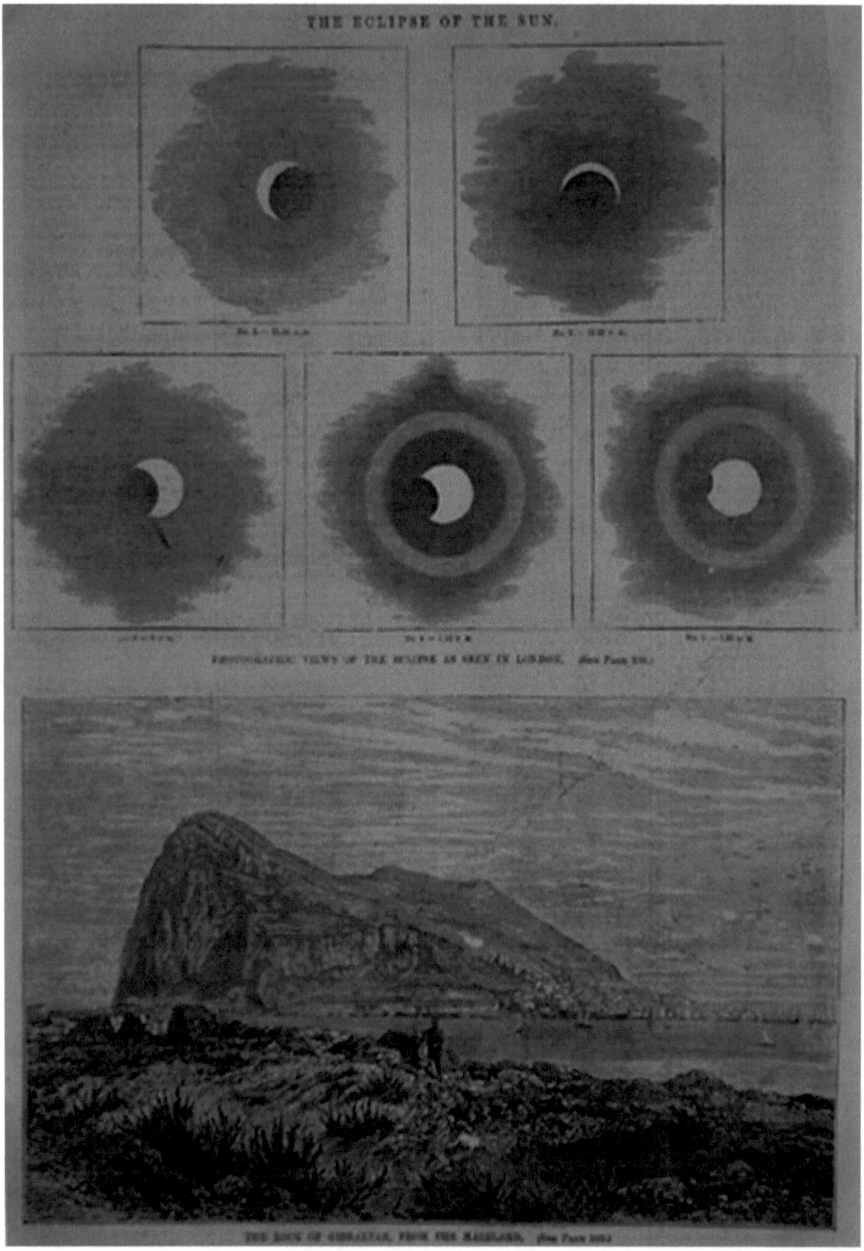

Fig. 4.9 The total solar eclipse of 1870 as seen at the Rock of Gibraltar (after *Every Saturday*, 4 February 1871)

Fig. 4.10 Lorenzo Respighi (courtesy: en.wikipedia.org)

(Mott 1938a, p. 35). On 14 August 1869 it printed "The Great Solar Eclipse", a personal account by an anonymous author who witnessed the solar eclipse of 1868 at sea on a voyage from Aden to Bombay.

The Galaxy. A Magazine of Entertaining Reading was an illustrated monthly, out of New York, published from 1866 to 1878. Among its contributors was Mark Twain (Mott 1938b, pp. 361–381). Subsequent to the successes achieved during the eclipse of 1869, there was an increased American presence in such scientific efforts on an international level. In January 1871 a short piece, "The Astronomical Expedition", was printed, informing the public of preparations for American involvement in the observations of the Mediterranean eclipse of 1870. Congress had allocated $29,000 for this purpose. Professor Peirce was to direct these efforts. Ironically it was only the Americans and the Italians who initially had demonstrated any interest in studying this eclipse. The British scientist and editor of *Nature*, J. Norman Lockyer, accepted Peirce's invitation to join his expedition to Sicily.

In August 1871 *The Galaxy* printed a 16-page article "The Mediterranean Solar Eclipse" describing the progress and ultimate results of the expeditions of 1870. Among the parties participating was Joseph Winlock's group of astronomers from Cambridge, Massachusetts, that had observed the total eclipse in Shelbyville, Kentucky, in 1869. They made their observations in Spain. Peirce's journey by boat and train to his party's site at a Benedictine monastery was described. There they met an English group whose yacht was wrecked outside the harbor at Catania. Fortunately all personnel and the scientific equipment of this group were saved. The interest in this eclipse was mostly centered on the solar corona, as other phenomena, in particular the solar prominences, were felt to be much understood by then. In 1868 the spectroscope had been put to use successfully on the prominences, less so on the corona. There was hope that studies of the eclipse of 1870 would lead to answers on whether this apparent envelope was entirely solar (Professor Young's supposition), was due to our terrestrial atmosphere (Lockyer's view) or was an effect of both (Professor Winlock's explanation). The weather was poor for scientific purposes at many of the observation sites, and some conclusions drawn were conflicting. Immediately following the eclipse Lockyer telegraphed London with the conclusion that there was "... surrounding the sun a mass of self-luminous gaseous matter, whose spectrum is characterized by the green 1474 line." (p. 190). It was felt that the lingering problem remained of determining the extent and composition of the corona. This article included four sketches of the shape of the corona as seen by four different individuals. It was apparent that Americans were now entrenched in the community of astronomical explorers (ibid., pp. 179–194).

In October 1871 *The Galaxy* printed an article "The Approaching Solar Eclipse", informing the American public of the next solar eclipse of interest, which would occur in 1871 and would be visible in India, Ceylon and Australia. Totality on the central line would last from about 2¼ min to about 4¼ min depending on the location. Spectroscopic and photographic studies would focus on the solar corona.

In July 1874 *The Galaxy* printed a short item on the encyclopedia of Ma-twan-lin, published in 1329 with several chapters on astronomy, including a list of solar eclipses that had been observed in China over more than 3,000 years (Chinese astronomy 1874).

Harper's Bazaar (spelled "Bazar" early in its history) was "... a ladies' *Harper's Weekly,* with the same type of English serials, double-page pictures, miscellany and humor, and with fashions and patterns taking the place of politics and public affairs." (Mott 1938b, pp. 388–390). "Sayings and Doings" was a regular feature of miscellany that frequently included items on scientific subjects. On 18 September 1869 there was a bit of speculation taken from *Scientific American* considering the relationship of solar eclipses to seasonal temperatures here on the Earth. On 25 February 1871 there was a paragraph in that same feature on the eclipse of 1870. Americans bore about half of the cost for the expeditions, and efforts were in large part frustrated by poor weather. On 3 August 1878 there was a report of Congressional appropriations for the eclipse of 29 July 1878. Favorable locations were listed and it was noted that distinguished European astronomers were expected to attend (Sayings and doings 1878a). On 10 August 1878 there was mention of Professor Henry Draper's private expedition to observe the eclipse at Rawlins, Wyoming Territory, and of a party of English astronomers who would observe from Pueblo, Colorado (Sayings and doings 1878b). On 31 August 1878 there was a report that two observers stated that they had seen the elusive intra-Mercurial planet (Sayings and doings 1878c).

From 1865 to 1870 Charles Scribner published a literary monthly *Hours at Home*. The editor was a Presbyterian clergyman and initially the journal had a religious flavor. This became less evident over time. *Hours ...* was discontinued in 1870 when Scribner focused on his new journal *Scribner's Monthly* (Mott 1938b, pp. 32–33). In October 1869 *Hours ...* printed a six-page article "The Total Eclipse of 1869". Here was given a little history of previous eclipses, followed by what was expected and accomplished during the American eclipse of 1869. The reader was told that the line of totality passed from Alaska eastward to North Carolina and was about 140 miles wide. It fell on no fixed observatory but temporary observatories were established at many sites along the route. The anonymous writer described his own experiences at the site of Burlington, Iowa. The sky was mostly clear. Venus, Mercury, Saturn and Arcturus were observed just before the corona appeared and there were two rosy prominences evident at totality. Photography and spectroscopy were employed to make records for later study. No intra-Mercurial planet was observed.

Lippincott's Magazine of Literature, Science and Education appeared from 1868 to 1871. Mott describes it as "… one of the best printed of American magazines …" (Mott 1938b, p. 396). After the first year it included woodcut illustrations. It published items from such distinguished contributors as Henry James, Emma Lazarus and Anthony Trollope. It introduced the American audience to Arthur Conan Doyle's Sherlock Holmes when it published "The Sign of the Four." Science was given significant attention (ibid., pp. 396–401). In May 1869 *Lippincott's* published a seven-page article "Spectrum Analysis". Here the history of spectroscopy was printed for the reader, beginning with Isaac Newton (1642–1727) and the prism, through Fraunhofer lines and Kirchhoff's explanation. The use of spectroscopy during the eclipse of 1868 was described, along with its importance in the understanding of the Sun and its constituents. Applications toward the understanding of other astronomical phenomena were mentioned (Morris 1869).

Littell's Living Age was a weekly magazine, published from 1844 until 1896, "… fostering taste in foreign literature …" (Mott 1938b, pp. 256 and 360). On 26 January 1867 an untitled article was published on the progress achieved in photographic techniques and the usefulness of these in art, science and commerce. Warren De la Rue applied photographic techniques during the total solar eclipse of 1860, which he viewed from Rivabellosa, Spain. He obtained more than 40 photographs, and concluded that prominences belonged to the Sun and not the Moon (Review 1 – No title 1867).

On 17 October 1868 *Littell's Living Age* published an account, originally from the London *Times*, "The Great Eclipse", by an unnamed German who observed the 1868 solar eclipse in Aden. His party's focus was on photographing the progress of the eclipse, equipped with a photographic telescope that could track its target by means of a clock-work mechanism. From their observations they concluded that the corona was an "… inflammable gas in a high state of combustion." (pp. 183–184).

On 25 September 1869 *Littell's Living Age* reprinted an article from *The Spectator* entitled "The Total Eclipse in America". Results obtained at the many eclipse sites available over most of the length of the track should provide fodder for study for years to come. The weather was generally fine, skies clear and temperature moderate. Of special significance would be the conclusions of the astronomers on the number and position of bright lines seen in the spectra of the solar prominences. Professor Winlock in Stubbville (*sic*), Kentucky detected no less than eleven bright lines. On 6 August 1870 "Recent Solar Researches", also from *The Spectator*, was printed. This recounted the progress in describing the various layers of the Sun from the solar prominences inward, mainly through the use of spectroscopy.

On 18 March 1871 *Littell's Living Age* reprinted "The Late Eclipse" from the *Cornhill Magazine*. An historical summary of previous eclipses would lead up to

Fig. 4.11 S.J. Perry (after *Popular Science Monthly*, 1896)

the questions to be addressed at this most recent eclipse of December 1870. In 1868 Tennant had determined the spectrum of the corona to be a continuous 'rainbow' with neither dark nor bright lines. In 1869 several Americans had found superimposed bright lines on this 'rainbow'. Professor Harkness had concluded that iron vapor was a constituent of the corona, while Professor Young concluded the corona to be a perpetual solar aurora. The English Jesuit astronomer, Father Stephen Joseph Perry (1833–1889; Fig. 4.11), head of the parties to Spain in 1870, confirmed the presence of bright lines in the coronal spectrum. Lockyer, in Sicily, also confirmed the American observations of 1869.

On 13 January 1872 *Littell's Living Age* reprinted Richard Proctor's article on "The December Eclipse" from the *Cornhill Magazine* which anticipated the results to come out of the recent eclipse of December 1871. For the non-scientific reader Proctor provided a lengthy description of how solar eclipses occur. He provided an historical summary of progress made since the eclipse of 1860, conclusions drawn and questions yet to be answered. At the time of the latest eclipse, most of the remaining questions surrounded the nature of the corona. Proctor concluded the article with two paragraphs stating the current position and the hopes for the recent eclipse. The current position was that the corona was a solar appendage, a portion of which was a self-luminous gas and a likely portion of its light being sunlight reflected from contained solid or liquid matter. It seemed probable that a portion of coronal light had an electrical origin similar to the aurorae. It was to be hoped that the astronomers in 1871, using photography, spectroscopy and polariscopy, might overcome the difficulties of the interference of the terrestrial atmosphere to obtain more satisfactory information regarding the components of the corona (Proctor 1872b).

On 24 February 1872 *Littell's Living Age* published a short item, "The Solar Eclipse," submitted by R.N. Taylor. This was simply a subjective account of "... a glorious and magnificent sight ..." (p. 490). On the same date, 24 February 1872, an article on "The Solar Eclipse" from *Nature,* by J. Norman Lockyer, was reprinted, regarding the eclipse of December 1871 which was observed in India and Ceylon. Lockyer listed the five sites and the main members of the respective parties and their particular tasks. He related his own observations during the relatively short time of totality, when he personally made observations visually, spectroscopically and with the polariscope. Months of work were ahead for all participants to compare and interpret their results.

After several years, on 1 June 1878, *Littell's Living Age* printed another article by Lockyer, also taken from *Nature*, "The Coming Total Solar Eclipse". The eclipse of July 1878 was to take place in parts of the Western United States and its territories. The many American expeditions welcomed the participation of European scientists. Questions regarding the corona still remained to be answered. Lockyer suggested a photographic record of the polarization of the corona might be attempted, certain of the radial polarization observed by himself in 1871. He wished the Americans well and anticipated success (Lockyer 1878a).

The regional farm journals existing during this time included the *Maine Farmer* which was published from 1833 until 1924. Farm journals would often deal with topics other than agriculture (Mott 1930, pp. 444–445). On 1 November 1866 a paragraph entitled "Great Shower of Meteors Expected" was printed, describing this upcoming event as next in grandeur only to that of a total solar eclipse. On 14 August 1869 "The Solar Eclipse" was printed, describing many of the dispatched results already received from various venues around the nation, including Vincennes, Indiana; Springfield, Illinois; Albany, New York; and White Sulphur Springs, West Virginia. No planetary bodies had yet been found within the orbit of Mercury. Bailey's (*sic*) beads were well observed at multiple sites. The nature of the corona was still in question.

On 13 August 1870 *Maine Farmer* published "Recent Solar Researches". This summarized conclusions drawn to date, mainly based on spectroscopic studies, regarding the constitution of the layers of the Sun, including the prominences, the chromosphere and the photosphere. On 17 August 1878 there was a paragraph within the article "August Skies at Night" whereby the reader was informed that both a solar and a lunar eclipse would be visible that month.

The Massachusetts Ploughman and New England Journal of Agriculture, another farm journal, was founded in 1840, in Boston, by the New England Agricultural Society and ran from 1842 to 1906 (Mott 1930, p. 442). On 7 August 1869, in a one-line filler, it made mention of the solar eclipse to occur on that same date (New way of raising beans). On 3 August 1878 it reported on various successes achieved during the eclipse of July of that year. According to reports so far received the planet Vulcan had not been found. Stated to be the most important result of Professor Young's observations was the fact that, at the moment of totality, the Fraunhofer lines were reversed. Other reports from other observing sites were cited (The solar eclipse – Successful observations at various points 1878).

Medical News was a successful journal founded by Dr Isaac Hays (1815–1879) in Philadelphia, running from 1843 to 1905 (Mott 1938a, p. 84). In February 1869 it printed the non-medical item "Composition of Solar Prominences", which dealt with the results obtained regarding these features subsequent to the spectroscopic work of Tennant, Herschel, Janssen and Lockyer.

The *Nassau Literary Magazine*, a student publication out of Princeton University, began in 1842 as the *Nassau Monthly*. It was in 1847 that the name was changed to *The Nassau Literary Magazine*. In 1930 it would become known simply as the *Nassau Lit* (http://etcweb.princeton.edu/CampusWWW/Companion/nassau_lit.html). This was one of the more long-lived college periodicals (Mott 1930, pp. 488–489; 1938a, p. 99; 1938b, p. 165). In October 1869 an article titled "The Eclipse Party" was published. It was a personal account of a member of the party at the Ottumwa (Iowa) site during the 1869 American eclipse. A temporary observatory was constructed to which local citizens were invited. At the time of the eclipse a large crowd gathered, prepared with pieces of smoked glass. The author underwent a religious experience with the return of sunlight. This article was signed "Kepler Copernicus."

The *New England Farmer and Horticultural Register* was one incarnation of an agricultural journal published from 1822 until 1890 (Mott 1930, pp. 317–319; 1938b, p. 152). On 15 June 1878 the short item "This Summer's Eclipse of the Sun" was printed. This related information about a pamphlet issued by the U.S. Nautical Office with instructions by Professor Newcomb for observing the eclipse of that year: "All the suggestions of the pamphlet are simple and may be carried out with little trouble by those who are favorably situated as to locality." (p. 4). On 13 July 1878 "The Total Solar Eclipse" was printed, containing a paragraph on Congressional appropriations and professional preparations being made for the eclipse. On 3 August 1878 some of the early reports of results obtained were printed. Successful spectroscopic observations were achieved at numerous sites. J. Norman Lockyer, who accompanied an American group to Separation Point, Wyoming Territory, was surprised to see only one faint prominence. Although most reports were negative, a dispatch from Rawlins, Wyoming Territory, reported Professor Watson's claim to have seen the intra-Mercurial planet Vulcan (The solar eclipse 1878). The search for Vulcan had been a goal of astronomers since the distinguished French astronomer Urbain Jean Joseph Le Verrier (1811–1877; Fig. 4.12) noted excess perihelion motion in the orbit of Mercury in 1859. Others claimed to have seen it. Yet others refuted the claims. It would not be until the famous German-born theoretical physicist Albert Einstein (1879–1955; Fig. 4.13) published his General Theory of Relativity in 1915 that astronomers would have a satisfactory explanation for the irregular motion of the planet Mercury (Marsden 1996).

The focused *New York Teacher and American Educational Monthly* was published from 1868 to 1869 (catalyst.library.jhu.edu). In December of 1868 there was the feature "Science and the Arts" covering aspects of the eclipse of 18 August 1868. Quoted were lines from Janssen to the French Minister of Public Instruction on the results of spectroscopic analysis of the solar prominences. Janssen determined that they were principally composed of hydrogen gas.

The *Overland Monthly and Out West Magazine* was founded by a San Francisco bookseller primarily as an outlet for regional writers, including Mark Twain, and was published from 1868 to 1935 (Mott 1938b, pp. 402–409). In June 1871 it

Fig. 4.12 Urbain Le Verrier (courtesy: en.wikipedia.org)

Fig. 4.13 Albert Einstein (courtesy: en.wikipedia.org)

published "The Eclipse in Siberia", the personal reminiscences of a participant in Professor Hall's expedition to the westernmost site selected for the eclipse of 1869. The author described the trip on the *Mohican* from San Francisco. Much of the content of the article dealt with his impressions of the aboriginal population encountered

Fig. 4.14 Albert S. Southworth, ca. 1845–1850 (courtesy: en.wikipedia.org)

en route. His description of the eclipse is non-scientific in nature, his observations of the prominences being "Three distinct lumps of something which had the appearance of heated iron …" (pp. 519–524).

The *Ohio Farmer* was established by Thomas Brown of Cleveland in 1848 and continued until 1906. As did many farm journals, it would often publish articles on topics other than agriculture (Mott 1930, pp. 89 and 444). On 18 April 1868 it published "Eclipse of the Sun," informing readers of the preparations for observing the eclipse of 18 August 1868. There would be French and English scientists present at various sites. On 18 November 1871 "The Approaching Solar Eclipse" was printed. Here British preparations to participate in expeditions to Ceylon and Australia were described. This article ended with the statement:

> We are sure that our American astronomers who earned great distinction by their observations during the last two solar eclipses will not be behind the European co-workers in doing all that can be done to aid the cause. (p. 727).

Despite its title *The Phrenological Journal and Science of Health* was a leading general magazine, and Walt Whitman was an important contributor (Mott 1938a, p. 43). In April of 1872 it published the article "History of Photography in America". After a general review of the history of the various methodologies, examples of its usefulness were given. The pioneering Boston photographers Albert Sands Southworth (1811–1894; Fig. 4.14) and Josiah Johnson Hawes (1808–1901; Fig. 4.15) daguerrotyped the Sun during the eclipse of 1842. This demonstrated only the solar disc. The first successful image of the corona was the daguerreotype (see Fig. 4.16) taken by a photographer named Berkowski at the Royal Observatory in Königsberg

Fig. 4.15 Josiah J. Hawes, ca. 1845–1850 (courtesy: en.wikipedia.org)

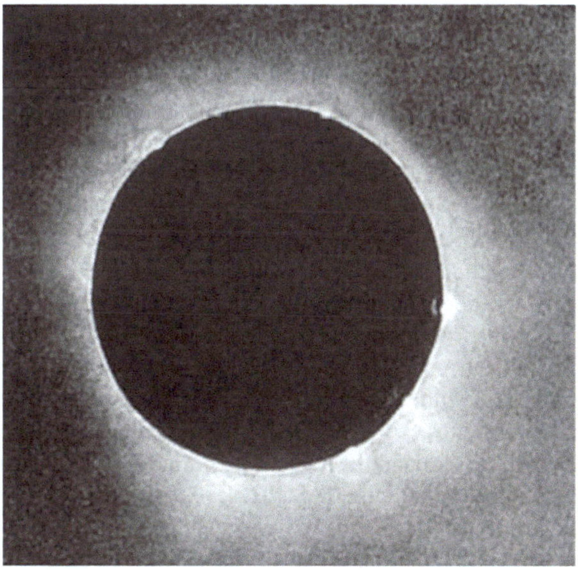

Fig. 4.16 A copy of Berkowski's photograph of the 28 July 1851 total solar eclipse, which not only shows the corona but also some prominences (courtesy: en.wikipedia.org)

(then in Prussia) in 1851 (Clerke 1902, p. 166). In 1850 the American inventor and photographer John Adams Whipple (1822–1891) daguerrotyped the Moon and then in 1852 some stars ranging down to magnitude 4. Mr Whipple, who accompanied Professor Winlock (incorrectly spelled "Menlock" in the article) to Shelbyville, Kentucky, for the 1869 eclipse, photographed and produced an excellent image of the solar corona.

In November of 1878, in its feature "Notes in Science and Agriculture," *The Phrenological Journal* printed a brief article "The Recent Solar Eclipse" referring to that of 29 July 1878. Many successes were enumerated in this short piece. Among others, Dr Henry Draper photographed the corona, Professor Young saw the bright bands of the corona and Professor Watson claimed to have discovered the elusive intra-Mercurial planet.

Putnam's Monthly Magazine was published from 1853 until 1910. An elite literary and political magazine, it succeeded in attracting many venerable contributors. Its first issue alone, printed in January 1853, included submitted works of Longfellow, Thoreau, Greeley and Dana (Mott 1938a, pp. 419–431). In February 1869 there appeared a summary of short items entitled "Science, Statistics, Exploration, etc." which included a paragraph on the successes of Jeussen (*sic*) and Lockyer in using spectroscopy to study solar 'protuberances' (i.e. prominences) outside of an eclipse. In September 1869 a short manuscript from the popular American writer James Fenimore Cooper (1789–1851) was printed posthumously, describing his recollection of the great eclipse of 1806, visible to him as a young sailor of 17 years on leave at his home in Otsego County, New York.

The *Round Table* was a weekly journal of opinion printed in New York, but it attempted to be national in character taking articles from correspondents in Boston, Philadelphia and Chicago. Articles were anonymous and frequently dealt with controversial issues of politics or religion (Mott 1938b, pp. 319–324). "Table-Talk" was a regular feature of miscellany. Occasionally an interested reader would be informed here of upcoming astronomical events. On 20 June 1868 there was a paragraph describing the preparations of the expeditions to observe the upcoming eclipse of 1868. Various European countries had divided the chosen sites increasing the likelihood of obtaining valuable data. The intent was to answer lingering questions, including those on the nature of the solar constituents and the existence of an undiscovered planet within the orbit of Mercury (Table-talk 1868a). On 1 August 1868 the importance of the upcoming eclipse is again addressed with regrets that this event will not be visible from America. However, Americans might take consolation in the fact that Encke's Comet would be visible to them (Table-talk 1868b). On 12 September 1868 this feature included mention of the meeting of the Germans, returning from eclipse sites, with the German archaeologist Johannes Dumichen (1833–1894) in Cairo, to expand their scientific horizons by examining and photographing hieroglyphics (Table-talk 1868c). On 30 January 1869 the *Round Table* printed a two-page article, "The Great Eclipse", intended to introduce to the readers the significance of the previous year's event to the scientific community. Here, some of the history of spectroscopy, so important for this eclipse, was described, along with the results obtained by its use.

The *Saturday Evening Post* has had numerous incarnations. Covering the period under consideration it was published from 1839 to 1885. When there were no Sunday editions of daily papers, Saturday journals became a popular reading source on Sundays. The *Post* could appeal to a wide variety of tastes and interests, as it included poetry, fiction, news items, editorials, birth and death announcements and advertisements. There also were some woodcut illustrations (Mott 1957, pp. 671–693). On 13 November 1869, referring to the recent American solar eclipse, the *Post* described some recently-associated phenomena believed to be observed by superstitious Indians and trappers (Wonderful appearances 1869). On 11 July 1874 there was a paragraph on Mr John William's possession of the encyclopedia of Ma-twan-lin, published in 1322, which included articles on Chinese astronomy and a list of observed solar eclipses (News of interest – Chinese astronomy 1874).

Scribner's Monthly was published from 1870 to 1881. Aimed at the middle-class public it included poetry, fiction and, frequently, illustrated articles on science, travel and art (Mott 1938b, pp. 457–480). In its feature of miscellany, "Memoranda", in March of 1875 there is note about photographs taken during the eclipse of 1871 showing signs of the presence of a comet in the corona of the Sun.

4.2.2 Religious Periodicals

Considered separately here are the reports from religious journals on astronomy. Religious journals were numerous during this period and interestingly had many articles on the events under consideration. They would frequently be lengthy and informative and usually were not weighed down with religious rhetoric.

The American Catholic Quarterly Review began publication in 1876 and lasted until 1924. "Religion and ecclesiastical policy were prominent in its contents; but literature, education, art, science, history and philosophy were not neglected." (Mott 1938b, pp. 68–69). In October 1878 an informative 14-page article, "The Total Solar Eclipse of July 29th, 1878", was printed. It enumerated many expeditions, but predominantly American ones. Some foreign astronomers, notably Lockyer, participated at their own expense. The Catholic Society of Jesus had an expedition that was based about one mile east of Denver. Specific data were provided in chart form, including the times, temperatures and relative humidity. The three stated goals for these efforts were specified as: (1) a better determination of the position of the Moon; (2) confirmation of the existence of an intra-Mercurial planet; and (3) a better understanding of the nature of the solar constituents, especially the corona. In his published guide for observers Professor Harkness encouraged documentation of the limits of the lunar shadow to better determine the position of the Moon. The existence of an intra-Mercurial planet was expected to explain perturbations in Mercury's orbit. Spectroscopic studies were to be a focus at these sites in order to increase understanding of the corona. Professor Young saw the coronal spectrum as mixed, partially due to gases reflecting solar light as seen in the production of Fraunhofer lines, and partially as a continuous spectrum due

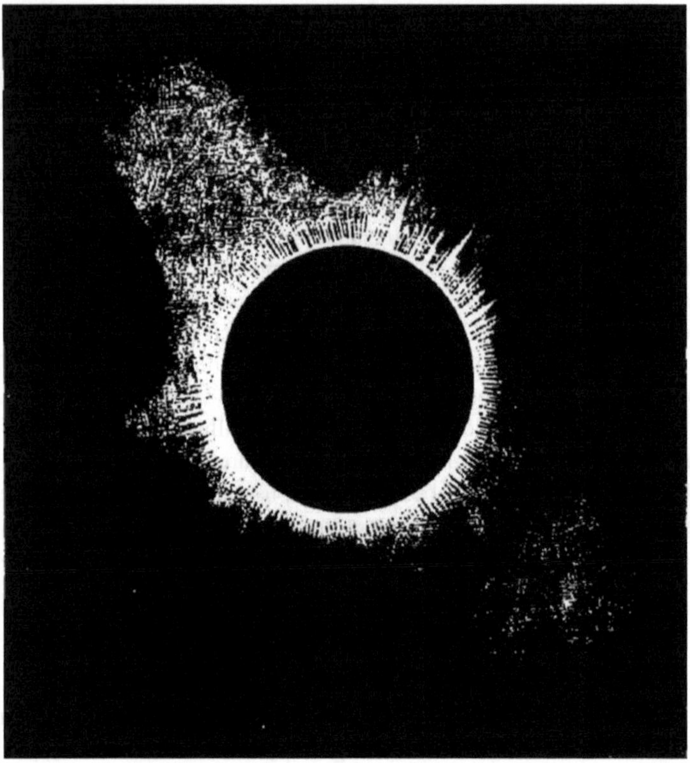

Fig. 4.17 The drawing of the total solar eclipse of 29 July 1878 by the Reverend Sestini (after *The American Catholic Quarterly Review*, October 1878, p. 641)

to the presence of solid or liquid matter. Polariscope studies seemed to indicate that the coronal light was partially polarized, indicating the presence of a substance that reflects light from the Sun. Spectroscopic studies also revealed a line at 1,474 Å, of unknown origin, which some associated with a substance they then called and is now universally accepted as 'helium', derived from the Greek word for the Sun, 'helios'. This article also included an illustration (see Fig. 4.17), a drawing of the event at totality by the Reverend Benedict Sestini S.J.

Catholic World, A Monthly Magazine of General Literature and Science was a popular magazine for Catholics, "… interesting to the general reader but designed to support doctrines of church and express Catholic views." (Mott 1938b, pp. 329–330). In February 1869 it printed "The Eclipse of the Sun of August 18, 1868", subtitled "A Report Addressed by M. Janssen to the Marshal of France, President of the Bureau of Longitudes", dealing with some specifics of Janssen's experiences in spectroscopic observations of two prominences. Since the spectra of these objects contained bright lines he concluded they were gaseous. There was a correspondence of some of these lines with those of the Sun, in particular the C and F lines, belonging to the element hydrogen. The prominence spectral lines were so brilliant that at this time Janssen (1869a, p. 700) speculated that they might be seen "… even when the sun was unobscured …" He confirmed this theory with spectroscopic studies the next morning. With these later observations he could state another characteristic of prominences which was that they can change form and position in a short period of time (ibid., pp. 697–703).

In October of 1869 *Catholic World* published "The Total Eclipse of August Seventh." This dealt with the American eclipse of that year, written by an anonymous author connected with the party at Shelbyville, Kentucky. At totality he saw Venus, Mercury, Saturn and several stars. He saw the bright red prominences with his naked eye. Professor Winlock saw 11 spectral lines in the great prominence. About 80 photographs were taken during the eclipse, several at totality.

The *Christian Advocate* was a Methodist Church weekly, published in New York (Mott 1938a, p. 66). On 22 October 1868 it published an article by Professor William Wells, "The Recent Solar Eclipse", within which he informed the reader of the rarity of eclipses, the value of the eclipse of that year due to its duration and location, and the intended goals of the various expeditions. All the results had not yet been reported but anticipated were those on the spectroscopic analysis of the prominences and the corona, along with the search for intra-Mercurial planets.

On 5 August 1869 the *Christian Advocate* published the article by A.J. Kalb, "The Eclipse – A Remarkable One". This article anticipated the results to be obtained at the many sites of the American eclipse of 1869. On 19 August 1869 there was an early report on the American eclipse from the site of Springfield, Ohio, which was under the leadership of Professor Peirce. Venus, Mercury, Lyra and Regulus were visible. Three rose-colored prominences were seen (The solar eclipse 1869).

On 31 January 1878, within the *Christian Advocate's* regular feature "Science", there was an item stating that there would be two solar eclipses that year, an annular event on 1 February and a total one on 29 July. On 20 June 1878, also within "Science", there was an item reporting that Professor Simon Newcomb had published particulars on the solar eclipse to take place on 29 July.

The *Christian Union* was a Baptist periodical established in New York in 1867. At first it consisted of only 16 pages but increased in size with advertising and publication of pieces by distinguished contributors such as Harriet Beecher Stowe and Louisa May Alcott (Mott 1938b, pp. 422–435). On 1 January 1870 it printed a communication from Professor J.D. Steele intended to allay popular fears about solar prominences. He stated that it had definitely been settled since the eclipses of 1868 and 1869 that these explosive prominences are not associated with the Moon, but with the more distant Sun.

On 24 December 1870, within its regular feature "Scientific & Sanitary", the *Christian Union* printed the short item "A Handsome Courtesy". Here the reader would learn that Congress had appropriated funds for the two solar eclipse expeditions that month, one to Sicily and one to Spain. As the English had not as yet provided such support for their own scientists, Professor Pierce (*sic*) generously invited a number of distinguished English astronomers to accompany these American expeditions.

On 1 February 1871 the *Christian Union* had advertisements for two books on popular astronomy to be released shortly by the publisher Scribner, Welford and Co.

One was by Richard A. Proctor, entitled *The Sun; Ruler, Fire, Light and Life of the Planetary System.* With illustrations drawn by Proctor himself, he would inform the reader of the nature of the Sun, its corona, its prominences and such events as eclipses. The second book was *The Heavens* by Amédée Guillemin. This was a new edition of a book previously well-received by reviewers, edited by Norman Lockyer and revised by Richard Proctor (Advertisement 21 – No title 1871).

On 15 February 1871, within "Scientific and Sanitary", the *Christian Union* printed the item "Astronomical Instruments". Referring to the eclipse of 1870, it explained that astronomers then needed more than telescopes to study the skies. The spectroscope and polariscope introduced new data for the study of astronomy. Within the article was a brief explanation for the layman of the use and significance of spectroscopy in solar studies. On 19 November 1873, also within "Scientific and Sanitary", "Constitution of the Sun" was printed. This was a summary of a paper read by Professor Young at a meeting of the American Association for the Advancement of Science on what was known, and on what was yet to be learned, about the nature of the Sun.

On 7 August 1878 there was the *Christian Union's* article, "The Solar Eclipse", about the most recent eclipse in 1878, visible in the Western Territories of the United States. Though by then it had been established that the corona belonged to the Sun, there was still the unanswered question of whether it was burning its own gas or whether it was cooler matter reflecting sunlight. Numerous parties were specifically identified, and the instruments their participants would utilize were described. Edison had fought the windy conditions and succeeded in concentrating coronal light upon the collecting opening to his tasimeter. His established scale was too narrow and the light collected surpassed the range of his instrument, the heat of the corona apparently being more than was anticipated. This knowledge was considered a successful outcome in itself. Results from most of the parties had yet to be received.

On 4 September 1878 the *Christian Union* printed, within "Science and Art", two short items relating to the eclipse of 1878. In "The Planet Vulcan" it was stated that the intra-Mercurial planet had likely been discovered by Professor Watson. In "The American Association" some presentations at the 27th Annual Meeting of the Association for the Advancement of Science were printed. Among other presentations was one by Thomas Edison, a new member, describing the results of his tasimeter measurements during the 1878 eclipse, and claiming success in detecting heat emitted by the solar corona.

The Friend. A Religious and Literary Journal was one of several periodicals for the Society of Friends, or Quakers. It was published from 1827 until 1906 and "… the eight pages of the quarto were filled weekly with brief literary and scientific articles, travel, original and selected poetry, pious biography, and current events, as well as doctrinal reading and articles dealing with the various 'concerns' of the Friends – peace, temperance, antimasonry, antislavery and colonization, and Indian education." (Mott 1930, pp. 562–565). On 24 October, 31 October and 7 November 1868 three parts of the article "Great Solar Eclipses" from the *Cornhill Magazine* were reprinted. These gave an historical perspective on eclipses from ancient records up to the current time, anticipating the results of the eclipse of that year. The first

Fig. 4.18 A pre-1846 engraving of François Arago by Alexandre Vincent Sixderniers (courtesy: en.wikipedia.org)

eclipse to lead to an actual planned expedition occurred in 1842, and in the article of 24 October some subjective impressions of François Jean Dominique Arago (1786–1853; Fig. 4.18) who was present at the event in Perpignan, France, were quoted. He shared that experience with "More than twenty thousand persons, with smoked glass in their hands …" (pp. 65–66). The second article began with a description of some of the environmental effects noted during eclipses, especially pertaining to animal behavior. It then briefly covered the next two eclipses that were of interest to the Europeans, those of 28 July 1851, visible in Sweden, and 18 July 1860, visible in Spain. The distinctness of prominences was noted in 1851 and drawings were made by numerous astronomers. These entities were still of considerable interest in 1860. The last of the three articles included a summary of some of the major expeditions that went to observe the 1868 eclipse and the anticipation of questions to be answered regarding the prominences when viewed with the spectroscope.

On 31 July 1869 *The Friend* printed "The Great Eclipse", this time anticipating the American eclipse of 1869. Specific details were provided of times and locations, most of these data being taken from the Supplement to the American Nautical Almanac. In the 11 March 1871 issue of *The Friend* there was the article "The Total Eclipse of 1870". In the Mediterranean this time, European efforts by the English and the French would be joined by American parties, including one under Professor Peirce in Sicily and one under Professor Winlock in Spain, as well as four observers from the U.S. Naval Observatory. Most parties experienced some frustration due to the uncooperative weather.

Fig. 4.19 Samuel Gummere (courtesy: www.freepages.genealogy.rootsweb.ancestry.com/~gomery/gummeresamueljames.html)

On 4 May 1878, within "Summary of Events", *The Friend* briefly described a party under Professors Young and Brackett who would be observing the eclipse of 29 July of that year. On 13 July 1878, also within the feature "Summary of Events", there was a mention of the Congressional appropriation of $8,000 for eclipse expeditions to the western territories later that year.

The *Friends' Intelligencer* was another instrument of the Society of Friends. It was published from 1853 until 1910 (Mott 1938a, p. 74). On 16 May 1868 under its feature "Items" was "The Coming Eclipse". In the short paragraph the reader was informed of an upcoming eclipse to occur on 18 August that would be visible for almost 7 min in several areas of Asia. The English and the French would be participating and, by using spectroscopy, would expect to learn more about the 'corono' (*sic*) and 'red flames'.

On 20 June 1868 the *Friends' Intelligencer* reprinted an article by Samuel James Gummere (1811–1874; Fig. 4.19), the President of Haverford College (at that time a Quaker institution), "The Solar Eclipse of 1869", anticipating the American eclipse of 1869 (1868a). An identical article was also printed in the *Friends' Review* (1868b).

On 19 September 1868, within the *Friends' Intelligencer's* "Items" were two paragraphs, entitled "The Solar Eclipse", covering some early results received from expedition sites of the eclipse of 1868. Though the weather was cloudy, Major Tennant, commander of the expedition of the Royal Astronomical Society of Great Britain, had achieved some success. The French astronomer Dr Janssen had used the spectroscope on the red 'protuberances'. His conclusions would follow based on whether the spectrum appeared as continuous, mostly continuous or only as bright lines. On 26 September 1868 within the same feature, "Photographing the Solar Eclipse" from the *New York Evening Post,* was reprinted. Mention was made of the British presence of Major Tennant and Lieutenant Herschel, as well as Mr Pogson who, as Director of the Madras Observatory, was already present in the area, and Dr Janssen for the French. To date, only Major Tennant had been heard from. Using a telescope made by Browning, he had achieved success in photographing the eclipse. On 8 May 1869, under "Items", were two paragraphs entitled "The Solar Eclipse", from the *New York Evening Post*, regarding the American eclipse expected that year. The path of the line of totality was to extend from Alaska to North Carolina. Professor Joseph Winlock of Harvard University would be participating at Shelbyville, Kentucky.

On 21 August 1869 the *Friends' Intelligencer* printed another article reprinted from the *New York Evening Post*, "Photographs of the Eclipse", within which J. Gardner shared his experiences in photographing the eclipse of 1869 at Bedford, Indiana. Coordinating efforts with Mr Baldwin and Mr Blackwell, 21 photographs were taken. One attempt during totality revealed the bright prominences.

On 13 November 1869 and 20 November 1869, the *Friends' Intelligencer* printed two parts of an article by Maria Mitchell titled "The Total Eclipse of 1869". Herein she explained how eclipses occur and gave some history of previous eclipses. Those of 1806 and 1834 were also visible from the United States. Several eclipse expeditions, starting with the first in 1842, were mentioned for their significance. Mitchell herself, observing the 1869 eclipse in Burlington, Iowa, was invited to use the grounds of the Burlington Collegiate Institute. She described the instruments used which included a 3-in. Alvan Clark telescope and a 4-in. Dolland equatorial. Descriptions were given of the coordination of efforts made during the progress of the eclipse. As darkness increased Venus, Mercury, Arcturus and Saturn were seen, and three prominences were noted.

On 2 December 1871 the *Friends' Intelligencer* printed an article, "The Late Aurora", dealing with the use of spectroscopy during an aurora. As such electrical disturbances in the Earth's atmosphere seem to be related to the activity of gases around the Sun, spectroscopy of solar prominences, now possible at times outside of total solar eclipses, will be of interest.

On 19 June 1875 within "Items" the *Friends' Intelligencer* printed two paragraphs announcing the untimely death of Professor Winlock, who had charge of solar eclipse expeditions to Shelbyville, Kentucky in 1869 and Spain in 1870. On 3 August 1878, within "Items", there was a short summary regarding the initial incoming reports from the various eclipse sites, which generally were satisfactory.

On 10 August 1878 there was the *Friends' Intelligencer*'s article, "The Late Solar Eclipse", with some early reports of that year's eclipse expeditions from Rawlins, Wyoming Territory. Dr Draper announced he had found the presence of oxygen in the Sun. Professor Lockyer noted a relationship between sunspot activity and brightness of the corona. Contradictory reports of the success of Edison's tasimeter, used to measure coronal heat, were received but its use during future eclipses seemed likely. The existence of the planet Vulcan seemed to be confirmed.

The *Friends' Review; a Religious, Literary and Miscellaneous Journal* was founded by the Quaker Enoch Lewis (1776–1856) when he was 70 years old and was edited by him until the time of his death 10 years later. As the Society of Friends were humanitarians, during the Civil War *Friends' Review* ... had frequent testimonies against slavery and published accounts expressing concerns for the Indians. "The *Friends' Review* was a well printed small quarto of sixteen pages." Besides its humanitarian concerns, "There were also many biographical accounts, journals, and so on, usually published serially. Poems of a thoughtful or religious cast were frequently reprinted, and there was a summary of foreign and domestic news." (Mott 1930, pp. 773–774). On 16 November 1867 there was "Work of the Spirit," a religious metaphor on the power of silent events, including solar eclipses (Phelps 1867).

On 6 June 1868 the *Friends' Review* ... printed an article by Samuel J. Gummere, President of Haverford College, on "The Solar Eclipse of 1869". The timing of this event had been calculated by Gummere 40 years earlier. He provided a little history of previous eclipses and some specifics on locations, times and the path of totality of the eclipse of 1869 (Gummere 1868b). This article had also previously been printed in the *Friends' Intelligencer* on 20 January 1868 (Gummere 1868a).

On 3 July 1869 and 10 July 1869 the *Friends' Review* ... printed the two parts of an article, also by Samuel Gummere, on "The Solar Eclipse of Eighth month 7th, 1869". Gummere provided an explanation of eclipse events and enumerated the locations for observation in 1869. He noted that the public could contribute, even without instruments, by noting the degree of darkness at their locations (Gummere 1869a, b). On 28 August 1869 there were two items of correspondence referring to the 1869 eclipse. Not surprisingly, one of these was from Samuel Gummere. He provided a subjective description of the eclipse which he had viewed from Oskaloosa, Iowa. He described the "mysterious" corona, noted the rose-colored flames (prominences) and the visibility of the planets Venus and Mercury, and the subsequent reappearance of the Sun (Gummere 1869c). In another letter Francis T. King describes the experience of 700 school children from the Friends' Normal School in Tyson at an open field on a nearby farm. There they witnessed the eclipse with smoked glass provided to them (King 1869).

On 24 December 1870 and 7 January 1871 the *Friends' Review* ... reprinted in two parts an article "The Sun's Corona" from the *Cornhill Magazine*. An historical account of eclipses, starting with that of 29 March 1652 and extending up to the most recent one in 1870, was given. Over most of these years observers would question whether the corona was appended to the Sun or was a terrestrial phenomenon. On 1 April 1871 "The Late Eclipse", also from the *Cornhill Magazine*, was reprinted. A short history of expeditions to observe eclipses was given, beginning with the first of 1842, and ending with the most recent in December 1870.

On 8 April 1871 there was the *Friends' Review* ... article of one paragraph, "Terrestrial Magnetism – The Eclipse", regarding the observation by the geophysicist Demetrio Emilio Diamilla-Muller that there was a relationship between the movement of a magnetic needle and the degree of totality of an eclipse. On 6 April 1872

Fig. 4.20 Julius Schmidt in 1884 (courtesy: en.wikipedia.org)

"Gossip about the Eclipse" was reprinted from *The Methodist.* Some personal observations of R.N. Taylor and Captain John P. Maclear (1838–1907) of the eclipse of 1871 were quoted. Scientific results from the event were still pending.

On 8 May 1875 the *Friends' Review* … reprinted from the *London Daily News*, "The Total Solar Eclipse", which referred to the event of 6 April of that year. Photography was especially useful this time. It was noted that during the most recent eclipses the Sun had more sunspots. Professor Young noticed that as the number of sunspots diminished the prominences became smaller and fainter.

On 10 February 1877 the *Friends' Review* … article, "Assyrian Records", was published. The Assyrians recorded eclipses and their records had been useful in establishing the dates of related historical events.

On 2 March 1878 the *Friends' Review* … feature "Scientific Notes" was printed, which included two items relevant to eclipses. One was a statement by the German-born Athens-based Professor Johann Freidrich Julius Schmidt (1825–1884; Fig. 4.20) on the existence of early records of observations of the solar corona, perhaps as early as 968. The other told of a new photographic telescope that Janssen would be using in the future to photograph the Sun (see Fig. 4.21).

On 3 August 1878 the *Friends' Review* … reprinted an article "The Eclipse of the Sun", taken from the *Chicago Tribune*, describing the locations chosen for observation and the goals of the parties for the eclipse of 1878. Reference was made to New York's Professor Draper's discovery of oxygen in the solar spectrum and the hope for confirmation. The article ended with a list of the solar eclipses expected for the remainder of the century, the locations for observation and the anticipated durations of totality.

On 14 September 1878 a letter received from the American scientist Pliny Earle Chase (1820–1886) was printed by the *Friends' Review* … Chase expressed his feeling

Fig. 4.21 Janssen's 'photographic revolver' used with a heliostat to photograph the Sun (after Flammarion 1875)

that Professor Watson's observation of the intra-Mercurial planet correlated to a prediction made by himself 5 years earlier. On 28 September 1878 "The Total Eclipse of 1878" was published. It was stated that the majority of readers would be most interested in the announcements of Watson and Swift that they had succeeded in finding two intra-Mercurial planets. Also, the conclusion regarding the corona, based on the very faint Fraunhofer lines observed in a continuous spectrum, was that coronal light was in part sunlight reflected off self-luminous particles. On 12 October 1878 "Astronomical Forecasts" was printed, describing how the study of the harmonic progression of the planets around the Sun could be used to provide a testable prediction of a theoretical orbit for the planet 'Vulcan'.

On 19 October 1878, within "Scientific Notes", the *Friends' Review* ... printed the item "Zodiacal Light and Sun-spots". At the most recent eclipse a relation was noted between the number of sunspots and the presence of meteoric action. As Giovanni Domenico Cassini (1625–1712) noted, zodiacal light was most brilliant at times when there were numerous sunspots, so the conclusion was drawn that meteoric activity was related to zodiacal light. On 15 February 1879, within "Scientific Notes", the item "New Asteroid Belt" was printed about Professor Swift's theory that the two intra-Mercurial planets he observed were different from those seen by Watson, and were actually part of an asteroid belt.

The Independent was a Congregationalist weekly newspaper published in New York, initially significant for its strong anti-slavery view. Over time it became more inter-denominational and appealed to a more general audience. It had a section

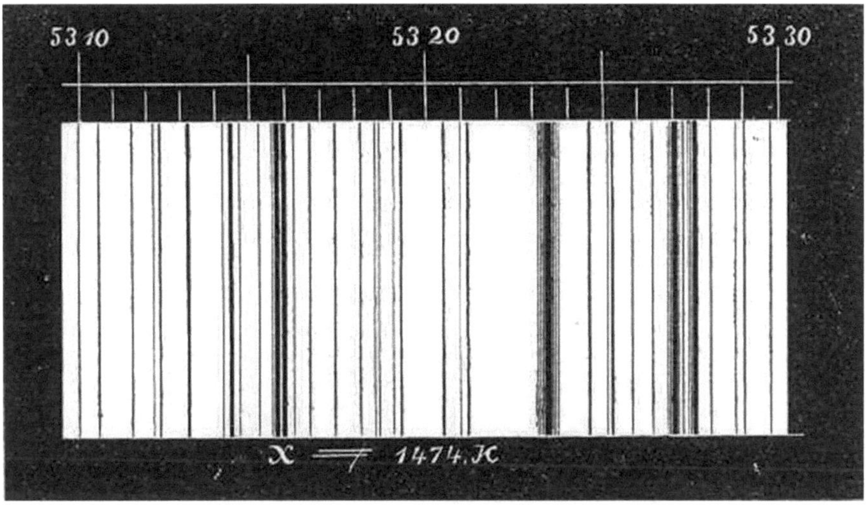

Fig. 4.22 The spectrum of the corona, showing a series of emission lines (in *black* here) superimposed on a fainter continuum. The scale along the top is in angstroms. The 'X' marks the coronal green line discovered by Young and Harkness during the 7 August 1869 eclipse and originally assigned to the new element 'coronium'; the '1474, **K**' refers to a scale derived earlier by Kirchhoff (after Young 1896, p. 258)

on science (Mott 1938a, pp. 367–379). On 16 July 1868 within a feature titled "Foreign Gleanings" there was a paragraph stating that various governments were "... fitting out ..." expeditions to eclipse sites in the East, and mentioning specifics about France and England (p. 2).

On 19 August 1869 *The Independent* printed an article "The Eclipse in Illinois," by Reverend Fred H. Wines (1838–1912), an early report from the Springfield, Illinois, site of observations of the eclipse of 1869. That site was under the leadership of Professor Peirce, Superintendent of the United States Coast and Geodetic Survey. Reverend Wines provided a non-scientific description of his personal impressions of the event.

On 3 November 1870, within *The Independent*'s feature "Scientific", there was a paragraph summarizing the question that was answered, i.e. that hydrogen is the main constituent of solar prominences, and the questions yet to be answered, i.e. those relating to the nature of the solar corona. On 19 January 1871, in that same feature, there was a paragraph with an early report from the Mediterranean eclipse site of 1870. The English found the weather unfavorable in Sicily but there was confirmation of earlier American observations of particular lines in the spectrum of the solar corona.

On 14 March 1872, within *The Independent*'s new feature, "Science", there were enumerated the results of the eclipse of December 1871. American observations of the self-luminosity of the corona and the existence of the 1474 line were confirmed (e.g. see Fig. 4.22). Young's observation of the reversal of the Fraunhofer lines at the solar limb during the event was also confirmed. Hydrogen was determined to be a

Fig. 4.23 Theodor von Oppolzer (courtesy: en.wikipedia.org)

major component of the corona. Janssen detected dark lines in the coronal spectrum, which indicated the presence of particulate matter that reflected sunlight. More than 20 good photographs were taken. There were still contradictory observations regarding polarization. On 26 August 1875 there was an item within "Science" regarding the new observatory to be built at Hamilton College. This will include two domes, one of which will house the excellent telescope that was ordered during the American eclipse of 1869.

On 16 May 1878, again within *The Independent's* "Science" section, there was a quote from J. Norman Lockyer, taken from the English journal *Nature,* regarding the next solar eclipse to occur, that would be visible in the western territories of the United States. He had very positive expectations of results to come out of the event, due to the venerable astronomers and excellent equipment available in this country. On 15 August 1878 the entire feature "Science" was subtitled "The Solar Eclipse" and covered the eclipse of 1878. It was felt that Professor Watson and Mr Swift had succeeded in finding the predicted intra-Mercurial planet. There was Professor Draper's report on his observations of the corona. He did note Fraunhofer lines in the corona, indicating that it is reflected light from the Sun. He stated that his party succeeded in actually getting photographs of the spectrum of the solar corona. Edison applied his invention, the tasimeter, to the determination of the presence of heat in the corona. Directed at the corona his "… index-beam of light went instantly and entirely off the scale." (p. 8). On 8 May 1879, within the same feature there was a summary of the opinion that the Czech-born Austrian astronomer Theodor von Oppolzer (1841–1886; Fig. 4.23) expressed in the journal *Astronomische*

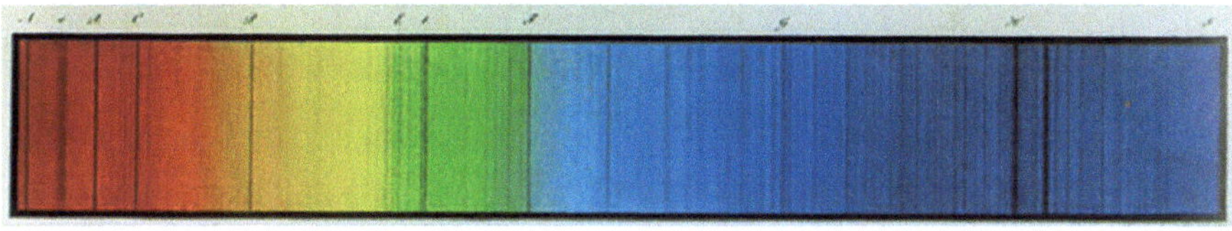

Fig. 4.24 In 1802 the British scientist William Hyde Wollaston reported the existence of dark (absorption) lines in the solar spectrum. Subsequently these were named the 'Fraunhofer lines' (courtesy: en.wikipedia.org)

Nachrichten. Although he agreed with Le Verrier on the existence of an intra-Mercurial planet, he did not believe that either of the objects seen by Watson was the planet Vulcan, for his calculations indicated its existence at a different position.

The Ladies' Repository; A Monthly Periodical Devoted to Literature, Art and Religion was published from 1841 to 1876. It originated with Methodist Samuel Williams (1786–1859) who felt there was a need for a more worthwhile journal for the feminine gender than was then available. "It contained sober, earnest and rather well-written essays of a moral character, plenty of poetry such as it was, some articles on historical and scientific matters, and book reviews." (Mott 1938a, pp. 301–305). In August of 1869 the lead item in the "Editor's Table" was "The Great Solar Eclipse". It described the movement of eclipse visibility across land from Siberia to Newbern, North Carolina, after which it ended in the Atlantic Ocean. It was stated, incorrectly, that there would not be another such event visible from the North American Continent until 28 May 1900.

In October 1869 two relevant articles were printed in *The Ladies' Repository* ... "What Has Been Discovered by Spectrum Analysis" was a five page article on the history of spectroscopy starting with Isaac Newton and the prism. There were descriptions of the discoveries of Wollaston (see Fig. 4.24), Fraunhofer and Kirchhoff relating to the solar spectrum. Father Secchi prepared a classification of stars based on their spectra. The tool of spectroscopy also was to be of use in determining the motion of stars towards or away from our planet. Most recently, spectrum analysis had been applied to the study of solar prominences and the corona. In that same issue was the article "The Great Solar Eclipse", which described the efforts that took place with Professor Winlock's group at Shelbyville, Kentucky. It was a personal description of the activity at that site and the author's own visual observation of the eclipse.

The Methodist Quarterly Review was published from 1841 until 1930 under a number of names (Mott 1938a, p. 66). In July 1869 a 19-page article, "The Application of Photography to Astronomy", was printed. Starting with some history, it recounted the invention of Daguerre first exhibited in 1839. During the total eclipse of 1860 De la Rue successfully photographed multiple stages of the event (see Fig. 4.25). An interesting outcome of these photographs was that they provided proof of the existence of prominences as extensions of the Sun.

The Monthly Religious Magazine of Boston was published from 1844 to 1874. It started with a Unitarian affiliation but was mostly a non-denominational periodical intended for pious homes (Mott 1930, p. 372). In February 1866 an article by the

Fig. 4.25 Warren De la Rue and his Kew photoheliograph in Northern Spain for the 18 July 1860 total solar eclipse (after De la Rue 1862, p. 31)

Boston theologian James Freeman Clarke (1810–1888), titled "The Summer Is Ended", was printed. It was an essay expounding on the beauty of nature provided by the Creator. Mention was made of the current understanding of solar eclipses which had "… once terrified impious nations." (p. 118).

The *New York Evangelist* was another Congregationalist weekly (Mott 1938a, pp. 18 and 140). On 2 November 1865 it published the sermon delivered by the noted clergyman Reverend William Adams (1807–1880) at the Madison Square Church in New York on 22 October 1865, the Sunday after a solar eclipse. The celestial event was the point on which he based his notion that there is a predictable constancy in the heavens, a "… uniformity of the works of God …" (p. 2).

On 6 August 1868, within the *New York Evangelist*'s regular feature "Scientific & Useful", there were several paragraphs enumerating some significant historical eclipses leading to the upcoming one of that month. It was noted that, though remote, the grand scale of this event would attract many scientific expeditions. Several countries and their particular sites were specified. Two of the most important questions these would hope to answer would relate to the nature of the 'red excrescences' and the possible existence of an intra-Mercurial planet. On 10 September 1868, again within "Scientific & Useful", there were the brief nostalgic

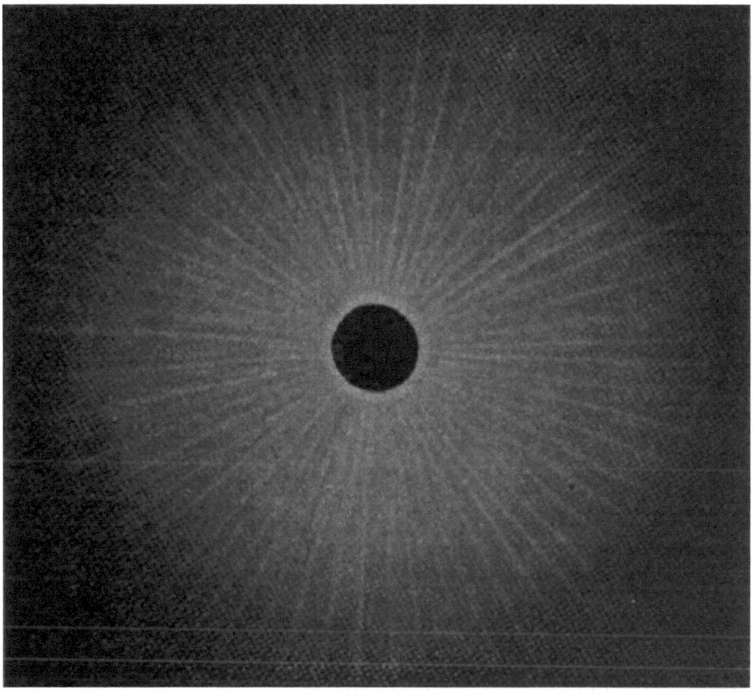

Fig. 4.26 A drawing of the 16 June 1806 total solar eclipse. It was José Joaquin de Ferrer who coined the term 'corona' for the outer atmosphere of the Sun (after Todd 1894)

recollections of an N.G. Carnes and another anonymous writer who observed the total solar eclipse in Boston in 1806 (see Fig. 4.26). Subjective memories included those of animal behavior, cold but clear weather and the number of stars visible during totality.

On 29 October 1868, within the *New York Evangelist*'s "Current Events", there was a paragraph on events in Australia which included a mild earthquake, a meteorite hitting a ship and killing a steersman, and the solar eclipse visible on 19 August.

In 1869 the *New York Evangelist* would have several mentions of the American eclipse to occur on 7 August. On 1 July under "Scientific & Useful", a short article was reprinted from the *Journal of Commerce*, "The Eclipse and Intra-Mercurial Planets". The author states the arguable point that the nature of the solar prominences and corona were solved in 1868, leaving little for consideration at the eclipse of 1869 except, perhaps, the search for an intra-Mercurial planet. On 22 July of that same year, again under "Scientific & Useful" there was the four paragraph item "Total Solar Eclipse in August". Here the path of observation across the United States was described, along with several expedition sites, from that of the U.S. Naval Observatory at the Behring Strait to those at many points eastward. The next week there was the announcement that Dr C.F.W. Peters, of the Litchfield Observatory at Hamilton College, had started on Monday for Des Moines, Iowa, in order to observe the eclipse. His companions were listed. His expenses were paid for through the generosity of the successful businessman Edwin Clark Litchfield (1815–1885) of Brooklyn (Hamilton College). On 12 August 1869, under "Current Events", there was the item "The Eclipse" describing the experiences of the local population in New York City. Newsboys and other speculators were selling varieties of colored glass. 'Ruby' glass was the most popular. Other observers might have pieces of smoked glass. Rooftops of high buildings along Broadway and elsewhere were occupied. Clouds interfered at the peak of obscuration. On 9 September 1869, under "Scientific and Useful", there was the short article "The Science Association of Salem". At this meeting, some early reports were given by Professor Price (*sic*) of the Coast and Geodetic Survey, Professors Hough, Murray and Young and Dr Gould. The article summarized some conclusions. The red prominences were determined to be mainly hydrogen gas and "The corona cannot be due to the solar atmosphere wholly; the interior probably is, but the exterior seems most probably an auroral phenomenon."

On 20 October 1870 there was the *New York Evangelist* article "The Approaching Eclipse of the Sun". Herein was reported the Congressional appropriation of $30,000 for expenses of the American expeditions to the Mediterranean eclipse of 22 December 1870. Professor Pierce (*sic*), Chief of the Coast and Geodetic Survey, would be supervisor. Professors Harkness, Hall and Eastman from the U.S. Naval Observatory would proceed to Sicily. Another party under Professor Newcomb would proceed to Gibraltar and Northern Africa.

On 26 January 1871 there was the *New York Evangelist* article, "Solar Discovery". Herein was a brief description of the 'auroral theory' of the corona proposed by Professor Norton.

On 9 February 1871, under its feature "Scientific and Useful", the *New York Evangelist* printed a letter from C.H.F. Peters of the Litchfield Observatory from Sicily, where he had observed the eclipse of 1870. He stated that parties there had all experienced some frustration due to weather but believed that "… when all the observations are combined, they will furnish very valuable results. Among other conclusions, it seems pretty certain that Young's green line in the corona exists." (Scientific and useful – Expedition of observation, p. 7). On 2 March 1871, under "Scientific and Useful", there was the item "English Astronomers versus American Astronomers", taken from the *Providence Journal*. One outcome of the 1870 eclipse was the admission by European astronomers to the validity of results obtained by Americans at the eclipse of 1869. Lockyer, previously dubious, graciously telegraphed "… the American observations have been confirmed." (p. 3).

On 25 January 1877, under "Scientific and Useful", there was the paragraph "Eclipses of the Sun and Moon," stating some generalities regarding upcoming lunar and partial solar eclipses.

The *New York Observer and Chronicle* was a Presbyterian weekly (Mott 1938b, p. 74). On 2 January 1868 it published "Great Eclipse of the Sun in 1868". This eclipse would be especially useful as it was of such long duration. Application of the new science of spectroscopy, already applied to other celestial bodies, was greatly anticipated for this event. On 10 September 1868 early reports on the eclipse were coming in. Major Tennant had experienced some uncooperative weather but

Fig. 4.27 King Rama IV who observed the 1869 total solar eclipse from Wah-Koa in conjunction with a team of French astronomers (King Rama IV Palace, Petchaburi; photograph by the second author)

still achieved some success (New York State agricultural fair 1868). On the following week the complete success of the Germans at their site in Aden in Arabia was announced (Foreign 1868). More specifics regarding their expedition were given 7 days later in "The Eclipse at Aden". Excellently equipped, they captured images of the Sun with a photographic telescope which tracked its target by means of a clock-work mechanism.

On 4 February 1869 in the *New York Observer and Chronicle* was the announcement of Congressional appropriations of $5,000 for the observation of the solar eclipse in August of that year (U.S. Congress 1869).

On 11 March 1869 the *New York Observer and Chronicle* printed the article "The Late King of Siam". The King (1804–1868; Fig. 4.27) had a strong interest in astronomy and demonstrated abilities in the calculation of such events as occultations and eclipses. He observed the eclipse of 1868 from the Gulf of Siam in the company of many foreign guests (see Orchiston and Soonthornthum 2014). He took ill at the site and ultimately died on 1 October (Wood 1869).

On 15 July 1869 the *New York Observer and Chronicle* published "The Great Solar Eclipse" with some early information for its readers on the various expedition sites and some details on the path of totality.

Fig. 4.28 George W. Hough, Director of the Dudley Observatory (courtesy: www.dudleyobservatory.org/History/history-hough.htm)

On 16 March 1876 the *New York Observer and Chronicle* mentioned a solar eclipse to occur on 17–18 September of that year. It would be visible only in remote areas for a short time, in fact for less than a minute of totality at Savage Island (discovered by Captain Cook in 1774), as well as at the Fiji Islands (Notes and queries 1876).

In the August 1878 issue of the *New York Observer and Chronicle* there were some early reports on the eclipse observed at the Western Territories of the United States in "The Eclipse". Here Professor Young gave his report on the reversal of the Fraunhofer lines at the Sun's edge at the moment of totality. In Denver, Professor George Washington Hough (1845–1909; Fig. 4.28) from the Dudley Observatory reported on his impression of the chromosphere and the prominences, while a member of his party, Professor L.F.M. Easterday (1839–1913) from Carthage, Illinois (Colbert 1878, p. 5; Hough 1878, p. 16), reported on the extent of the corona. From Pike's Peak, S.P. Langley reported on his impression of the corona, its extent and appearance. Henry Draper reported from Rawlins, Wyoming Territory, that four photographs were taken of the corona and that Edison's tasimeter demonstrated heat in the corona. On 29 August 1878 "The Planet Vulcan" was printed quoting Professor Watson's claim that he had observed the elusive planet at his site. He described it as a 'disk' and stated that it was not on any star chart.

Old and New was established by Edward Everett Hale (1822–1909), a Unitarian minister. He intended to create a magazine like the *Atlantic* with more religious emphasis. Contributions included serial pieces by Harriet Beecher Stowe and Anthony Trollope (Mott 1938b, pp. 436–439). In May of 1871 it published "The Eclipse of

Fig. 4.29 *H.M.S. Caledonia* and two steamers attempting to tow *H.M.S. Psyche* off a sunken rock on the coast of Sicily (after *The Illustrated London News* 1871)

1870" by Professor Edward C. Pickering from the Massachusetts Institute of Technology. Pickering opened his article with a general description of eclipses. After mention of the significance of the eclipses of 1868 and 1869 he described some preparations made and subsequent results obtained from the eclipse expeditions of 1870. Congress appropriated $29,000 for American efforts. Professor Peirce organized two expeditions, one to be led by himself to Sicily, the other by Professor Winlock to Spain. A third American party was organized for the U.S. Naval Observatory in Washington and would include Professors Eastman, Hall, Harkness and Newcomb. The English were encouraged to participate and would send observers to Spain, Algeria and Sicily. One of their boats, the steamer *Psyche*, crashed on rocks (see Fig. 4.29) but all survived and made their observations at Catania. Also to participate was one of Ireland's leading astronomers, Lord Lindsay, the 26th Earl of Crawford (see Fig. 4.30). The French sent Janssen to Algeria by balloon where he had no success due to cloudy skies. American observations were of five types: general observations, photographic, photometric, polariscopic and spectroscopic. Spectroscopist Professor Young, in Sicily, and Professor Winlock, in Spain, both saw the green line in the coronal spectrum that had been observed by Americans in 1869 and was doubted by Europeans at that time. The degree of polarization of the corona was determined to decide whether the corona was self-luminescent (un-polarized) or reflected sunlight (polarized). Results were inconsistent but seemed to lean toward some degree of polarization. Photographers focused on obtaining images of the corona and there were several successes. Ross used a photometer to determine the amount of light during totality. Several English observers would conclude that the corona had two parts, the inner being considered brighter. In general, observations were relatively poor as cloud cover interfered at most sites. An outcome deemed important was the agreement by the English that the green line observed by the Americans in 1869 was indeed present.

Fig. 4.30 A caricature of Lord Lindsay, painted by Leslie Ward, alias 'Spy' (after *Vanity Fair* 1878)

The *Princeton Review*, published from 1878 to 1884, had appeared under earlier titles of *Biblical Repertory* and *Presbyterian Quarterly and Princeton Review*. Under this most recent title its content became less theological and more general (Mott 1930, pp. 529–535). In the issue of July-December 1878 there was an excellent 24-page article by Professor Charles A. Young, "The Recent Solar Eclipse". This described the results obtained from observations at a number of eclipse sites from the first true eclipse expedition of 1842 to the most recent, and ended with a table describing sites and times of seven more upcoming eclipses. At the expedition of 1842 the polariscope was first applied and demonstrated polarization in the solar corona, but the main objects of interest were the rose-colored projections from the Sun (i.e. the prominences). In the Scandinavian eclipse of 1851 these prominences were again the main objects of interest. There was an uncertainty as to their source. Were they solar, lunar or were they optical illusions? In the eclipse of 1860 photography was first employed and the solar character of the prominences was established. In 1868 the spectroscope was first used, and Lockyer and Janssen soon afterwards demonstrated that one could study solar prominences outside of eclipses. Again polariscope observations seemed to demonstrate the polarized nature of the corona though there were negative reports as well. In 1869 an eclipse was visible in the United States. The first satisfactory photograph of the corona was taken at Shelbyville, Kentucky. Polariscope observations generally seemed to demonstrate that coronal light was polarized in radial planes, as expected if it was caused by particulate reflection of sunlight. Spectroscopes were used on the corona and the

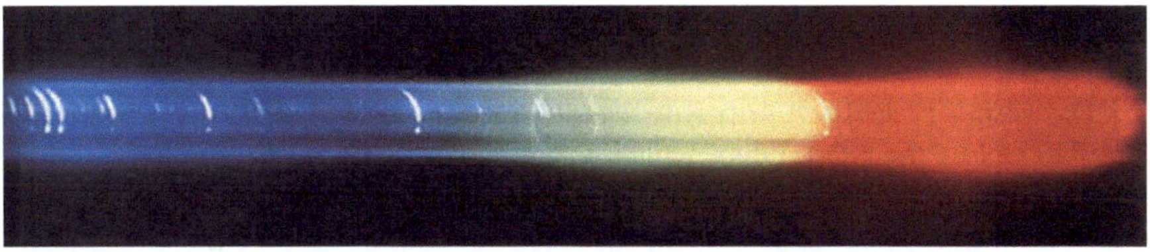

Fig. 4.31 A recent photograph of a solar eclipse showing the spectrum of the 'reversing layer' when for a few seconds just before totality the dark Fraunhofer lines are seen in emission (courtesy: http://history.nasa.gov/SP-402/ch2.htm)

unrecognized 1474 line (initially associated with the element 'coronium') was found. In the Mediterranean eclipse of 1870 spectroscopists detected the 'reversing layer' of the Sun's atmosphere. This occurs when the Moon just covers the Sun and the solar atmosphere alone is just projecting. At this point the dark Fraunhofer lines are transferred to bright lines (see Fig. 4.31). In 1871 there was an eclipse in India where Janssen observed some dark lines in the coronal spectrum. At the 1874 eclipse in South Africa there were no major original discoveries but confirmation of previous results. The Siamese eclipse expeditions of 1875 met with little success due to poor weather. After this summary of previous eclipse expedition results, Young provided a description of a number of parties scattered across the western territories to observe the eclipse of 1878. These had a variety of goals that were listed by the author. There were observations of precision (especially referring to the position of the Moon); general observations (especially naked eye observations of the corona); photographic, spectroscopic and polariscopic observations; some using miscellaneous apparatus such as photometers or Edison's tasimeter; and the search for intra-Mercurial planets.

In the January-June 1880 issue there was another article by Professor Young, on "Recent Progress in Solar Astronomy". After a summary of progress in establishing an accurate value for the solar parallax, there was a discussion of methods applied to better understand the nature of the Sun, itself, i.e. its temperature, its chemical constitution and the nature of the corona.

The *Reformed Church Messenger* was a weekly periodical out of Philadelphia published from 1867 until 1875 (chroniclingamerica.loc.gov). On 14 July 1869 under its feature title "Scientific and Useful" the article "Total Solar Eclipse in August" was printed. This provided particulars on various government parties to observe the American eclipse of 7 August 1869. On 18 February 1874 it printed the article "Professor Appel's Lectures". Professor Theodore Appel (1823–1907) from Marshall College had been delivering lectures at various venues on astronomy. It was related that projected on the screen for the audience were illustrations that were very satisfying, in particular, a photographic view obtained by a government party at the total eclipse of 1869.

The *Shaker Manifesto* was a monthly periodical for the benefit of the small community of the Shaker sect, and was published from 1878 to 1883. In October 1879 there was an article by Lucy Bowers, entitled "The Stars", which waxed philosophic about the grandeur of nature and natural phenomena. Long before the Christians, there were schools and disciples contributing to human knowledge, including philosophers who could calculate accurately the time of a solar eclipse.

The *Western Christian Advocate* was a Methodist weekly published from 1834 until 1883 in Cincinnati, Ohio, which was then considered western United States (Mott 1938a, p. 66). On 4 August 1869 it published the article "Total Solar Eclipse in August" informing readers of the arrangements to be made by the Government and others to observe the American eclipse of 1869.

The Youth's Companion was an offshoot of the *Boston Congregational Recorder*, intended to be instructive and entertaining to youth but of interest to the whole family (Mott 1938a, pp. 262–274). On 19 November 1868 it published "A Talk about Eclipses". This presented, in dialog form, an explanation of eclipses, total, partial and annular, with specific mention of the one to occur in the United States the following year. On 26 July 1877 there was an item written by Maria Mitchell, "The Planet Jupiter". Among other things, she described how eclipses, solar and lunar, would appear on Jupiter. Illustrations were provided. On 25 July 1878 the timely article "The Coming Solar Eclipse" was printed, which referred to the western American eclipse of 29 July 1878. Included were a diagram that demonstrated how solar eclipses occur and a picture representing the appearance of the eclipsed Sun at Des Moines, Iowa, during the 1869 eclipse (Austin 1878).

Zion's Herald was a Methodist periodical published in Boston (Mott 1938a, pp. 67 and 140). This publication made occasional mention of eclipses in passing but did not publish any extensive pieces on the topic. On 9 September 1869 there was an article on four young men participating in "The Boat Race" on the Thames (see Fig. 4.32), and the fact that "The sun's eclipse had caused far less excitement."

Fig. 4.32 The 'Great International Boat Race' between Harvard and Oxford Universities, on the Thames River near London on 27 August 1869, attracted an audience of about 750,000 (courtesy: www.rowinghistory.net/1869.htm)

Though this boat race was not at a venue where the eclipse would be visible, the article indicated that it generated comparable excitement (The boat race 1869, p. 427). On 14 October 1869 there was the article "The Storm – Have We a Prophet among Us?" speculating about the possible relationship between eclipses and atmospheric disturbances on Earth. On 13 May 1875 within the feature "Educational" the following sentence was published: "It is calculated that no solar eclipse of the present century will equal in length of totality, that which occurs on April 5, except that of April, 1893." (p. 147). On 30 August 1877 some excerpts from a lecture titled "The Telescope" by the theologian-astronomer Enoch Fitch Burr (1818–1907), given at Chautauqua, were published, describing briefly what can be learned by use of a telescope and further, in the case of an eclipse, the use of a telescope armed with a spectroscope (From our exchanges – The Sun 1877). On 16 May 1878, within "Educational," was a piece about the Princeton party going to Denver to witness the solar eclipse.

It is apparent that the many religious journals of the decades of the middle 1800s provided interesting and secular information on such subjects as astronomy.

4.2.3 *Scientific Periodicals*

Popular scientific periodicals were not common during this period. The first popular journal of astronomy, Ormsby MacKnight Mitchel's *Sidereal Messenger*, lasted only a little more than 2 years. Two periodicals of general science that had frequent articles on astronomy that could be appreciated by the general public were the *American Journal of Science and Arts* and *Scientific American.*

The *American Journal of Science and Arts,* published from 1818, is America's oldest scientific journal, and now is titled simply the *American Journal of Science.* Early editors included Professor Benjamin Silliman of Yale College and his son Benjamin, Jr. (1816–1885). In its early years it was commonly referred to as "Silliman's journal". It was never subsidized by Yale or any professional society. It was not intended to be a magazine of popular science, but its wide variety of content and literary quality made it of interest to all. Among other interesting contributions was Eli Whitney's description of the cotton gin (Mott 1930, pp. 302–304; 1938a, pp. 78–79). In March 1867 it published "Researches in Solar Physics" by Warren De la Rue, then President of the Royal Astronomical Society, the Scottish physicist Balfour Stewart (1828–1887) and Benjamin Loewy (1831–1892), dealing primarily with studies they made on the nature of sunspots. Regarding the corona it was stated: "... to this very day it has not been finally decided whether this luminosity proceeds from the sun's solid body, or from an envelope which surrounds it." (p. 179). The methods of observation were naked eye and photographic. The definitions of umbra and penumbra, and a description of gradations of luminosity were given. Among other conclusions drawn from their studies were the facts that the umbra is deeper into the Sun than the penumbra, and that a sunspot, as a whole, is a phenomenon taking place below the Sun's photosphere (ibid., pp. 179–192).

In January 1871 the *American Journal of Science and Arts* published "On the Corona Seen in Total Eclipses of the Sun" by Professor W.A. Norton of Delaware College, Newark (Delaware). Here Professor Norton discusses his auroral theory of the solar corona. He felt the irregular shape of the corona and the phenomenon of zodiacal light were best explained by such an electrical effect. He observed the total eclipse of 1869 in Des Moines, Iowa, and there observed the corona's "... great inequality in the extent of its outstreaming in different directions." (page 11; e.g. see Fig. 4.33). Polariscope observations indicated that the corona was unpolarized and thus self-luminous. Spectroscopic observations of the corona demonstrated a continuous spectrum with one or more bright lines, the latter which also indicated self-luminosity. Norton states that Professor Young observed in the coronal spectrum a bright line corresponding quite closely to a line recently discovered by Professor Winlock in the aurora borealis. Whatever theory for the origin of the corona might eventually be proven correct, it was most certainly a solar phenomenon, since photography of the 1869 eclipse showed "... as the moon advanced the corona was progressively covered." (p. 15).

In February 1871 the *American Journal of Science and Arts* printed excerpts of several letters announcing results of 1870 solar eclipse expeditions. From Professor B.A. Gould, dated 9 November 1870 in Cordoba, there was a letter to the editor

Fig. 4.33 The corona on 7 August 1869 seen with the aid of a 4-in. telescope (after Sands 1870)

regarding his experiences preparing for the eclipse site there. From General E. Abbot, in Sicily, there was the statement that success was limited due to cloud cover. He noted that the expedition had been divided into five parties. There were good sketches of the corona. Polariscopy results obtained indicated the coronal light was polarized in radial planes. Attempts at spectroscopy mostly failed although the English were able to confirm the American observations of the previous year. From Professor Peirce, also in Sicily, came the report that his party was successful in sketching the corona and obtaining polariscopy results. Professor Harkness reported from Sicily that the coronal light was strongly polarized indicating it shined by reflected light. Spectroscopic findings included a clear green line of unknown source. Professor Young was in Spain and at his site one good photograph was obtained during intermittently cloudy skies. His most interesting spectroscopic result was the reversal of the spectrum observed at the base of the chromosphere at the instant of totality's commencement (Appendix 1871).

In the *American Journal of Science and Arts* of July 1871 "On the Application of Photography to the Determination of Astronomical Data" by Asaph Hall was published. Hall describes the value of photography as a record of events as well as a means for accurate measurement. In 1860 Warren De la Rue first applied photography to determine the contact times during an eclipse.

In August 1875 the *American Journal of Science and Arts* published Edward S. Holden's article "Historical Note on the Observation of the Corona and Red Prominences of the Sun." He described the early attempts by the late George Phillips Bond (1825–1865), former Director of the Harvard College Observatory, to observe the prominences in the year 1851 by occulting all but its very edge, outside of the time of an eclipse. His attempts were unsuccessful but Bond's concluding words, transcribed from his diary, were: "On the whole, I am more than ever sure that the experiment can be made, and I think will be by someone more fortunate than I." (p. 83).

The September 1878 issue of the *American Journal of Science and Arts* printed Professor Henry Draper's "The Solar Eclipse of July 29th, 1878". He expressed his interest in solar spectroscopy and stated that the main point of recent solar eclipse expeditions was to determine "… whether the corona was an incandescent gas shining by its own light, or whether it shone by reflected sunlight." (p. 227). He identified the significant participants, their instruments and their tasks at his expedition site of Rawlins, Wyoming Territory, in 1878. His observations regarding what he considered to be the most important question led to the conclusion that the corona had a somewhat dynamic nature. At this event Professor Norton's work with the polariscope showed the corona to be reflected sunlight. Previously there had been evidence that the corona

> … had been infiltrated with materials thrown up from the chromosphere, notably with the 1474 matter and hydrogen. As the chromosphere is now quiescent this infiltration has taken place to a scarcely perceptible degree recently. (p. 228).

In October 1878 the *American Journal of Science and Arts* printed a contribution by Professor C.A. Young (1878b), "Observations upon the Solar Eclipse of July 29,

1878, by the Princeton Eclipse Expedition." His expedition party of ten persons, supported by the estate of the late John C. Green, was at Cherry Creek (locality 6 in Fig. 3.40), two and a half miles southeast of Denver. Their "… principle object was to investigate the spectrum of the corona and chromosphere – not only the visible portion, but also, and especially, the invisible portion below the red and above the violet." (p. 279). Young described the equipment used, including a tasimeter loaned to his party by Mr Edison. Unfortunately, it was found that this instrument needed such modifications as could not be done in a timely manner and so, although they were grateful for Edison's generosity, the device was useless to them. The weather was almost perfect and spectroscopic studies were carried out. Here Young describes the reversal of spectroscopic lines noted by him at this eclipse:

> At the moment when totality began, the field was filled with bright lines, which came into brightness, not instantaneously, but brightening perhaps half a second, remained steady nearly a second and then faded out and disappeared, not all together but successively, most of them being lost within two seconds of their first appearance, while some lasted three or even four seconds. (p. 286).

This remarkable effect would not be documented with photography until 1896 and again by a Lick Observatory party at a solar eclipse in Jeur, India, in 1898 (see Orchiston and Pearson, 2011; Pearson et al. 2011).

In the 1879 January-June issue of the *American Journal of Science and Arts* there was a letter from Thomas Edison to Professor Draper regarding the use of the tasimeter at Rawlins, Wyoming. The instrument had been created only 2 days before the expedition onset and Edison had been unable to test it adequately. The tasimeter was designed to detect the level of infrared radiation in the corona. The radiation would be focused on a hard rubber rod which expanded with the heat and pressed against a carbon button. Sensitive to the pressure, there would be a change in electric resistivity which would register on a galvanometer (see Israel 1998, pp. 160–162). The calibrated range was too narrow as the heat of the corona was too great. With proposed modifications Edison felt he could achieve success at a future event. Illustrations of this instrument were included in this article, one of which is shown here in Fig. 4.34 (and it is interesting to compare this with the very similar but clearer image presented in Fig. 3.43, in Section 3.1.4.3).

In February 1879 the *American Journal of Science and Arts* printed the letter from George Frederick Barker (1835–1910; Fig. 4.35), Professor of Physics at the University of Pennsylvania and one of Professor Henry Draper's party, "On the Results of the Spectroscopic Observation of the Solar Eclipse of July 29th, 1878". He described the results he obtained performing the tasks assigned to him. He observed no bright lines in the solar corona, concluding that it was not composed of an incandescent gas, but he did observe Fraunhofer lines, showing

> … conclusively the presence of reflected sunlight in the light of the corona and goes to establish the theory, long ago suggested, that masses of meteoric matter raining down upon the solar surface from all directions reflected to us the light of the sun. (p. 124).

By the time of these eclipses *Scientific American* was a well-established periodical covering many subjects in popular science. When started in 1845 it was a weekly publication and it remained so until 1921 (Mott 1938a, pp. 316–324). Being a

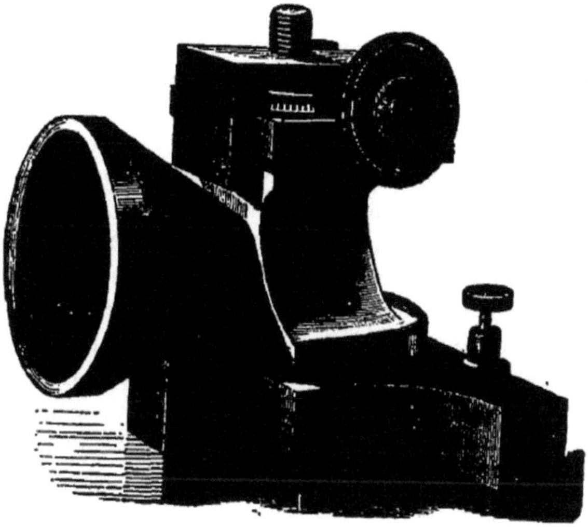

Fig. 4.34 The tasimeter (after Edison 1879, p. 52)

Fig. 4.35 Professor G.F. Barker (courtesy: en.wikipedia.org)

weekly publication, usually with relatively short items, there were frequent references to eclipses and related topics. On 27 June 1868 there was the article "The Latest Novelty in Photography". It described the unique invention of photographs visible only in the dark due to the use of phosphorus compounds. Speculation is made on the presence of such phosphorescence in nature, perhaps explaining some of the appearances of celestial bodies. Some of these might have been visible partly due to reflected sunlight and partly due to their own phosphorescence.

On 23 September 1868 there was an untitled article in the *Scientific American* with a paragraph entitled "The Astronomers in Luck". Telegraphs had been arriving from various sites reporting great successes in photography and spectroscopy at 1868 eclipse expedition sites (Article 11 – No title 1868). On 21 October 1868 there was the article "Total Eclipse in 1869" announcing that this next total eclipse would be visible in the United States. On 4 November 1868 "Accurate Observation" was printed. This was a generic article on the profits to be gained through scientific enterprise. In particular, valuable data might be expected from results to be obtained at the 1868 eclipse expedition sites. On 11 November 1868 there was a paragraph on an upcoming 'eclipse' (transit) of Mercury, visible only to those with telescopes, and the total eclipse of the Sun in 1868, to be visible by naked eye, in the United States (Article 3 – No title 1868). On 25 November 1868 "Lunar Assistance" was printed, covering the significance of eclipses in establishing dates of historical events. An example provided was that of the Thales' eclipse, an event which interrupted the war between the Lydians and Medes, ultimately resulting in peace. Professor Airy calculated the date of the eclipse and thus the date attributed to the cessation of hostilities. On 1 January 1869 "The Eventful Year of Our Lord 1868" was published. This article enumerated significant scientific events of the year including the eclipse. The tone of this item would demonstrate that religious undertones might introduce themselves even into secular publications. On 13 February 1869 there was a statement about the almost simultaneous discovery by M Jenssen (*sic*) and Mr Lockyer that spectroscopic studies might be made of the solar prominences outside of the time of eclipses (Article 8 – No title 1869).

On 14 August 1869 the *Scientific American* printed the item "How to Observe the Eclipse and Save Your Eyes" describing the use of pinhole cards and projection to safely observe eclipses. A week later it published the full-page illustrated article "The Great Solar Eclipse" referring to the one just witnessed in the United States. Herein was printed quite a bit of the history of observations made during other eclipses, going back to 1706. A map (see Fig. 4.36) was provided showing the central track and geographical limits of visibility of the recent event. A diagram of the fully-eclipsed Sun showed the corona, and another diagram showed three stages of the partial eclipse as it was seen in New York.

On 28 August 1869 the reader could learn of some early reports. Many sites reported fair weather and successful spectroscopic observations. So far, no observations of intra-Mercurial planets had been reported. (The solar eclipse 1869). On 4 September 1869 there was a speculative essay, "Is the Weather Affected by Solar Eclipses?" It ends with a request for data from its readership supporting either conclusion. On 11 September 1869 there was the report of G.H. Knight, "Observations of the Eclipse as Seen at Westport, KY". The writer gave some visual impressions of the event. In particular, he described three very different prominences, the largest as emanating from a concentric explosion, another resembling stag's antlers, and the last a horse's tail. For these he provided an illustration (see Fig. 4.37).

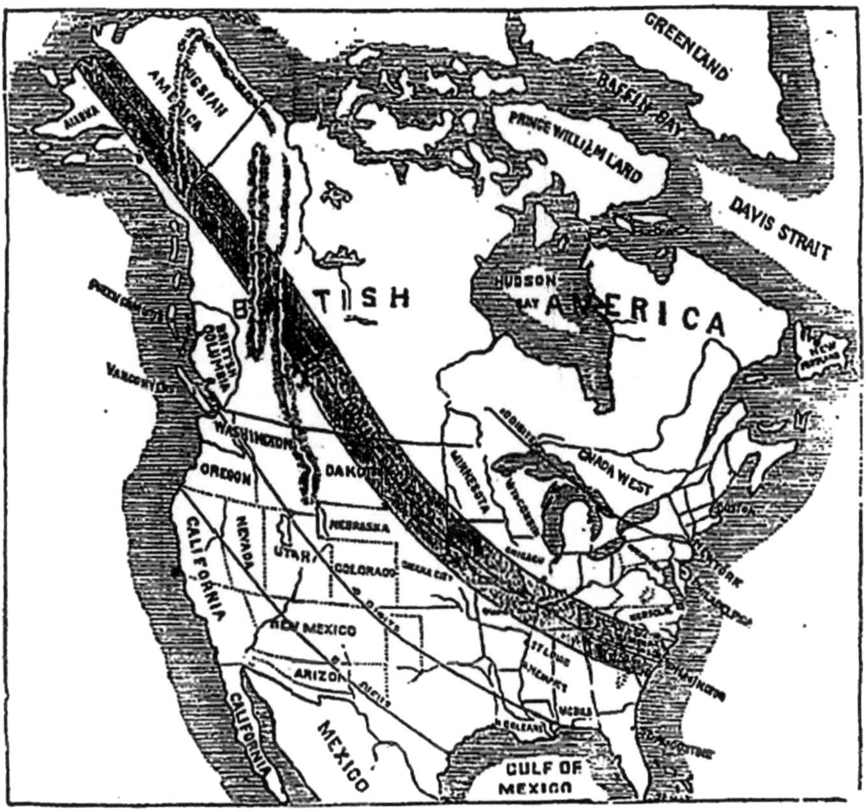

PATH OF THE ECLIPSE.

Fig. 4.36 The solar eclipse of 1869 (after *Scientific American*, 21 August 1869, p. 117)

On 1 January 1871 the *Scientific American* printed "The Solar Eclipse of 22nd December, 1870". While Congress provided \$30,000 for American efforts in this Mediterranean event, the British provided \$10,000 and one ship. The article provided a sketchy history of eclipse observations starting with the eclipse of 1706 where the prominences were first seen to the most recent event of 1870. Reports were still anticipated on polariscopic studies that should determine whether the corona is a luminous gas or particulate matter reflecting sunlight. On 21 October 1871 a paragraph entitled "The Approaching Solar Eclipse" was printed. The opinion was expressed that the Americans would continue to earn distinction with their contributions to solar physics during this eclipse visible in India, Ceylon and Australia as they had in the two previous events. On 3 February 1872 there was a report on this eclipse, "The Solar Eclipse, December, 1871". European astronomers confirmed Professor Young's report of a reversal of spectral lines from dark to bright at the commencement of totality. On 9 March 1872 "American Solar Eclipse Observations" was printed. This was a brief history of such observations since that of 1806, the first visible in the new nation of the United States. In the same issue

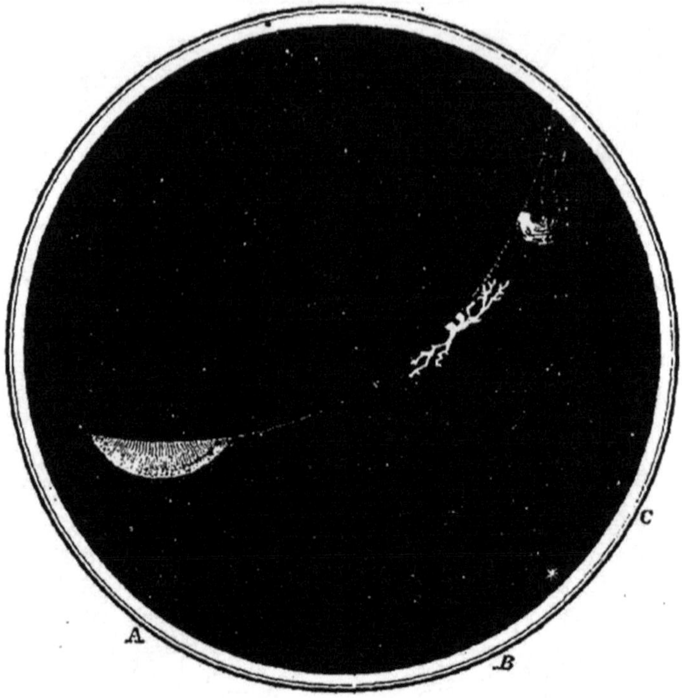

Fig. 4.37 The types of prominences G.H. Knight reported during the total solar eclipse of 1869 (after *Scientific American,* 11 September 1869, p. 165)

there followed a more recent report, "The eclipse of December, 1870." This time the United States sent two expeditions, one to Spain and one to Sicily. Spectroscopy and polariscopy were employed. There was confirmation of radial polarization of the corona.

On 16 March 1872 there was a letter in the *Scientific American* by John H. Leach of Dartmouth College who observed solar prominences on 16 February outside of an eclipse. He noted and provided illustrations of how these changed over time (Leach 1872a). On 6 April 1872 and 13 April 1872 there were two parts of an article by the same writer, John H. Leach of Dartmouth College titled "The Discovery and History of the Chromosphere". He described the chromosphere, the solar envelope outside of the photosphere, and the source of the solar prominences. He provided illustrations of the engravings of De la Rue from the 1869 eclipse when the chromosphere was discovered. The spectrum of bright lines was illustrated and Janssen's success in observing these lines the day after an eclipse was described (Leach 1872b, c).

On 30 August 1873 there was a *Scientific American* feature permitting readers to ask questions and receive responses. One questioner asked how solar and lunar eclipses were calculated. A response was given with reference to Burritt's *Geography of the Heavens* and Chauvenet's *Spherical and Practical Astronomy*

(Article 17 – No title 1873). On 11 July 1874 "The Solar Eclipse of April 16" was printed. Her Majesty's Astronomer at the Cape of Good Hope, Edward James Stone (1831–1897), was present for this event at the Cape and confirmed the reversal of the Fraunhofer lines at the instant of totality (The solar eclipse of April 16, 1874). On 6 March 1875 there was an article "Progress and Prospects of Solar Chemistry". It was not until what were then recent decades that this topic could be broached. The progress in these studies, starting with photography in 1860 and spectroscopy in 1868, was outlined. On 22 May 1875 the brief item "Recent Astronomical Studies" was printed which, among other things, announced the success of an English party in Bangkok in photographing the spectrum of the chromosphere (Scientific and practical information ..., 1875).

On 4 September 1875 the *Scientific American* printed an obituary, "Professor Winlock", on the death of Joseph Winlock of Harvard University. His contributions to improvements in astronomical instrumentation and his successes at the eclipse expeditions of 1869 in Shelbyville, Kentucky, and 1870 in Spain were recounted. On 11 March 1876 there was an article "The Solar Eclipse of March 25, 1876", anticipating an annular eclipse in the North Pacific Ocean that might achieve the rare occurrence of attaining totality for a few seconds. On 25 March 1876 there was the article "The Total Solar Eclipse of September 17–18, 1876", describing the track of a total eclipse in the South Pacific Ocean. On 14 April 1877 "An Observatory on Etna" was printed expressing the wish of Professor Pietro Tacchini (1838–1905) to establish an observatory on Mt. Etna where English and American expeditions had observed the total solar eclipse of December 1870.

On 25 May 1878 the *Scientific American* published "The Transit of Mercury and the Inter-Mercurial Body." Studies of the orbit of Mercury would simplify the calculation of the orbit of the proposed intra-Mercurial planet. The total eclipse of 29 July might prove to be an opportunity to discover this planet. On 13 July 1878 there was the short article "Where to Observe the Solar Eclipse of July 29th". General Albert James Myer (1828–1880; Fig. 4.38), the Chief Signal Officer of the Army, provided a list of favorable sites within the path of totality of the upcoming eclipse as well as a number of sites showing a partial eclipse of more than 90 %. On 2 November 1878 within the item "Astronomical Notes" there was a paragraph on the discovery of intra-Mercurial planets during the eclipse of 1878 by Professor James C. Watson, and confirmed by Lewis Swift of New York. It was believed by Watson that Lescarbault's predicted planet 'Vulcan' was seen as well as a second new planet (Wright 1878, p. 277). On 28 December 1878 "The National Observatory" was printed describing work in progress. Data reduction had been ongoing on recent transits of Venus and Mercury as well as the total solar eclipse of 1878.

On 8 February 1879 the *Scientific American* printed the article "The Solar Eclipse of 1880", describing the track of the shadow and stating the possibility that totality might be visible briefly in California near sunset.

Scientific American was an important source of information for the public on scientific topics during these events and continues to be so in the present day.

The number, content and quality of articles on the subject of the total solar eclipses under consideration here would indicate a real and growing interest

Fig. 4.38 General A.J. Myer (courtesy: en.wikipedia.org)

among the general public in astronomical subjects. The dramatic nature of solar eclipses had been described for the vicarious appreciation of the events as well as a scientific understanding.

4.3 *The New York Times*

The New York Times was a daily publication, usually only eight pages in length, at the time of these various eclipses. The first issue published as the *New York Daily Times* came out on 18 September 1851 at the cost of one cent, showing its intention of being "… the paper for the masses." (Emery and Emery 1984, p. 152). "*Daily*" was dropped from its name in 1857. The first page was usually devoted to national and international news. The next two pages would be devoted to book reviews and general articles. The fourth page was frequently editorial and might include late telegraphic news. The next three pages would be local news and advertisements and the last page would be financial news and advertising (Davis 1921, pp. 21–27, 36–38, 46, 62–64, 156, 168–169). The placement and amount of space delegated to articles of astronomical topics in such a publication would give an indication of anticipated public interest.

In 1867 preliminary to our first eclipse under consideration, that of 1868, this newspaper made little mention of such events. There was a reprint of an article from the *Boston Traveller* on an annular eclipse to take place on 6 March in Europe.

In this article there was also anticipatory mention of the next total solar eclipse to take place in the United States, in 1869 (Central eclipse of the Sun 1867). Later that year, on 19 August, there was an article on a total solar eclipse that would be visible across much of South America (Minor topics 1867).

4.3.1 The Solar Eclipse of 18 August 1868

1868 was a significant year in eclipse observations as this year was the first where the science of spectroscopy was applied to the study of the solar corona. *The New York Times,* however, made little mention of this important advance in astronomy. On 26 April of that year the *Times* reprinted an article from the *Mechanics Magazine* on the preparations to photograph the eclipse in India (Preparations to photograph the great eclipse 1868). On 2 August there was an article dealing with the likelihood of finding an intra-Mercurial planet during the event (The coming eclipse 1868). Later that month there was a paragraph reporting that the eclipse was observed satisfactorily in much of India, except at Bombay where the weather was uncooperative (Telegrams – India – The total eclipse of the Sun 1868). On September 20 a letter to the editor signed "A.G." was printed describing the conditions and successes of visual observations made at Aden (A.G., 1868). A week later there were some findings on the nature of the prominences seen during eclipses, confirming recent conclusions that these were indeed associated with the Sun and not the Moon, as once thought (Foreign news by mail … 1868).

It was not until 6 December 1868 that mention was made in the *Times* of the usefulness of spectroscopy in solar studies during an eclipse. Here were discussed the successes of a "French savant" (i.e. Janssen) and the English "Mr Lockyer" in discovering that the prominences consisted of burning hydrogen gas. In the concluding paragraph it was stated, "… science has taken another step, and one which is worthy of popular recognition." (The latest advance in solar science 1868). A mention is made of this eclipse on 1 January of the following year when it is listed as one of the significant events of the year (The old year … 1868). On 18 July 1870, almost 2 years after the eclipse, the *Times* printed its most extensive recognition of the importance of the eclipse. Here was briefly summarized what had been learned to date regarding the constitution of the Sun based on the methodology of spectroscopy. It was further speculated that "… there really seems to be no reason for believing that we have as yet reached the limits of the knowledge which spectroscopic analysis is capable of supplying." (Recent solar researches 1870).

4.3.2 The Solar Eclipse of 7 August 1869

In 1869 articles in *The New York Times* would reflect a growing interest in solar eclipses by the American public. This time reports would enthusiastically be shared among those observing from metropolitan areas as well as from more remote expedition sites, where there were observers hoping to witness a total solar eclipse.

Fig. 4.39 A caricature of Sir William Crookes, painted by Leslie Ward, alias 'Spy' (after *Vanity Fair* 1903)

On 19 May 1869 it was reported that the Government made preparations to establish a 'meridian line' in Springfield, Illinois, where the eclipse would be seen as total: "… this will be the most complete and interesting eclipse that will occur in this country during the century." (A new meridian line 1869). Four days later an article was reprinted from the *Springfield Journal* of 17 May with more specifics on the path of totality (The solar eclipse in August 1869).

On 28 May a Letter to the Editor was published with an excerpt from a report by the British physicist William (later Sir William) Crookes (1832–1919; Fig. 4.39), Editor of the *Chemical News* and Co-editor of the *Quarterly Science Review*. Crookes recommended the use of multiple photographic methods during the event. Both the daguerreotype and dry albumin processes were capable of yielding more detail than the wet colloidal process and also had the logistical advantage of allowing some preparation beforehand and development afterwards, thereby permitting the photographer to expose a maximal number of plates. He also recommended a system of lenses to magnify the image prior to development. Crookes concluded from the experiences of previous European expeditions that photographers of talent were necessary to guarantee success (The approaching total eclipse of the Sun … 1869).

On 17 June 1869 there was a paragraph describing the sites chosen and equipment to be used by the party of Professors Harkness and Newcomb from the U.S. Naval Observatory. At a site outside of Des Moines, Iowa, they would use three telescopes as well as photographic and spectroscopic equipment (Washington – The solar eclipse in August 1869).

On 6 July 1869 the suggestions of Commodore B.F. Sands, Superintendent of the U.S. Naval Observatory, were printed for useful meteorological data to be obtained during the eclipse. The path of totality was to be approximately 140 miles wide and was to pass through a number of cities and towns that he listed. He described equipment to be used and a schedule of data collection for all participants. All were encouraged to share their observations with the Observatory (The solar eclipse – Arrangements by Commodore Sands 1869). On the next day a paragraph was printed about Dr Edward Curtis (1837–1912) of the United States Army who would join the U.S. Naval Observatory party in Des Moines to photograph the progress of the eclipse (The coming solar eclipse 1869).

On 23 July two parties sent from Cambridge, Massachusetts, to observe the eclipse were described. Joseph Aimlock (*sic*, a misspelling of Winlock) would lead a party to Shelbyville, Illinois (*sic*, another mistake as the site was Shelbyville, Kentucky). Winlock's party included, among others, Charles Pierce, John Adams Whipple, Arthur Searle and Messrs. Clarke (*sic*, should be Clark), "… the mathematical instrument makers …" of Cambridge (Massachusetts – Observers of the eclipse 1869). They

> … are as finely equipped as any party of observation that ever went out, either in this country or in Europe, having a dozen powerful telescopes and every other instrument that could be serviceable to them which the country affords (ibid.).

The other party mentioned was that of the members of the United States Coast and Geodetic Survey, which included Professor Benjamin Pierce (*sic*) and photographer J.W. Black, who would be going to Springfield, Illinois (ibid.).

On 28 July a reprint of a letter by C.H.F. Peters to the Editor of the *Utica Herald* was published on 27 July. Peters, of the Litchfield Observatory (see Fig. 4.40), noted that the eclipse would achieve approximately 7/8 totality at that site. He would be taking a party outside of Des Moines, Iowa, where there would be totality lasting approximately 2 min and 45 s. He stated:

> While observations for precising the times of beginning and end will not be neglected (though nowadays of comparatively little use) the principle object will be the investigation of certain questions regarding the physical constitution of the sun and its envelopes. (Peters 1869).

The newer methodology of spectroscopy would be important at this expedition site (ibid.).

A more remote expedition site was the subject of an article on 31 July. The party of Professor Davidson of the United States Coast and Geodetic Survey had arrived safely at Sitka, *en route* to the Chilakahat's country in Alaska where they would observe totality (The Pacific Coast … 1869).

On 2 August 1869 *The New York Times* published an extract from a paper written earlier by the popular American author James Fenimore Cooper (1789–1851; Fig. 4.41),

Fig. 4.40 Stereo images of the Litchfield Observatory (courtesy: en.wikipedia.org)

Fig. 4.41 James Fenimore Cooper in about 1850 (courtesy: en.wikipedia.org)

on the eclipse of 1806. This eclipse was visible in parts of the United States, but prior to the scientific advantages of photography and spectroscopy. Born in 1789, Cooper was just a teenager in 1806, but his talents would later bring to reality this event of his youth. He described the sky, the weather and the surrounding responses of the animals and humans. He concluded:

> I shall only say that I have passed a varied and eventful life, that it has been my fortune to see earth, heavens, ocean, and man in most of these aspects; but never have I beheld any

> spectacle which so plainly manifested the majesty of the Creator, or so forcibly taught the lesson of humanity to man, as a total eclipse of the sun. (The eclipse of 1806, 1869).

In early August the *Times* printed articles to prepare the public for the eclipse event. On the 5th the public was apprised of the best locations and the times for observing. They were also informed of some associated phenomena such as the corona, Bailey's (*sic*) beads and protuberances (i.e. prominences). The history of some of Janssen's spectroscopic successes was described. Some interesting comments were made regarding the use of calculated eclipse times in establishing the dates of associated historical events. The article ended with a list of times when the 1869 eclipse might be observed from different U.S. cities; in some cases the eclipse would only be partial (The great solar eclipse – Where it begins and ends 1869).

On 7 August, the actual date of the eclipse, there was an article with more specifics informing the public on how to observe it. Two methods were suggested: using smoked glass or a card with a small round hole in the center that would permit passage of the image of the Sun onto a selected background screen. Mention was made of the fact that Professor Harkness at the Des Moines site would be using the spectroscopic method first employed during the eclipse expeditions the previous year in India. Included in the content of this article were the somewhat disparaged words of a citizen who predicted an earthquake with the eclipse, based on the coincidental eclipse/earthquake events in South America the previous year, a potential topic of discussion for the intrigued public (The eclipse to-day 1869). On that same day a program was published of the usual Saturday musical concert to take place in Central Park, but this time to be accompanied by a partial solar eclipse (The eclipse and music in Central Park today 1869)!

On 8 August the *Times* printed their longest article regarding this eclipse. In New York the public was "… made aware in ample time to prepare for the show …" by the newspapers. Young street vendors had been selling colored glass for days. In Central Park the concert-goers, having been provided with pieces of colored glass, had their eyes turned to the sky. Representatives of the press went to observe the eclipse from the meteorological observatory set up in Central Park by Daniel Draper (1841–1931). The report of Professor A.E. Thatcher of New York, who made his observations from the top of the Astor House, was made here. He noted the times of contact and the meteorological conditions. Telegraphic reports from the sites of Washington D.C., Des Moines, Shelbyville, San Francisco, Cincinnati and Wilmington, North Carolina were printed. An interesting comment made here was that this well-predicted event did much to re-establish the public's confidence in the astronomical community. The public had been much disappointed a year or two previously when a predicted great meteor shower did not occur as promised (The eclipse – How it was seen in New-York and elsewhere 1869).

Reports of results continued to be printed during the month. In its summary of the event as experienced in New York, on 9 August the *Times* noted, "… the exhibition amply gratified the curiosity of the spectators, and has doubtless led many to study, with a greater deal of interest than heretofore, the movements of celestial bodies …" This was in spite of the fact that the event was obscured much of the time by clouds. In this article some results from other sites also were printed. Professor Pearce (*sic*) of Harvard had charge of the observations at Springfield, Illinois.

One hundred photographs were taken by Mr Black of Boston and exceptional observations were made of the corona, prominences and Baily's beads. Members of the public filled the streets and rooftops in Wilmington, North Carolina. Also printed were the scientific observations of Winlock at Shelbyville and the various parties at Des Moines. Briefer reports were printed from various other sites across the country and from Montreal in Canada (The great solar eclipse – City scenes 1869).

On 11 August there was an article of two paragraphs submitted by 'Our Own Correspondent' as a witness at the University Grove site in Chicago. The skies were ideal and the times provided precise. He mused: "But when shall science penetrate the mystery of the sun itself, and tell us the source of its light and heat?" (Chicago – The eclipse 1869). On that same date there was an article submitted by Daniel Draper of his observations from his meteorological observatory in Central Park. He reported on the variations in temperature, barometric pressure and various wind parameters as noticed during the eclipse. He anticipated making a future report of long-term effects several days later (The eclipse – Its meteorological effects 1869). On the following day a more extensive article from 'Our Own Correspondent' from Springfield, Illinois, was printed. Here, after brief mention of the scientific parties present, he attempted to describe the experience of the "… unscientific observer …", i.e. that, "It was the grandest sight of a lifetime." He noted planets and stars observable by name and the effects on the local fauna (The eclipse – Observations at Springfield and Illinois 1869).

As the days passed more reports from further sites would come in to be printed. On 21 August there was a short article of various news items from Canada. Included was the fact that Commander Edward David Ashe of the Royal Navy participated in observations in Iowa with great success (Canada – Arrival of Prince Arthur's horses 1869). On 26 August an anecdote was recounted, that would now undoubtedly be considered politically incorrect, of the amusement of Dr Peters at the reaction of an old Negro's observations of his chickens during the eclipse (The late eclipse – An old Negro and his chickens 1869). In 1867 Secretary of State William Henry Seward (1801–1872; Fig. 4.42) had arranged the purchase of the future state of Alaska from the Russians for the sum of $7,200,000. On 28 August 1869 the *Times* reported of his subsequent visit to Professor Davidson's camp in Alaska, where, during breaks in cloud cover, all were able to see the various phases of the eclipse as well as the phenomena of "… the rose-colored flames and the corona …" (The Pacific Coast – Arrival and reception of Mr. Seward 1869). On the next day there was a report of Professor Davidson's serendipitous discovery of a mountain range of iron ore in Alaska during his eclipse expedition (The Pacific Coast – Discovery of iron in Alaska 1869).

On 2 September 1869 a piece was published speculating on the possible relationship between four shipwrecks and the eclipse, all occurring on the same date (Article 1 – No title 1869). On 5 September there was another account of the eclipse in Alaska and the visit of then Governor Seward to the expedition site. Comment was made of the fear experienced by the local Inuit people during the event (The eclipse in Alaska 1869). Later that month, on the 14th, an article was reprinted from the San Francisco *Alla California*, yet another account of Seward's experiences in Alaska.

Fig. 4.42 William H. Seward (courtesy: en.wikipedia.org)

The article began with a description of his reception upon arrival on the *Active* to the town of Sitka. He then reboarded to continue to the eclipse site at the Indian village of Chilcat. It was actually during the trip up river that Seward and his party observed the eclipse. The subsequent reception and dinner parties were described, stylish but oftentimes tedious due to the complications of communication. Two interpreters were always needed to translate among the English, Russian and Indian languages. A few days later Seward and his party returned to San Francisco (Mr Seward in Alaska 1869). On 23 September there was a report of the total failure of the Siberian expedition led by Professors Hall and Rogers of the U.S. Naval Observatory (The Pacific Coast – The astronomical expedition to Siberia unsuccessful 1869).

On 3 October 1869 Professor Hall's response to previous reports of failure was printed. There were three stations and there were indeed some successes in seeing the eclipse distinctly through the clouds (The eclipse – Observations at Behring Straits 1869). The following day there was a front page article relating experiences in Siberia from Rear Admiral Thomas Turner who had transported Hall and Rogers to their sites. He noted predominantly overcast skies with occasional periods of good visibility. In his opinion he did "… not think that the weather was sufficiently clear to make the observations of the astronomer of value." (The eclipse in Siberia 1869). On 8 November 1869 the readers were informed of the return of the Behring Strait expedition party to San Francisco the previous day (The Pacific Coast – Return of an eclipse-observation party 1869).

The following year interest in solar eclipses seemed to wane as evidenced by the decreased number of articles on the subject in *The New York Times*. More than a year after the eclipse on 22 August 1870 there was an article stating that there were still reports to be heard from some expeditions to sites of the 1869 solar eclipse. Of the three Government parties, those from the U.S. Naval Observatory, the Coast and Geodetic Survey and the Nautical Almanac, only the U.S. Naval Observatory had published its results (Washington – The last great eclipse 1870). The nature of the U.S. Naval Observatory publication was described on 7 October (Washington – Astronomical observations 1870).

Some mention of anticipation of the eclipse of 22 December 1870, best visible in parts of southern Europe, would be made. On 20 November there was an article explaining how spectroscopy had solved the mystery of the nature of solar prominences. It was expected that the upcoming eclipse might clarify the nature of the corona (The coming eclipse of the Sun 1870). Although this eclipse did not seem to intrigue the American public as much as the one that had just occurred at home, their scientists would send some expeditions. This was related in an article published on the day of the eclipse (Today's eclipse 1870). On 13 November 1870 it was reported that the American parties had invited eminent English physicists to join them (Foreign items 1870). Professor Pierce (*sic*), joined by Mr Lockyer, led the main expedition to Syracuse, but unfortunately had little success due to the poor weather conditions, as reported on 26 and 27 December. However, it was reported on the second of these dates that another expedition in Cadiz, under the leadership of S.J. Perry, obtained satisfactory photographs of the corona (General European news … 1870; The recent eclipse. 27 December 1870).

4.3.3 The Solar Eclipse of 29 July 1878

Starting approximately 2 months before the event *The New York Times* would mention plans for the upcoming expeditions to observe the eclipse of 1878. Public interest should be piqued as most of these sites would be within the boundaries of the United States and its territories. On 30 May 1878 there was a special dispatch from Princeton College on preparations taking place there including those of Professors Charles A. Young and Cyrus Fogg Brackett (1833–1915) who were taking an expedition to Denver (Princeton College – The approaching commencement 1878). On the 16th of the following month in another special dispatch Princeton would reveal the plans of these professors who would leave for the expedition shortly after commencement. Another expedition of un-specified membership had already left the previous week (The Princeton commencement 1878). On 11 July the public was notified of the plans of Professor Langley of the Allegheny Observatory (see Fig. 4.43) to observe at Pike's Peak (The coming eclipse of the Sun 1878) and on 14 July of New York Professor Henry Draper's plans to observe at Rawlins in the Wyoming Territory (Eclipse expedition from New-York 1878). On the 15th of that month the public would be reminded of the plans of Young and Draper as the leading item in the regular feature "Scientific Gossip."

On 18 July 1878 the *Times* printed a special dispatch on the progress of the Draper expedition which had just passed through Omaha, Nebraska. Among others

Fig. 4.43 A model of the original Allegheny Observatory, as it appeared at the time of the 1878 solar eclipse (courtesy: en.wikipedia.org)

in the party would be Professor Edison who would use his tasimeter to measure the infrared radiation from the Sun during the eclipse and who would also test his new 'carbon telephone' between Omaha and the site at Rawlins (The Draper eclipse expedition 1878).

On 19 July 1878 there was a summary paragraph stating that the upcoming eclipse would be the last visible in the United States in the century and listing parts of the country where it could be seen, lasting on average 2 min and 50 s (Topics of the season 1878). On 21 July there was an article by 'Our Own Correspondent' with an extensive list of expedition parties, their sites and some considerations regarding the likelihood of satisfactory weather conditions. On the same page of that issue there was a short paragraph relating the various locations of Professors Newcomb, Todd, Hall, Harkness and Holden, all representing Government interests (The eclipse of the Sun, 21 July 1878; Untitled, 21 July 1878). On the 26th there was another progress report on the Draper expedition. After safely arriving at Rawlins the members were preparing their telescopes and equipment for the event. Specifics were provided on photographic equipment for those interested. Professor Draper and his wife were to do the photographic work. Professor G.F. Barker would operate the analyzing spectroscope and Stevens Institute of Technology President, Professor Henry Jackson Morton (1836–1902; Fig. 4.44), the polariscope. Mr Edison would use his tasimeter to measure the heat of the corona (Draper's eclipse expedition).

Fig. 4.44 Professor H.J. Morton (after Reich 1900, p. 430)

On the day before the actual eclipse *The New York Times* printed a nice summary of plans and expectations for the following day. As the "… earth is indebted to the sun for almost all of the manifestations of force which we find here… [one] … should seek to learn as much as possible about the sun …" It was total in western areas of the United States and its territories but a partial eclipse would be visible throughout the rest of the country. The approximate times for the beginning and the end were listed for numerous large cities. Scientists would be using photography, spectroscopy and polariscopy to gain an increased understanding of the corona. They would search for a theorized intra-Mercurial planet. "Never before have preparations so elaborate been made to observe an eclipse." (The coming total eclipse, 28 July 1878).

On the day of the eclipse the *Times* printed a description by General Henry E. Oliver of the eclipse of 1806 (The eclipse of 1806, 29 July 1878).

The day after the eclipse the *Times* printed several articles related to the event. A brief untitled article stated that telegraphic reports were coming in indicating that all observers had good atmospheric conditions for their observations (Editorial article 1 – No title, 30 July 1878). On that same day a tongue-in-cheek item was printed theorizing on some of the mysteries of the Sun. As the heat of summer brings flies on Earth, an observer outside of our planet might see this fly-sphere as a homogeneous covering of the surface. A brushfire or hailstorm might cause movement within the fly-sphere giving the appearance of a spot. The writer offers this as an explanation for sunspots (Solar mysteries 1878). Also on this date the interesting anecdote was published that at an international baseball championship in Utica, New York, a game ended with the first half of tenth inning owing to the darkness caused by clouds and the eclipse. However, the effects of the latter may have been exaggerated as the site was quite far from the line of totality (Base-ball 1878). Also on the 30th there was an extensive article with reports from various venues of observation. It led with the dispatch from New York's own Henry Draper:

> Totality just over. I have splendid photographs. Two are spectra of the corona, taken with a grating, and show the spectrum to be continuous. Two photographs of the corona are full of details. Edison's tasimeter showed the heat of the corona.

It was reported that Professor Watson, at the site of Separation Point, discovered an 'extra-Mercurial planet'. In the city of Denver, the populace stood on their rooftops with smoked glass to best see the event. Astronomers there made satisfactory spectroscopic observations but failed to find Watson's extra-Mercurial planet. At Idaho Springs, Professor Darwin Groves Eaton (b. 1822) from the Packer Institute and S.V. White from Brooklyn, New York, had cloudless skies and they made numerous sketches of the corona as did Professors Holden and Compton at Central City. Brief reports of success and/or visibility were also received from 20 other sites (The eclipse of the Sun – Successful observations in the West 1878).

Over the next several weeks summaries of reports would continue to be printed as received. On 31 July 1878 there were three reports of success from Virginia City in Montana, Havana and Quebec (Monday's eclipse of the Sun 1878). On 1 August there was the brief report from La Junta, Colorado, "Rogers' photographs developed last night; they are excellent. Structure of corona well shown. – Hall" (The solar eclipse 1878). The following day there was a report from Denver where astronomers from numerous parties met. Professor Watson felt certain he had discovered the intra-Mercurial planet Vulcan and Professor Draper got a photograph of the coronal spectrum (The solar eclipse 1878). On 3 August an article was printed expressing the opinion that most astronomers were hiding an awful truth about the Sun. Only Professor Lockyer would admit that some noted changes in the Sun must have serious results. There had been a decrease in the number of sunspots, and in recent eclipse studies a disappearance of hydrogen from the solar corona was noted. It was argued that this must mean there was a decrease in solar activity and heat. Decreases in solar heat meant more heat on Earth which seemed to reflect the weather pattern noted over the previous several years (The truth about the Sun 1878).

On 4 August 1878 there was an educational item from the *Providence Journal* for readers with an interest in astronomy. It was noted that the Moon was a 'star' in the sky during this lunar cycle. Starting with the solar eclipse, she would within 2 weeks also be party to a partial lunar eclipse. She would further demonstrate a conjunction with the planets Jupiter, Mars, Uranus and Mercury (The August moon 1878). On the same date there was the special dispatch that Professor Swift of Rochester had certainly discovered the intra-Mercurial planet, agreeing with Professor Watson's data (The recent solar eclipse 1878).

On 8 August there were three mentions of the solar eclipse. On this date there was a special dispatch from Professor Watson of Ann Arbor stating his success in discovering Vulcan (The discovery of Vulcan 1878). Reports on New York's own Professor Henry Draper were printed with results of his private party expedition to Rawlins, Wyoming Territory. In his interview for this article, he clarified the misconception that had been printed in numerous accounts that he was looking for oxygen in the Sun's corona, which he knew could not be present. His studies did, however, center on the corona, by methods photographic, spectroscopic and polariscopic, which led him to the conclusion that coronal light was reflected sunlight. Party member Mr Edison used his very sensitive tasimeter to measure coronal heat (The examiners of the Sun 1878). Reprinted from the *Denver Tribune* was the story of the popular actor Joseph Jefferson (1829–1905), who was taking a rest in Colorado, fishing during the solar eclipse (Joe Jefferson goes a-fishing 1878).

On 11 August there was an article on Professor Lockyer and the nebular hypothesis as it accounted for the origin of worlds. Having just returned from Colorado, where he observed the eclipse, he stated that nothing learned there would negate his theory (The nebular hypothesis 1878).

On 16 August there was a submission from Professor C.A. Young of Princeton on the event. At this time he was willing to accept the conclusions of Watson and Swift of the existence of the planet Vulcan. He felt that this and the apparent relationship of sunspot activity to the coronal atmosphere were the two significant outcomes of the various expeditions for this eclipse (Vulcan and the corona 1878). The next day a human interest story from the *Denver Tribune* was reprinted on the response of the Pueblo Indians of New Mexico to the solar eclipse. At the Pueblo village at Taos, New Mexico, it was an unexpected event and the darkening sky caused fear among the inhabitants. Their governor stated that it was due to a grievous sin of one of their own and if proper penance was not paid, crops would fail and certain death would result. The women of the community made penance by running naked on the fast-track until such time as light grew stronger (Their wrathful deity 1878).

Noted in *The New York Times* "Scientific Gossip" on 15 September was the conclusion by Jean Baptiste Aimable Gaillot (1834–1821) from the Paris Observatory that Professor Watson's intra-Mercurial object closely followed one of the potential paths predicted by Leverrier.

On 8 November 1878 there was an article on the meeting of the National Academy of Science at Columbia College. Among those presenting papers was Professor Draper, who had made significant observations in astronomy, most recently at the solar eclipse (Science on the platform 1878). The next day there was an article on the last day of the same meeting. Professor Stephen Alexander (1806–1883; Fig. 4.45) of Princeton College presented a paper requesting that observers of the solar eclipse coming in January 1880 direct their attention to the confirmation of the existence of the planet Vulcan (Miscellaneous city news – National Academy of Science 1878).

On 24 December there was an article on the many successes of Princeton College that included Professor Young's expedition to observe the solar eclipse of that year, "... one of the best equipped and one of the most successful." (Princeton College 1878).

There was little specific mention of the 1878 solar eclipse the following year but articles in *The New York Times* would demonstrate that there was still a lingering interest among the public in astronomical topics. On 1 March 1879 there was an

Fig. 4.45 Professor S. Alexander (courtesy: paw.princeton.edu)

article on lectures given by John Lockwood on astronomical subjects ranging from the "… wonderful guesses of Kepler …" to the recent use of spectroscopy (Astronomical researches 1879). On 11 and 21 November 1879 there were two items describing a series of lectures on the progress of astronomy, given at Chickering Hall in New York, by England's ever-popular Richard Proctor to "… a cultured and refined audience, not composed of astronomers." Proctor discussed topics of interest to the public, such as the habitability of the moon and other planets. He discussed the use of spectroscopy as a tool to understand the nature of the Sun and stars, presenting as examples photographs of several spectra collected by Henry Draper. He predicted a great future for the use of spectroscopy including "… how it might be applied to the determination of the question whether a star is advancing or receding." (Progress of astronomy, 11 November 1879; Worlds beyond the skies, 21 November 1879).

On 30 April 1879 there was a notice that Lewis Swift, a hardware dealer and amateur astronomer, had been elected as a fellow of the Royal Astronomical Society of England. Swift had been one of the individuals who "… discovered …" an intra-Mercurial planet during his observations of the 1878 eclipse in Denver (Astronomical researches rewarded 1879).

There were a number of articles referring to the conflicting conclusions of Mr Lockyer and Professor Draper on their interpretations of spectra collected during the solar eclipse of 1878. On 7 January 1879 there was an article on a paper presented by Lockyer at a meeting of the Royal Society in London. Lockyer's comments were preceded by a paragraph on the basics of spectroscopy for the benefit of the interested reader. He explained to his audience how spectroscopy might be used to determine a number of parameters, such as the presence and quantity of particular elements or compounds, as well as the temperature of a star (Nature of the elements 1879). On 14 January 1879 the *Times* printed an article reporting on the response of some scientists, Professor Henry Draper in particular, to conclusions drawn by Lockyer from spectra collected at the last eclipse. Lockyer had concluded that "… so-called elements are really compound bodies." He inferred from his observations that the hottest stars have the simple spectra of thick hydrogen lines with very few metallic lines while cooler stars produce lines of "… compounds of metals with non-metals and of non-metals in a state of isolation." He proposed that heat led to the decomposition of the compounds. Draper did not dispute Lockyer's theory but he felt it was at yet unproven (The new solar theory 1879). On 27 January 1879 there was an article describing Professor Draper's discovery of oxygen in the Sun. Lockyer's response to this varied from skepticism to the suggestion that he, himself, had found such spectral lines prior to Draper's revelation. Asked for documentation of this claim, Lockyer claimed to have lost his photographs in the metro tunnel in London. At the time of this article, Draper was working on providing larger photographs that would make his claims indisputable. Within the article the interesting notion is expressed that science in America had become a force to contend with. A distinguished American scientist was quoted: "Germany to-day fears the physical experimentalists of the United States more than she fears those of any other country, and has long since practically adopted a protection tariff in the department of scientific discovery." (Prof. Draper's discovery 1879). Later that year, on 28 August 1879, there is a report on the meeting of the American Association for the Advancement of Science. Here the President of the meeting, Dr George F. Barker, expressed his concurrence with Draper's opinion on the presence of oxygen in the Sun (Scientists at Saratoga 1879).

Late that year there was some discussion of Edison's tasimeter. On 3 December, Professor Charles A. Young of Princeton delivered an address at a meeting of the New York Academy of Science. He discussed various difficulties met during observations of the solar corona from 1869 to 1878, these bracketing years being those where eclipses could be observed on American soil. Preceding some final photographs of his staff at work at the 1878 eclipse he discussed the "… failure …" of Edison's tasimeter (Problems of solar eclipses 1879). The next day there was an untitled editorial discussing the difference in opinion between Professor Young and Mr Proctor on the usefulness of the tasimeter. Proctor suggested its application as a detector of icebergs! It was implied that Proctor was practicing "… platform astronomy …" (Editorial article 7 – No title 1879). The *Times* printed Proctor's response 4 days later. Proctor was very gracious in his comments regarding Professor Young. He agreed that the tasimeter was a failure in 1878. This new instrument was set under the supposition that only delicate adjustments would be necessary. In fact, it was so sensitive that the indicator overshot the expected range. This was a matter of calibration and the instrument's sensitivity should indeed be able to be set such that the tasimeter might indicate the nearing of a ship to an iceberg (Edison's tasimeter 1879).

There were other articles that year on eclipses that dealt less with the science of astronomy. Two articles discussed the use of eclipses, whose dates could be precisely calculated, as markers for the dates of historical events. Mr Hind calculated an eclipse on 15 June 763 BC and thus confirmed the year 721 BC when the Assyrians took Samaria (Assyrian and Biblical history, 11 May 1879). On 3 August there was an article dealing with several historical events associated with "… sun-darkenings …", frequently solar eclipses in various stages of totality (Historical Sun-darkenings 1879).

Other articles in 1879 mentioned the eclipse in passing. On 7 March 1879 an article described the facilities of the United States Signal Service Station at Pike's Peak. "The best and most complete report of the last total eclipse of the sun received at Washington was the report of Prof. Loud, of Colorado, from observations taken at Pike's Peak." (The highest inhabited point 1879). On 30 August Professor Stephen Langley of the Allegheny Observatory in Pennsylvania spoke on the study of the Sun. He discussed the significance of photography and spectroscopy and the contributions of several named individuals. He also spoke of the future of these studies. He referred to the possibility of meteorological predictions and its significance for farmers. He predicted the future of a 'solar engine' which man might employ "… to better advantage than we now use steam power." (The study of the Sun 30 August 1879).

It is apparent from *The New York Times* articles printed about the eclipses during the years discussed here that there was frequently an assumption of some scientific knowledge on the part of the readers. However, on occasion the publication would take it upon itself to educate or re-educate the public with some informational articles to assure an understanding of the astronomical topics covered.

Chapter 5
Popular Astronomy and the Transits of Venus of 1874 and 1882

5.1 The Transit of Venus of December 1874

After the Civil War the United States enjoyed some decades of peace, industrial growth and relative prosperity. Though still primarily an agrarian nation, availability of jobs in the new industries was drawing people to urban areas. America had a growing middle class with more free time to take an interest in world events. The transit of Venus of 1874 was the first such transit to occur after the creation of this growing nation.

5.1.1 A Sampling of Popular Periodicals

During this period there were a number of publications, reflecting a variety of interests, designed to appeal to the general public. These would often have articles on the transit, including scientific explanations, periodic updates and relevant human interest stories. Descriptions of the journals that have been provided in the previous chapter on solar eclipses will not be repeated here. Once again the periodicals are grouped into three thematic categories.

5.1.1.1 General Periodicals

The 3 January 1874 issue of *Appleton's Journal of Literature, Science and Art* had two relevant articles within the feature "Science". One was a summary of the distribution of planned transit stations up to that date. Another page described the amusing suggestion of Émile-Hortensius-Charles Cros (1842–1888; Fig. 5.1) from the French Academy of Science that the approaching transit of Venus might yield an

S. Cottam, W. Orchiston, *Eclipses, Transits, and Comets of the Nineteenth Century*, Astrophysics and Space Science Library 406,
DOI 10.1007/978-3-319-08341-4_5

Fig. 5.1 Charles Cros (courtesy: en.wikipedia.org)

opportunity for communication with dwellers of that planet. As the hostile nature of conditions on Venus was unknown at this time, the suggestion of life there was not uncommon. In the issue of 7 February 1874, within the same feature "Science", a summary of a lecture by Richard A. Proctor (see Fig. 5.2) from the Royal Astronomical Society was given. Lectures by this international popularizer of astronomy were very well attended. He discussed the significance of the transit and the techniques that would be employed to determine the mean distance from the Earth to the Sun.

On 11 April 1874, in its new feature, "Science and Invention", *Appleton's Journal* ... had a brief description of the French plans for the transit. On 4 July 1874, in the same feature, the parties for the English and French expeditions were listed. The English had three parties in the Southern Hemisphere, one at Rodriguez and two at Kerguelen, all in the Indian Ocean. The American Professor Peters would also be at Kerguelen, and joint efforts were anticipated. The French would have parties in Yokohama, Japan; Peking, China; and Saigon, Cochin-China in the Northern Hemisphere, and at St. Paul, Campbell Island (see Fig. 5.3) and Noumen (*sic*—it should be Noumea), New Caledonia (see Fig. 5.4) in the Southern Hemisphere. Amounts of funds allocated by the major participating governments were listed, with the Russians spending the most at $240,000. The dates and times for the next transits through to the year 2125 were listed. On 18 July 1874, in

Fig. 5.2 A caricature of Richard Proctor, painted by Leslie Ward, alias 'Spy' (after *Vanity Fair* 1883)

Fig. 5.3 The French transit of Venus station on Campbell Island (courtesy: www.transitofvenus.nl/history.html)

Fig. 5.4 A drawing by Albert Tissandier of the French transit of Venus station at Noumea, New Caledonia (courtesy: transitofvenus.nl/wp/past-transits/1874-december-9/)

"Science and Invention", the leading two paragraphs described the flora and fauna of the almost barren Kerguelen's Island where both an English and an American party were to observe the transit of Venus. Moss and cabbage-plants supplied what little color there was. Wild ducks were good eating, and the marching King Penguins provided entertainment.

On 8 August 1874 *Appleton's Journal* … reprinted a five-page article contributed by the French astronomer and mathematician Rodolphe Radau (1835–1911) describing for the public the scientific theory behind the various observational methods to be employed, as well as an historical overview. Radau shared some human interest stories of triumph and tragedy of earlier transits that would further stimulate public interest. Since the previous transits of 1761 and 1769 scientific theories and technologies had improved such that it was worthwhile to again send well-prepared and well-equipped expeditions to suitable observing sites in a bid to obtain a better value for the solar parallax.

The Chautauquan; A Weekly Magazine was a regional publication originally tied to the prescribed reading of the New York Chautauqua Literary and Scientific Circle (Mott 1938b, pp. 544–547). In the issue of January 1883 was the obituary of Professor Henry Draper, physician by training, whose legacy resulted more from his interest in astronomy. He had discontinued his medical practice after 2 years and concentrated on his new duties as a Professor at the City University in New York, having been elected to the Chair of Physiology. His later years were devoted to the study of chemistry, astronomy and electricity. He constructed his own reflecting telescope of aperture 28 in. in his private observatory at Hastings-on-the-Hudson (see Fig. 5.5). His most significant contributions were in astronomical photography. He discovered

Fig. 5.5 The equatorial room at Dr Henry Draper's observatory (courtesy: old.transitofvenus.org/mercury-draper_obs.jpg)

the gelatino-bromide dry process of photography, significant for its use in astronomy. He was the superintendent of the photographic department for the American contingents to observe the 1874 transit of Venus (Professor Henry Draper 1883).

In May of 1869 there was a relevant article previously cited in Chapter 4, from *The Eclectic Magazine of Foreign Literature,* "Perceiving without Seeing – A Romance in Astronomy", by Charles Pritchard. It included some history of Venus transit expeditions to determine the distance from the Earth to the Sun, anticipating that another such opportunity would occur in the year 1874. In July 1875 the *Eclectic Magazine ...* published a 13-page article by Richard Proctor. The article, "The Success of the Transit Expeditions", described some relevant scientific theory and summarized the varying degrees of success of the parties of all nationalities. In December 1876 the periodical published another article by Proctor, "Astronomy in America", where he described the development of astronomical research in America, beginning with the proposal for a meridian for the United States in 1810, through the growth of observatories, and the part that America played in the transit of Venus expeditions. Proctor found America appealing and, in fact, he emigrated to Florida 1 year after the 1874 transit (Sheehan and Westfall 2004, p. 233).

On 21 February 1874 *Every Saturday* published "America as Seen from Europe", a transcription of an address delivered by the Reverend Joseph F. Thompson in the American Chapel of Berlin on 27 November 1873. Not exactly reflecting the point-of-view implied by the title, Thompson was an American describing to his European audience the nature of America. He mentions the lack of uniform laws of mandatory education throughout the country though there was no lack of desire for knowledge. He said:

> The tone of our popular education is seen first in the enormous demand among what Mr. Lincoln called 'the plain people' for the best results of science and art. Every popular newspaper and magazine must now have its special departments of science and art, fitly edited and well illustrated; the *New York Tribune* sells large editions of Extras with *verbatim* reports of scientific lectures and of the meetings of scientific bodies; and such authors as Herbert Spencer, Darwin, Tyndall, Huxley, Helmholz and Mayer have even a wider reading in the United States than in their own countries. Editors and publishers know the demand of the market, and they find a taste for true science wide-spread among the people. (p. 203).

Furthermore, he states that this scientific inclination of the populace is reflected in their own Government, as demonstrated by "… the liberal measures for observing the transit of Venus of 1874." (p. 204).

Occasionally periodicals with more focused topics might digress to respond to the public interest in the transit. *Forest and Stream* was a New York weekly dedicated to game conservation, and was instrumental in forming the Audubon Society (Mott 1938b, p. 210). On 19 February 1874 there was a paragraph enumerating upcoming astronomical events of the year including two solar eclipses and two lunar eclipses, but

> … the strangest phenomenon will be the transit of Venus over the sun's disk, December 8, which will be visible only in parts of Asia, Africa, Australia and the Sandwich Islands. The last occurrence of this phenomenon was in 1796. (Article 1 – No title: 18 1874).

This last sentence has a typographical error, as the last transit took place in 1769 not 1796.

The Galaxy: A Magazine of Entertaining Reading in January 1874 published "Interplanetary Signalling", an article dealing with the interesting notion of the afore-mentioned Frenchman Charles Cros that the inhabitants of Venus might rec-

Fig. 5.6 A caricature of Sir George Airy, painted by Carlo Pellegrini, alias 'Ape' (after *Vanity Fair* 1875)

ognize scientists' efforts during the transit and attempt to communicate with them. The journal printed the tongue-in-cheek response of Stanislas Meunier of the French Academy that likewise the Martians might be looking for signals from Earth when they observe a transit of Earth across the face of the Sun. After the 1874 transit of Venus, in August 1875 *The Galaxy* printed a summary of Simon Newcomb's report on the success of the expeditions, the number of photographs obtained and the accuracy to be expected from the subsequent calculations (The transit expeditions 1875).

On 7 February 1874 *Harper's Bazaar* printed a paragraph, in its regular feature "Sayings and Doings", about Richard Proctor's lecture at the Brooklyn Academy of Music on his preferred methodology for the expeditions during the transit of Venus. His methods were ultimately selected over those preferred by the Astronomer Royal, Sir George Airy (see Fig. 5.6). Then on 11 April 1874, also within "Sayings and Doings", there was an item on "… a French astronomer …", (unnamed) who saw the coming transit as potentially leading to communication with inhabitants of Venus (p. 239). On 20 June 1874 "Sayings and Doings" had a lead item dealing with the upcoming transit plans. There would likely be more that 80 stations comprised of astronomers from around the world. It was reported that the *S.S. Swatara* would be carrying eight parties of Americans to their sites of observation, five to the Southern Hemisphere and three to the Northern Hemisphere. In actuality the *S.S. Swatara* only carried those to observe in the Southern Hemisphere. The significance of the results in determining the distance of the Sun was emphasized. Aristarchus of Samos (c. 310–230) had calculated the Earth-Sun distance to be only 18–20 times that of the Earth-Moon. In Kepler's time the distance was believed to

Fig. 5.7 J.E. Hilgard (after *Popular Science Monthly* 1875)

be only about 26,000,000 miles, but by the time this article was written, in 1874, it was known that the distance was much greater. On 11 July of that year, in the same feature, there was a brief introduction to the instruments to be used. There would be transit instruments to establish time and latitude, and equatorial instruments useful in establishing longitude by various means, as well as equipment for the observation and photographic recording of the transit.

The *International Review* was initially a bi-monthly, later a monthly, publication of varied contents from international authors (Mott 1938b, p. 35). In March 1874 it published a 13-page article on the transit of Venus by Professor Julius Erasmus Hilgard (1825–1890; Fig. 5.7) of the United States Coast and Geodetic Survey. Hilgard described plans for the upcoming transit, and how those of the Americans would differ from those of other countries.

Littell's Living Age on 24 April 1874 printed an article, "Modern Astronomy", taken from the *London Quarterly Review,* which described the results of efforts by participants in the transit of Venus expeditions to determine an accurate value for the distance to the Sun. On 22 December 1877 it published a 3-page article, taken from *Nature,* describing Sir George Airy's report on the 1874 British transit expeditions. The Astronomer Royal enumerated the expedition sites and their respective degrees of success (The Sun's distance 1877).

Maine Farmer on 12 December 1874 printed a one-paragraph summary of the recent Venus transit expeditions. It noted that the United States sent out eight expeditions, five to the Southern Hemisphere and three to Russia. The last statement was inaccurate as only one of the three Northern Hemisphere sites was in Russia, that at

Fig. 5.8 The American transit of Venus station at Nagasaki. Features from left to right are: the 'transit house'; the weight drive and adjacent pier holding the heliostat; the wooden framework leading towards to the photographic house; the 'equatorial house'; and part of the 'photographic house' (courtesy: transitofvenus.nl/wp/past-transits/1874-december-9/)

Vladivostok under Asaph Hall. It mentioned that the next transit would be visible in the Atlantic States in 1882 and after that the event would not occur again until 2004. The importance as a means to more accurately determining the Earth-Sun distance was stated (Article 4 – No title 1874). On 26 December 1874 there was a longer article summarizing results learned by telegraph of the degree of success of the various expeditions, which on the whole achieved satisfactory results. Particular mention was made of a dispatch from Professor Davidson, at Nagasaki, Japan (see Fig. 5.8), of results obtained. Weather precluded good timing of the first contact but the others were noted with better accuracy, and "... about two hundred micrometric surveys and measurements were taken, besides about sixty excellent photographs of the transit." (The transit of Venus 1874, p. 2). The importance of this event was the establishment of the distance of the Sun which would provide a basis for measuring the distance to other celestial bodies. There would also be benefits for the science of navigation (ibid.).

The farm journal *Massachusetts Ploughman and New England Journal of Agriculture* on 26 September 1874 had in its regular feature "Read and Run", a brief one-sentence article mentioning the American expedition on the *S.S. Swatara*, which had reached its destination at the Cape of Good Hope.

The *Massachusetts Teacher* was an important educational journal, although it was short-lived, lasting only from 1873 to 1874 (Mott 1938a, p. 99). In January 1874, included in its reviews of valuable reading, was mention of the December issue of *Popular Science Monthly*, which among other articles had "Preparations for the Coming Transit of Venus." (Review 3 – No title 1874).

In December 1882 *The Medical and Surgical Reporter* printed an obituary of Professor Henry Draper, a physician by profession but a most enthusiastic astronomer. It was for his work in astronomy that he would be best remembered. He was appointed by the Government Commission to be the superintendent of the photographic department sent to observe the transit of Venus in 1874. His work was so successful that Congress authorized the striking of a special gold coin in his honor, "… the first such public recognition … given by Congress to a scientist." (Obituary notices: Henry Draper 1874, p. 644).

The farm journal *Ohio Farmer* had an article published on 13 June 1874, "Transit of Venus", anticipating the scientific value of the upcoming event and listing the likely number of parties from various countries. This short item included the inaccurate statement that there would not be another transit for more than 100 years. On 19 December 1874, there was a more extensive item titled "The Transit of Venus". It conveyed the interesting story, taken from a children's periodical, of how Jeremiah Horrocks was the first man to predict, observe and document a transit of Venus, on 4 December 1639.

In one of the regular features of the *Overland Monthly and Out West Magazine*, "Scientific Notes", there was a brief discussion of recent scientific topics. In February 1874, in this feature, readers were notified that:

> Professor Davidson, of the United States Coast Survey, who had been appointed to lead one of the expeditions to Japan for the observation of the transit of Venus this year, has announced that he will read a paper to the Academy on the subject of the transit before his departure. (p. 190).

California had only become one of the United States in 1850, but it, too, had an Academy and important scientists lecturing to a public with an interest in science.

The Phrenological Journal and Science of Health, in its publication of October 1874 provided a summary of upcoming transit expeditions and an illustrated explanation of Halley's and Delisle's methods (The transit of Venus 1874).

There were a number of journals written specifically for juveniles that might include relevant and timely articles on science. *St. Nicholas: An Illustrated Magazine for Young Folks* was a monthly publication for juveniles, originally 48 pages in length, but later increased to 96 pages. It contained stories, poems and articles (Mott 1938b, pp. 500–505). The British popularizer of astronomy, Richard Proctor, would contribute a monthly star register for a number of issues in later years (New publications, *The New York Times*, 1877). In the issue of November 1874 there was a 3-page article on the transit of Venus. It described the concept of parallax, with text and illustrations, as well as the preparations for the international expeditions (Smiley 1874). The following month's issue related the story of Jeremiah Horrocks and his prediction and observation of the transit of Venus of 1639. This included the inspirational story of how Horrocks left his observations, not having yet seen the transit, to fulfill his duties of faith when he heard the church bells. He was gratified on his return to his observations to find the planet Venus already on the face of the Sun (Butterworth 1874).

In the October 1874 issue of *Scribner's Monthly* there was the one-paragraph article "The Transit of Venus", which gave a description for the uninitiated of the progressive appearance of the planet transiting the Sun. In December 1874 there was a more extensive article by David Murray describing preparations that had taken place at the Greenwich Observatory as early as 1871 for the event. It was noted that the Americans had also been conducting similar preparations. In the United States, a Commission of distinguished astronomers, supported by Congressional funds, would apply the discoveries of Kepler to establish the parallax of the Sun. Included were diagrams that accompanied an explanation of Halley's method using Kepler's law. Pictures demonstrating the difficulties caused by the black drop 'ligament' effect were provided. Professor Young's proposal to use the spectroscope to eliminate this problem was described. He would watch a particular line of the solar spectrum, perhaps the C line, and note the time of its extinction at the solar surface caused by the interposition of the planet Venus. This would define the time of first contact. The importance of photography was also described as it would eliminate much of the subjectivity in specific determinations and provide a permanent record for study. In the same month there was a short piece about the transit, among a number of other scientific items. It stated that the transit would be observed from about 75 stations. Its significance was recognized to be more than that of determining the distance of the Sun. Longitudes worldwide would be more accurately determined and meteorology would be better understood (The transit of Venus 1874).

5.1.1.2 Religious Periodicals

There were many journals which reflected the great variety of religious faiths of the nation during this period and these also included frequent articles on scientific topics. Though usually positive in his comments about the American people and their accomplishments, Richard Proctor (1876, p. 687) disdained the part that he felt religion played in their science:

> … religion (so-called) attacks and denounces discoveries inconsistent with the views which the orthodox have been accustomed to advocate … [until] … when there is no longer any choice, the orthodox quietly accept these discoveries as established facts, expressing a naïve astonishment that they should ever have been thought in the least degree inconsistent with received opinions.

In light of these comments, it is interesting to see how religious periodicals treated the topic of the transit of Venus.

In May 1874 *Catholic World, A Monthly Magazine of General Literature and Science* published the article "The Coming Transit of Venus". It described very well the theory and methods involved. It related some history of astronomy starting with Aristarchus, on through Kepler and Halley, and the transits prior to the upcoming one of 1874. This 16-page article was relatively free of references to religion until the last page where the author reminds us that the heavens "… declare the glory of God …" (p. 162).

The Methodist *Christian Advocate,* in an article of 7 May 1874, briefly named the participants of the southern expeditions that would soon leave on board the refitted warship, the *S.S. Swatara* (The transit of Venus 1874). The next week it carried an article, "The Approaching Transit of Venus", which described some early determinations of the Earth-Sun distance and some of the upcoming expeditions, English and American. After the transit, on 31 December 1874, the publication carried a longer article on the history and methodology, followed by the results of some of the expeditions. Again, it is not until the last paragraph that the reader is reminded of the 'glory of God.' (The transit of Venus 1874). On 31 January 1878, the feature "Science" reported that British Astronomer Royal, Sir George Airy, determined the Earth-Sun distance, based on data collected at the British transit of Venus expeditions in 1874, to be preliminarily 93,300,000 miles, but further reduction of photographs was pending.

The Baptist *Christian Union,* on 7 January 1874, within their regular feature "Scientific and Sanitary", sketched the progress of astronomy during the previous year. It concluded that much of the preparation for the expeditions to observe the transit of Venus was accomplished during this period (Scientific and Sanitary – Astronomy in 1873, 1874). On 17 June 1874 there was a paragraph enumerating the various American parties that would be leaving on the *S.S. Swatara*, well-equipped, and considered to be professionally comparable to the more experienced European scientists (Article 2 – No title 1874). On 9 December 1874, the day after the transit, the *Christian Union* published a background article on astronomical history and the methods of Halley and Delisle (The transit of Venus 1874). On 17 March 1875 there was the report of a journal entry by Captain Charles Walker Raymond (1842–1913), who was in Tasmania for the transit of Venus (e.g. see Orchiston and Buchanan 1993, 2004). Through a discouraging weather situation of intermittent clouds and rain his party stood steadfast at their positions, taking advantage of all breaks in inclemency to perform their individual tasks and ultimately to obtain 132 photographs and observe one contact (Scientific and sanitary – The transit in Tasmania 1875). On 17 April 1878 the journal printed a favorable review by Professor Maria Mitchell of Vassar College of a book by Professor Simon Newcomb, *Popular Astronomy*, which included a section on the transit of Venus.

The Quaker publication *The Friend: a Religious and Literary Journal* printed some extracts from a report in the *New York Tribune.* A *Tribune* correspondent related the happenings at the American expedition site at Nagasaki, Japan. He listed the main participants and their particular tasks. He described the overall procedure at the photographic tent and the degree of success achieved throughout the event due to the various degrees of cloud cover. He summarized:

> The whole of the party may, therefore, be summed up as follows: – First contact observed by Davidson and Tittman; second contact observed well by Davidson and Tittman; third contact observed by Davidson and Tittman; 190 micrometric and meridian measurements taken; 60 available photographs taken. (The American transit expedition 1875, pp. 245–246).

The Quaker *Friends' Intelligencer* on 20 June 1874, within their regular feature "Items", had a paragraph on the departure of the *S.S. Swatara* from Brooklyn with 26 scientists and 200 officers and crew. They would stop at Rio Janerio (*sic*) and Cape Town on their way to observe the transit of Venus at various sites, the first being Crozet Island. The next month on the 18th, also within "Items", there was a short article on the amateur astronomer Lord Lindsay who had established an

Fig. 5.9 The transit of Venus observing station that Lord Lindsay set up at Belmont, on Mauritius (courtesy: Mauritius Transit of Venus 2004 Committee)

observing station on Mauritius (see Fig. 5.9). While waiting to observe the Venus transit he planned to observe the minor planet Juno. He hoped by this additional means to obtain an independent value for the solar parallax that approached in accuracy that expected from the observations of the transit. On 19 December 1874 there was an article, "What the 'Transit' Means", and it described the transit's usefulness in determining the Earth-Sun distance, and it also enumerated the American expeditions. On 27 February 1875 there was the preliminary account from Professor George Davidson reporting the degree of success of his expedition at Nagasaki. This was taken from *The New York Times* (Preliminary account by Professor Davidson of his observations 1875).

Printed on 19 December 1874 in the Quaker *Friends' Review; a Religious, Literary and Miscellaneous Journal,* was the writer's anticipation of the transit to take place that evening, on 8 December, when "... a hundred telescopes, stationed at widely separated places on the other side of the globe, will be directed to a small spot on the face of the sun." (The transit of Venus 1874, p. 286). The locations of the American expeditions were stated. Their use of excellent photographic equipment was described as important in achieving consistency and accuracy in determining the distance to the Sun, the yardstick by which other celestial distances might be determined (ibid., pp. 286–287). An early report after the event was given in January 1875. Within 3 days reports had already been received from 37 of the parties, 16 of which considered themselves "… wholly successful …" The parties under Captain Alney (*sic*), M Janssen and Sir G. Arey (*sic*) all gave negative reports on the observation of the expected 'black drop' effect (The transit of Venus 1874, p. 326). The *Friends' Review* published a three-page article on 20 February 1875, an account by a *New York Tribune* correspondent who accompanied Professor Davidson's transit expedition to Nagasaki, Japan. There were astronomers and photographers well prepared for their particular tasks, which they performed to the best of their ability considering the mostly cloudy conditions. Longitude had been established by telegraphic connection to Vladivostok which was, in turn, connected to St. Petersburg. Times for the first three contacts were noted with some discrepancy by Davidson and Mr Tittman. In the end, 60 useful photographs out of 100 were obtained (The day of the transit 1875).

The Congregationalist weekly, *The Independent,* published on 4 June 1874 a summary of the plans in place for the eight American expeditions to observe the transit of Venus. The sites were listed along with the names of their leaders. The *S.S. Swatara* would transport the five southern expedition groups. The three northern groups would be met at Yokohama where they would be transported to their sites by the *Lackawanna* of the U.S. Navy. The parties would each be composed of a chief astronomer, an assistant, a chief photographer and two assistant photographers. All parties would carry the same astronomical and photographic equipment (Science). On 4 March 1875 there was the report of Professor C.A. Young on the transit of Venus expedition to Peking. Leaving from Chicago on 15 July, they arrived at their ultimate destination on 16 September. The instruments, assigned duties for various party members and preliminary work were described. The first and second contacts were plagued by intermittent cloud cover and the third and fourth by hazy conditions. However, some useful photographs were obtained, about two thirds of the 99 printed. At the time of this writing the party was still unsure of how they were to get back home to the United States as their original transportation was cancelled by Admiral Pennock for fear of the unsafe waters of the season (Young 1875a). On 20 July 1876, in its feature "Science", there was an article by C.A. Young, "American Contributions to Astronomy". Among other things, Young contended that the American expeditions for the 1874 transit of Venus compared favorably to all others "… in completeness and accuracy." (p. 8).

Another Congregationalist weekly, *New York Evangelist*, under its feature "Current Events", on 28 May 1874, enumerated all the transit party locations and their leaders. On 13 August 1874 there was a single untitled paragraph on some of Jules Janssen's apparatus for photographing the event:

> Fifty photographs can be taken in as many seconds on different parts of the same plate, with the apparatus for photographing the transit of Venus, invented by Janssen, and perfected by Mr Christie, the first assistant at the Royal Observatory, Greenwich, England. (Article 1 – No title 1874, p. 4).

On 1 October 1874 the *New York Evangelist* printed a contribution by the Presbyterian mission administrator Frank Field Ellinwood (1826–1908) on his thoughts during his travels from San Francisco to Japan. After expressing his impression of the moral growth of the city of San Francisco under Christian influence,

Fig. 5.10 The *S.S. Alaska* (courtesy: www.murrayofstanhope.coom/images/alaska.gif)

he advanced his feelings on the prejudicial treatment of the Chinese immigrants working there. Among his companions on his voyage of 20-days across the Pacific on the *S.S. Alaska* (see Fig. 5.10) were two scientific parties on their way to observe the transit of Venus. One party was under the leadership of Professor Watson, on its way to Peking and the other, under Professor Hall, was bound for Wladiwostock. Ellinwood then briefly described the composition of the other six parties, all to observe in the Eastern Hemisphere at "… points as remote as possible from each other in latitudes north and south of the Equator" in order to determine the parallax of the Sun, "… a basis of measurement which must enter into nearly all astronomical computations." (p. 1). It was anticipated that the advantages of these parties due to technical advances should make their results much improved over those of the previous transit of 1769.

On 3 December 1874 the *New York Evangelist* recounted the story of Jeremiah Horrocks, taken from the juvenile periodical *St. Nicholas* (The boy astronomer 1874). After the transit, on 18 February 1875, the success of the expedition under Professor Christian Heinrich Freidrich Peters (1813–1890, Fig. 5.11) at Queenstown, New Zealand, was reported (Scientific and useful 1875). On 25 February 1875 there was an article by Professor Peters describing his experience (for recent accounts see Dick et al. 1998; Orchiston et al. 2000). The participants had prepared well for their remote site (see Fig. 5.12) to the point of building a little aqueduct to move water from a stream 600 ft away for the use of the photographers. Of the various expeditions to New Zealand he felt his was the only one to be favored with cooperative skies (for details see Orchiston 2004b). Following a period of discouraging cloud cover, the first two contacts were observed with no difficulty. Observation of the latter part of the transit was disrupted by occasional cloud cover but photographs were efficiently taken at all intervening opportunities. In the end they had obtained 239 pictures.

On 15 April 1875 the *New York Evangelist* printed a letter from S.N. Castle, then a resident of the Sandwich Islands (Hawaii) expressing an appreciation for the many scientific advances in recent years that had made communication with other parts of the world so expeditious. Present during the time of the Venus transit, Castle briefly described the activities there organized by Captain George Lyon Tupman (1838–1922)

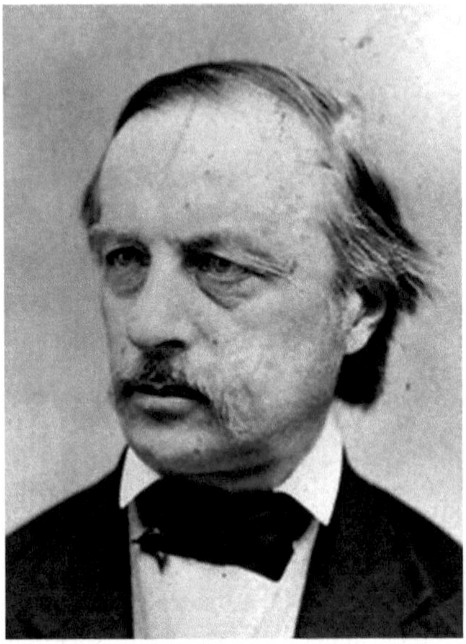

Fig. 5.11 C.F.H. Peters (courtesy: en.wikipedia.org)

with the authority of Airey (*sic*), Astronomer Royal of the Greenwich Observatory. Tupman, himself (see Fig. 5.13), was in charge of expeditions on three of the Sandwich Islands, Hawaii, Kauai and Oahu (for details see Chauvin 2004). Tupman told Castle that final results from the observations could not be expected for a number of years.

The Presbyterian weekly *New York Observer and Chronicle* on 7 January 1875 related an account of a British party at the Hawaiian Islands, which should be "… of great interest, even to the unscientific." (The transit of Venus 1875, p. 6).

The Methodist *Western Christian Advocate* on 18 February 1874 in their regular feature of short items, "Gleanings", enumerated the significant astronomical events of 1874, including eclipses and the transit of Venus. On 22 July 1874 there was an article, "The Transit of Venus", describing briefly the upcoming expeditions, the concept of parallax and the methodology used to determine it. This article concluded: "It is a subject of congratulations that our Government is willing to assist science for its own sake, and by so much to increase the knowledge of the world." (p. 228).

The Youth's Companion on 12 November 1874, in anticipation of the transit, which was less than a month away, published a very nice article with diagrams describing the concept of the transit and the methodologies to be employed by the expeditions (The transit of Venus 1874).

The Methodist *Zion's Herald* briefly reviewed, as interesting, the paper by Professor Hilgard on the transit published in the *International Review* (Article 8 – No title 1874). On 23 July 1874, under its feature "Miscellaneous News", there was a paragraph announcing the departure of Professor Watson and his party from Omaha,

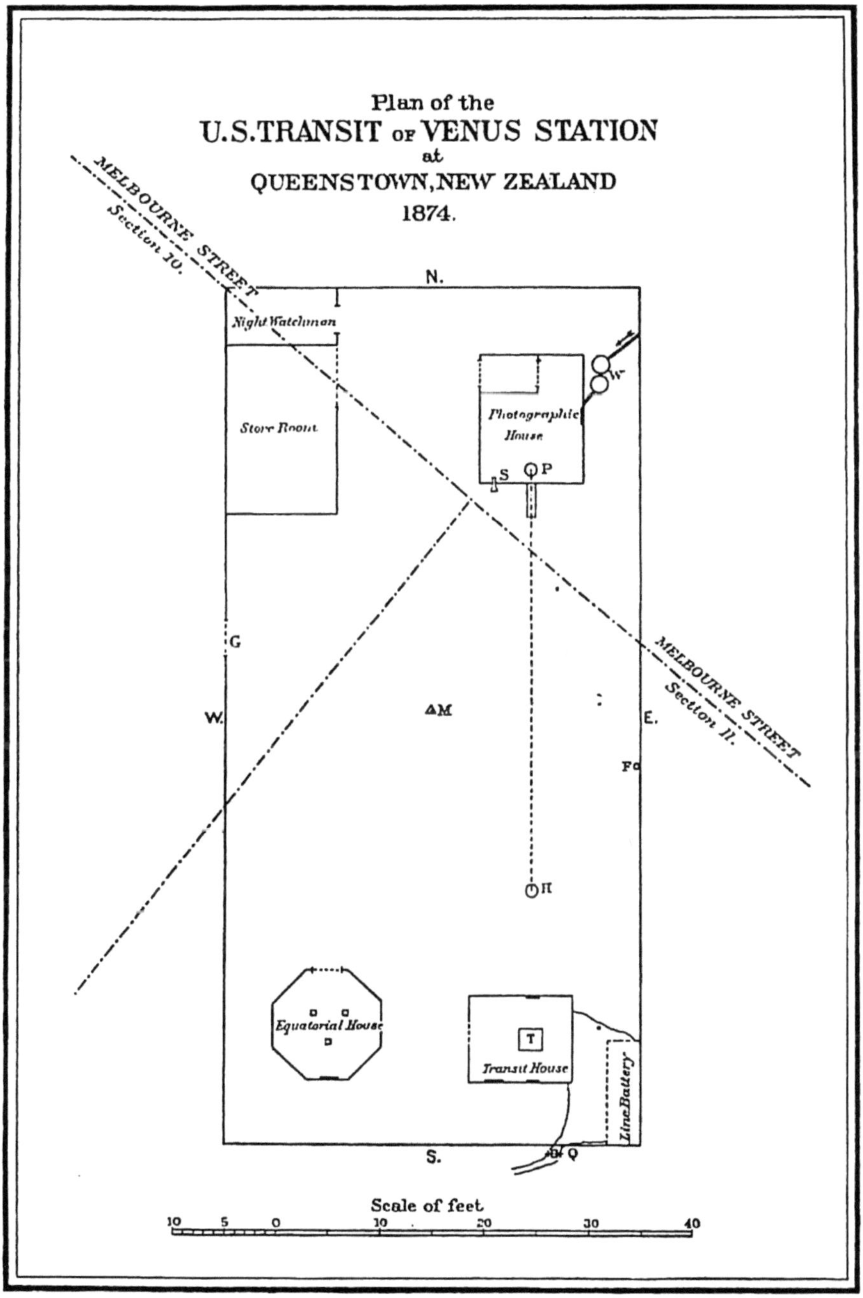

Fig. 5.12 Plan of the Queenstown transit station (after Peters 1881). Key: *T* transit telescope, *H* heliostat, *P* photographic plate-holder

Fig. 5.13 Captain G.L. Tupman (courtesy: en.wikipedia.org)

Nebraska, for the China and Japan expeditions to observe the transit of Venus. "Four car-loads of astronomical instruments preceded them." (p. 237).

It seems that these religious periodicals, in general, did a very good job of describing the transit history, methodology and expeditions to the general public. Surprisingly, very little religious dogma was injected.

5.1.1.3 Scientific Periodicals

The previously-mentioned journals of popular science were also valuable in conveying information about the transit during this period.

The American Journal of Science and Arts in July 1874 published an article by Professor Charles Francis Himes (1838–1918) from Dickinson College on the proper preparation of photographic dry-plates in daylight by the manipulation of silver compounds. He suggested the use of this method at expedition sites to observe the transit of Venus that year.

The pages of *Scientific American* contained frequent references to the transit. As early as 7 February 1874 there was a Letter to the Editor anticipating what might be learned about Venus itself subsequent to the transit (Vander Weyde 1874). On 21 February 1874 there was an article about the popular British lecturer Richard Proctor, with reference to his controversy with the Astronomer Royal regarding the methodology to be applied at the transit of Venus observing sites (Professor R.A. Proctor 1874).

On 9 May 1874 *Scientific American* printed an account of a meeting of the National Academy of Science, including a report on the preparations for the American transit expeditions (The meeting of the National Academy of Science 1874).

On 22 August 1874 an article summarized the techniques to be used during the transit, and included diagrams. A couple of minor errors in detail were noted in this article. For instance, it was stated that the estimate for the Earth-Sun distance was previously calculated by Encke to be about 92 million miles, while his figure actually was greater than 95 million miles. Also, there was mention of an English expedition to Alexandria in Northern India, whereas Alexandria is in Egypt (The transit of Venus 1874).

The following week the *Scientific American* made reference to the 'black drop' problem in transit observations (Recent advances in science, 29 August 1874). On 5 December 1874, in the regular feature "Astronomical Notes", there was a reminder of the upcoming transit and reference to previous transits seen.

By 26 December 1874, some results had come in. There were reports from the American and British parties of their degrees of success in observing the different contacts and in obtaining photographs (The transit of Venus 1874). On 20 March 1875 there was a favorable comparison of the work of the American astronomers to that of the British during these transit expeditions (The English and American transit campaigns compared 1875).

On 22 May 1875 the determination of the solar parallax as 8.879″ was published, based on data collected by the French party at Peking (Scientific and practical information – Recent astronomical discoveries 1875).

On 28 August 1875 *Scientific American* noted a proposal made that there might be value in observing Venus as it passes behind the Sun rather than across the face of it (Transits of Venus behind the Sun 1875). On 21 October 1876 *Scientific American* quoted a section of a speech by Sir William Thomson (1824–1907) as Chair of the Physical Science Section at the British Association in Glasgow. Thomson was impressed with the status of American science. He referred specifically to the photographic instrumentation used during their astronomical expeditions (Science in America 1876). On 27 July 1878 readers were told of an upcoming eclipse expedition and the anticipated use of the same photoheliograph that had been used in the transit expeditions of 1874 (The total eclipse of the Sun 1878). On 28 December 1878 one learned that Professor Harkness was measuring the 221 photographs taken by the American transit of Venus parties to correct for such distortion as colloidal shrinkage (The National Observatory 1878). In the publication of 2 December 1882 there was the obituary of Henry Draper, who had been Superintendent of the Photographic Department of the Transit Commission (Henry Draper 1882).

5.1.2 The New York Times

Due to the numerous articles on the transits in *The New York Times*, a well-read citizen could expect to have considerable understanding of the transit expeditions, their goals and their successes. Also, in the classified section one might find announcements of lectures on the subject. These were generally well-attended by both men and women. Public astronomy lectures were a popular form of entertainment in the 1800s (Inkster 1978, 1980). See, for example, the notice for the lecture by Henry M. Parkhurst, "The Coming Transit of Venus", on 3 December 1874, or the notice for the lecture by Leonard Waldo (1853–1929), "An Astronomer's Voyage around the World", on 25 April 1876. The latter was to be "… brilliantly illustrated by the Magic Lantern." The magic lantern was a popular projection device, a precursor to the modern cinema.

Probably the most significant of the many articles on the transit published by *The New York Times* was that of 8 December 1874, the day of the transit. On this date the reader would find a 6-column article defining the transit and describing the two primary observing methods, all with diagrams (see Fig. 5.14). The instrumentation was discussed briefly. Practical difficulties that might be encountered were described. Preparations to be made by the various nations were summarized (The transit of Venus 1874). Preceding this article were a number of others during the year that brought this event to the attention of the public.

There were human interest stories of past transits that would intrigue the reader and increase his or her interest in the current ones. On 25 July and 30 August 1874 articles were published about the trials of Le Gentil de la Galasière, as previously described (The transit of Venus and Le Gentil's observation 1874; An astronomer's evil star 1874).

Also preliminary to the 1874 transit there were references to useful writings in other periodicals, or in books (New publications – *Quadrature of the Circle,* 14 March 1874; New publications – *Harper's Magazine*, 23 September 1874; New publications – *St. Nicholas*, 22 October 1874). There was a summary of an informative lecture given by Richard Proctor at Cooper Union "… before a large and attentive audience …" (Prof. Proctor's lecture on the transit of Venus 1874). There was an opinion column published on 19 April 1874 expressing concern about America's slow start to involve itself in the transit expeditions (The astronomical event of the century 1874).

There was as much interest in the foreign transit expeditions as in the American ones. Readers would learn the plans of those who had more experience in such matters. Although there was ostensibly a spirit of cooperation among the scientists of all nations, there was also an undercurrent of competition and national pride.

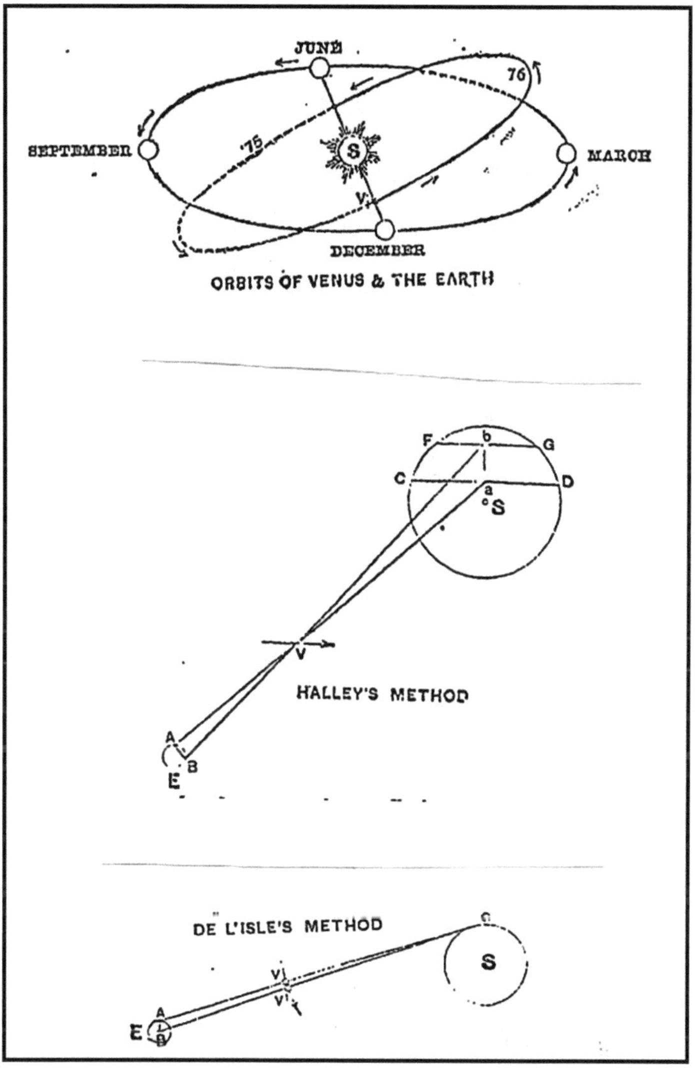

Fig. 5.14 Diagrams taken from *The New York Times* article of 8 December 1874

As the United States was such a young country, its citizens were not very far removed from their roots and might find some satisfaction in the accomplishments of those of their ancestry. A regular reader of *The New York Times* would have a good understanding of the level of participation of the many nations around the world.

On 11 April 1874 there was a summary of the major countries participating and the number of stations and locations for each (Foreign Notes 1874). On 15 and 27 July 1874 there were two articles regarding Richard Proctor's recommendations for

Fig. 5.15 John B.N. Hennessy (1829–1910) from the Trigonometrical Survey of India successfully observed the transit from Masoorie in the Shivalik Hills (photograph after Phillimore 1968)

the British stations (Transit of Venus: Prof. Proctor makes a suggestion 1874; The transit of Venus 1874).

After the transit, results started to come in. They were frequently reported in the *Times* as short fillers of only a line or two. On 10 December 1874 there was the 9 December report via London that observations at Nagasaki were "… completely successful." (The transit of Venus and Observation in Japan successful 1874). On the same date there was a report from the Greenwich Observatory of British success in India (see Kapoor 2014; and Figs. 5.15 and 5.16). More than 100 photographs were obtained (The transit of Venus telegrams received at the Greenwich Observatory 1874). In the same issue, on the front page, was another item announcing perfect success by the British in Egypt (Washington 1874). On 11 December 1874 in "The Latest News by Cable" the British announced telegram reports that several Siberian sites had failures but there was success at Yokohama. One line was printed on 13 December 1874 from a London report of 12 December 1874: "The transit was successfully observed at Beyrout [*sic*], Syria." (The transit of Venus; Successful observations at Beyrout, Syria 1874).

On 30 December 1874 the *Times* had a reprint from the *Honolulu Commercial Advertiser* of 12 December. It related the story of the British expedition in Hawaii (see Fig. 5.17). Besides the details about the level of success of the transit observations, there were some interesting details about the natives who prepared themselves to experience the transit with smoked "… broken panes of glass and bottles."

Fig. 5.16 John F. Tennant (1829–1915), also from the Trigonometrical Survey of India, successfully observed the transit from Roorkee (photograph courtesy Indian Institute of Astrophysics Archives)

Fig. 5.17 The British transit of Venus station at Honolulu, Hawaii (after Chauvin 1993)

Fig. 5.18 The French transit of Venus station at Kobe, Japan (courtesy: www.transitofvenus.nl/history.html)

(Observations in the Hawaiian Islands 1874). On 31 December 1874 the *Times* published a lengthy front-page article from C.A. Young. He mostly described the experiences of the Americans at Peking, China, but there was also mention of a Russian party nearby, as well as a French party which was struck with tragedy when one of their party was mysteriously paralyzed (The transit of Venus and Letter from Prof. Young at Peking 1874).

On 23 January 1875 the *Times* mentioned multiple successes of nations at Japanese sites. Besides the Americans there were the French (see Fig. 5.18), Germans, Russians, Mexicans (see Fig. 5.19) and the Japanese themselves. The Emperor of Japan observed the event on his own grounds with American instruments (The Far East 1875). On 31 January 1875 there was a reprint of an article from the *Japan Herald* of 21 December describing the activities of various parties from Germany, Mexico, and Austria, as well as the Japanese: "… smoked glass had been all the rage, and in every street eager gazers might have been seen scanning the sun." (The transit of Venus; Successful observations in Japan 1875).

Several years later, on 16 December 1880, the *Times* printed the obituary of Lord Lindsay, the Earl of Crawford, and

> Among his contributions to the cause of science was the equipment and preparation, entirely at his own expense, of an expedition which proceeded to the Mauritius to take observations on the transit of Venus in December, 1874. (Obituary: The Earl of Crawford 1880).

Americans with interest in the expeditions of their own country might have followed closely a series of articles in *The New York Times* authored by "From Our Own Correspondent", a participant in the travels of the southern expeditions. This unnamed individual started his series with an explanation of the transit's

Fig. 5.19 The two Mexican observing stations at Yokohama, one at Nogeyama (*left*) and the other at The Bluff (*right*) (courtesy: www.transitofvenus.nl/history.html)

significance and the funding for its observation, and he then continued with other reminiscences throughout the months of adventurous travels to exotic places. Subsequently, he discussed the results of the many parties.

On 21 May 1874, the writer gave some background on the anticipated expeditions. He stated the reasons for the interest in the Venus transits and described the initial activity of the Government of the United States in providing funding and creating the Commission which would oversee the activities of the various parties. The equipment and observing methods to be used were described. The reader learned that the *S.S. Swatara* was designated as the mode of transportation for all southern expeditions, and would leave on the first of June. Mention was made of the northern parties, which would not leave for Yokohama until a later date (The transit of Venus; Preparations made for its observation 1874).

On 19 September 1874 *The New York Times* printed the correspondent's report on the progress of the *S.S. Swatara.* The parties had by then reached Bahia, Brazil. This article did not deal with the transit. It was rather a description of local flora and fauna and the native culture, in itself of great interest to the general public (The transit of Venus; Arrival at Bahia 1874). On 22 September 1874 readers learned of the entertainment of Captain Chandler from the *S.S. Swatara* and the chiefs of the scientific parties at Cape Town (The transit of Venus; Entertainment of Captain Chandler 1874).

On the front page of the issue of 6 December 1874 the *Times* printed the correspondent's latest report, dated 17 August 1874. By this time they had reached the Cape of Good Hope. Museums and gardens were described. Curious details were noted: "… an American cannot fail to notice that carriages passing each other always turn to the left instead of the right." There were reminiscences of astronomical accomplishments that had taken place here by Abbé Lacaille and John Herschel, and "The most important scientific establishment now existing at the Cape is, without doubt, the Royal Observatory." Ostrich farming was described as an interesting local industry. The reader learned of a meeting with the British party that was also heading for Kerguelen Island, of social events and of local entertainment (The transit of Venus; The expedition at the Cape of Good Hope 1874).

On 7 December 1874 the report of 10 October 1874 was printed. The *S.S. Swatara* had reached its first expedition destination. The "… bleak and desolate …" Crozet Islands were approached. The correspondent shared some local color as he related his conversation with a man who used to be a member of a sealing party in years past. Ultimately bad weather conditions led to the decision to abandon the Crozet Islands as an observation site and the party moved on (The transit of Venus; The *Swatara* on her way from Capetown 1874).

On 30 December 1874 there was the report of 9 November 1874. The first party was deposited at Malloy Point at Kerguelen Island. It was a rocky place with little flora and only birds, ducks and penguins for fauna (The transit of Venus: Voyage of the *Swatara* from Crozet 1874).

On 9 February 1875 the reader received the report of 17 December 1874, 9 days after the transit. Here the author went into great detail on the activities of Professor Harkness' group at Hobart in Tasmania. After the usual description of the local geography, the correspondent described the set-up of buildings and instruments for the astronomers at the site (see Fig. 5.20). He described the transit itself. It was "… a sad disappointment …" Observations were periodically interrupted by clouds and heavy rain. Some photos were obtained. It was noted that Captain Raymond's party at Campbell Town, originally to have been positioned at the Crozet Islands, had similar bad luck with weather, although they did obtain about 125 photos (The transit of Venus; Observations at Hobart Town 1875).

On 23 February 1875 there was the report dated 13 January 1875. Here the correspondent brought the reader up-to-date on the last two southern expeditions, which went to Queenstown and Chatham Island in New Zealand. Queenstown, an inland village of about 780 inhabitants, though inconvenient to reach (40 miles by railroad, 60 by stage and finally 20 miles by steamboat) proved to be an ideal site with excellent visibility. The party saw most of the transit and obtained more than 200 photographs. Finally the Chatham Island party had been delivered to their site on 19 October 1874. They had cloudy weather during most of the transit and obtained only 13 photographs. This installment related the story of how Captain Chandler of the *S.S. Swatara* responded to a request by the German Council at Melbourne to find a German expedition based on Auckland Island (see Fig. 5.21) that had not been heard from for some time. Chandler found the group. They had encountered some bad weather but were safe (The transit of Venus; Experiences of the Observing Parties 1875).

On 29 March 1875 *The New York Times* printed the final installment of the saga of 'Our Own Correspondent', which was dated 9 February 1875. It described the last days of the expedition members at Chatham Island and their return to Hobart

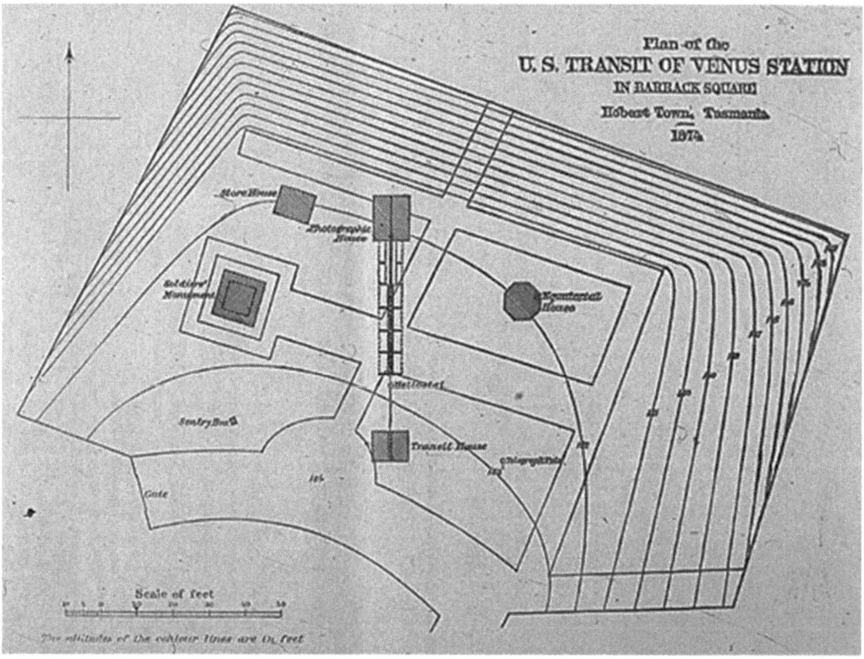

Fig. 5.20 The American transit station at Barrack Square, Hobart, showing (left to right) the soldiers' monument, a store house, the 'Transit House', and above it the photoheliograph, and the 'Equatorial House' (after Dick 2003)

Town, from which they would return to the United States (The transit of Venus; Homeward bound 1875).

'Our Own Correspondent's' follow-up article on 5 August 1875 summarized the successes of the American parties in catching the contacts of the transit, as provided by Professor Harkness. Only the Peking party observed all four. A description of the advantages of the use of Mars oppositions to determine the parallax was made. However, it was admitted that the rare transits of Venus could not be ignored (The astronomers' work 1875).

Interspersed among the many segments of 'Our Own Correspondent' were many articles updating the public on the preparations relating to the transit.

As early as 12 February 1874 *The New York Times* printed a Letter to the Editor applauding Congress's financial support for the upcoming expeditions. It was signed "A Citizen of New York." (Aid to the Centennial 1874).

On 30 April 1874 there was an article listing the various expeditions, and the schedule of the *S.S. Swatara* also was described in detail (The transit of Venus and United States observing expeditions 1874). Similar information was repeated in articles on 18 and 22 May 1874 (The transit of Venus and Government preparations… 1874; The transit of Venus and List of the persons… 1874). The latter, a special dispatch to *The New York Times*, listed all personnel of each party for the anticipated expedition sites. On 28 and 31 May 1874 there were articles describing

Fig. 5.21 The German transit station at Terror Cove, Auckland Island (courtesy: www.transitofvenus.nl/history.html)

the *S.S.* S*watara*, its principle personnel and its schedule of destinations (The *Swatara* and Destinations… 1874; The *Swatara* 1874).

On 1 June 1874, readers learned that the steamer *Colorado* set sail a day earlier for Yokohama with the instruments for Professor Davidson's Venus transit expedition (The Pacific Slope… 1874).

On 4 June 1874 there was an updated list of astronomers and assistants who would set sail on the *S.S. Swatara* (The transit of Venus 1874). On the following day readers would have seen a list of all the scientific instruments that the *S.S. Swatara* would carry. Readers would also have learned that plans for the northern expeditions had advanced and that they would probably be departing on the *Lackawanna* (The transit of Venus and description of the scientific instruments 1874).

On 21 June 1874 *The New York Times* published the fact that the Senate had approved appropriations to the amount of $25,000 to the Secretary of the Navy to complete the observations of the transit of Venus (The Sundry Civil Appropriation Bill 1874). Through *The New York Times* members of the public were able to follow the scientists on their adventures, as they traveled to the far corners of the Earth.

On 30 August 1874 the newspaper reported that Professor Davidson and party had sailed to Nagasaki the previous day on the steamship *Japan* (Departure of Prof.

Davidson for Nagasaki 1874). On 5 September 1874 readers learned that the southern groups had arrived at Cape Town on 5 August 1874 on their way to their ultimate sites and then on 22 September of the local entertainment enjoyed by these groups (The transit of Venus and The American party at Cape Town… 1874; The transit of Venus expedition 1874). On 21 November 1874 the *Times* published the arrival of Professor Davidson's party in Japan, where they would be stationed at Nagasaki. Readers were informed that telegraphic operations were to be made to establish longitude, and a list of party members was given (The Davidson transit of Venus party 1874). On 2 December 1874 readers learned of a steamer in preparation to bring the parties of the southern expeditions home (Latest news by cable 1874).

On 31 December 1874 there was the aforementioned letter from Professor Young where he described some of the international groups. The main body of his letter dealt with the travels of the three northern expedition parties. The Vladivostok and Peking parties left together from San Francisco on 25 July 1874. Professor Davidson's party to Japan left about a month later. Professor Young also described the observation sites, the equipment used and the preparations made (The transit of Venus and letter from Prof. Young at Peking 1874).

Over the next several months *The New York Times* would respond to the public's curiosity about the success of these expeditions.

Thus, on 1 February 1875 the *Times* printed a brief report:

> Drs. Peters and Harkness, of the American expedition, were the only successful observers of the transit of Venus in New Zealand. The observations at Hobart Town and Campbelltown, Tasmania, were obstructed by clouds, and the first part of the transit was lost. The German party at the Auckland Islands had a successful observation. (The transit of Venus 1875).

On 11 February 1875 there was Professor Davidson's preliminary account. His party did obtain some data and photographs of the second contact and felt they had "… more than average satisfactory results." (Transit of Venus; Preliminary account by Prof. Davidson 1875). On 12 February 1875, Dr Peters gave a report on his observations at Queenstown, New Zealand. Except for occasional clouds, the transit was observed almost in its entirety, and 239 photographs were taken (The transit of Venus; Report of Dr. Peters 1875). On 25 February 1875 the *Times* printed two relevant items, a lengthy report on the Peking expedition, submitted by Professor Young who was first assistant at the site, and a short summary by Professor Watson who was chief of the party there. It was deemed a qualified success. Although frequently cloudy, data were collected at all four contacts, and 99 photographs were taken of which approximately two thirds seemed useful for measurement. Professor Watson confirmed their relative success in a separate article on the same page (Young 1875b; A letter from Prof. Watson 1875).

On 3 March 1875 there was an article on a lecture given by Professor C.A. Young, in Hoboken, New Jersey, on the transit expedition to Peking. He spoke of the work of the party, gave a little travelogue on some Chinese excursions and expressed pride in the comparable accomplishments of the Americans (The transit of Venus; Lecture by Prof. C.A. Young 1875).

On 2 May 1875 readers learned of the partial success of the Chatham Island party in New Zealand. Included in this article, which originated from the *San Francisco Bulletin,* was mention of a number of photographs that were obtained, although most of the transit was obscured due to poor weather (The transit of Venus; Results of the observations at Chatham Island 1875).

On 1 June 1875 the *Times* related the return of the *S.S. Swatara* to New York, bringing back all the ship's personnel (of which all officers were listed), expedition members and "… Australian animals and botanical curiosities …" (Local miscellany – Arrival of the *Swatara* 1875).

On 3 June 1876 there was a short piece reprinted from the *Boston Globe* saying that the German Minister at Washington forwarded a letter to Captain Chandler of the *S.S. Swatara* from Prince Bismarck expressing thanks to the Captain for aid rendered to the German expedition party in distress on Auckland Island (Thanks to Captain Chandler 1876).

An indication of sustained interest in astronomical subjects tangential to that of the Venus transit may be reflected in what was published in *The New York Times* over the following years.

On 19 November 1875 there was an item on other celestial events as interesting as the transit of Venus, and mentioned were the upcoming lunar occultations of the Pleiades and Saturn, neither of which would require the use of smoked glass for observation (The occultation season 1875).

On 14 December 1875 there was the obituary of Henry James Anderson (1799–1875), retired Chair of Mathematics and Astronomy at Columbia College. Included was the fact that, at greater than 75 years of age, he traveled to Australia to witness the transit of Venus in 1874 (Obituary: Henry James Anderson 1875).

On 4 October 1876 readers learned that the astronomers at the U.S. Naval Observatory used the same powerful instruments as those used in observing the transit of Venus to search for the planet Vulcan, which theoretically was supposed to travel within the orbit of Mercury. This search was to no avail (The planet Vulcan 1876).

On 16 July 1877 the *Times* published an opinion that an upcoming opposition of Mars might provide a preferable method for determining the Earth-Sun distance. Two variations on the use of this event to determine parallax were proposed here (The opposition of Mars 1877). On 3 May 1878 another alternative was published, proposed by Mr Lawrence Slater Benson. His method would entail observing the difference between the time of light travel between the Earth and one of Jupiter's satellites, when closest and when furthest from the Earth (Letters to the editor – The tall tower and the transit 1878). On the same date the *Times* printed an article on a transit of Mercury with comparisons to the previous transit of Venus (The transit of Mercury 1878).

On 19 August 1880 one would find the tongue-in-cheek opinion that science is a slow process. The writer gave as an example the transit of Venus expeditions which involved months of travel to "… pagan places …" followed by a much longer period to reduce and interpret the data collected (A powerful report 1880).

5.2 The Transit of Venus of December 1882

The United States had not grown much in the 8 years since the last transit. To the 37 states of the Union, only Colorado had been added, in 1876. These years were a period of continuing urbanization and industrialization.

5.2.1 A Sampling of Popular Periodicals

5.2.1.1 General Periodicals

For the 1882 Venus transit the public would again be reminded of its rarity and significance.

In *The Chautauquan; A Weekly Magazine*, shortly following the Venus transit of 1882, two relevant items were published. In January 1883 there was an article in response to a reader asking for "... a description of the transit of Venus ..." The response was taken from a meeting of the New York Academy of Sciences where Professor John Krom Rees (1851–1907) of Columbia College (see Fig. 5.22) spoke of his own observations (The transit of Venus 1882, p. 239). In the issue of May 1883 readers found a short item by Bishop Warren describing the difficulties of obtaining an accurate Earth-Sun distance (Bishop 1883).

Fig. 5.22 Professor John Rees observed the transit from the Columbia University Observatory, which is shown here with the green dome near the center of the picture, on the top of the law building (courtesy: www.transitofvenus.nl/history.html)

Fig. 5.23 The Halsted Observatory at Princeton (courtesy: www.transitofvenus.nl/history.html)

The Critic was a literary monthly. It had regular departments of reviews, drama and fine arts, and Walt Whitman was a leading contributor (Mott 1938b, p. 430). *The Critic* would digress from its usual content with such news items as that submitted by C.A. Young on the transit of Venus. On 7 October 1882 *The Critic* published Young's article "The Coming Transit of Venus", in which he gave the public a synopsis of transit history, methodologies and upcoming expeditions (Young 1882a). On 16 December 1882, some days after the event, he presented another item, "The Transit of Venus" (Young 1882c), in which he described he own success in observing the event with the 23-in. Alvan Clark refractor at the Halsted Observatory at Princeton (see Fig. 5.23).

Frank Leslie's Popular Monthly was a literary monthly, typically of 128 pages, and beautifully illustrated with black-and-white prints (Mott 1938b, pp. 510–512). In the issue of May 1883 there was the article "Transits of Venus". It contained 11 illustrations, including a mythological image of Venus crossing the Sun (see Fig. 5.24), pictures of equipment used at sites occupied by various parties and some of the individuals involved. The text included a history of the earlier transits and their varying degrees of success, including the most recent transit of 1882.

Fig. 5.24 Venus Crossing the Sun (after: *Frank Leslie's Popular Monthly*, May 1883, p. 628)

Littell's Living Age published an article on 3 November 1883, "The Distance of the Sun". It related how the famous Irish astronomer and popularizer, Professor (later Sir) Robert Stawell Ball (1840–1913; Fig. 5.25; see Jones 2005), described lucidly for the layman the different methods by which one might determine the distance to the Sun.

The Massachusetts Ploughman and New England Journal of Agriculture departed from its typical subject matter in an article on 11 November 1882 titled "The Planet Venus". With reference to a piece written by the popularizer Richard Proctor, it contained speculation on the likelihood of life on the planet Venus.

Potter's American Monthly was a well-illustrated journal dealing with history, literature, science and art (Mott 1938b, pp. 260–261). In June 1882 it published an article by Richard Proctor (1882b, p. 636) describing some of the superstitions that

Fig. 5.25 A caricature of Robert Ball, painted by Leslie Ward, alias 'Spy' (after *Vanity Fair* 1905)

accompanied astronomical events over the centuries. Wise men would attach meaning, in the form of prophecies of disaster, to the appearance of meteors, comets and other coincidental phenomena, "… and absolutely on December 6th Venus transits the sun's disc! Something will surely come of this, if only we live to see it."

Puck was a comic weekly of political satire and social commentary (Mott 1938b, pp. 518–532). On 12 December 1882 it published a futuristic account of the 2004 transit of Venus. The writer pointed out that the tools of the trade had much advanced since the transits of the nineteenth century. In 2004 there were no clouds in the sky. There are telegraph and telephone wires connecting Earth to Venus. One could travel to the second planet in less than 4 h. The Great Venusian Anchoring Company had plans to anchor Venus to the Sun such that seasons would be uniform throughout the year. Transits would "… be a thing of the past …" (2004, Transit of Venus 1882, p. 227).

In anticipation of the event *The Saturday Evening Post* presented the tale of Jeremiah Horrocks, who had calculated and seen the first documented Venus transit in 1639 (Across the Sun, 20 August 1881).

5.2.1.2 Religious Periodicals

Again, religious periodicals would have a generous number of articles on the subject of the transit of Venus.

Fig. 5.26 The Brazilian transit station near Punta Arenas, Chile (courtesy: www.transitofvenus.nl/history.html)

The American Catholic Quarterly Review in October of 1882 published an informative 19-page article, "The Coming Transit of Venus". It described the methodologies of Halley and Delisle, supplemented with diagrams. It summarized the successes of the 1874 expeditions and the preparations for those of 1882.

On 28 September 1882 the Methodist *Christian Advocate* had a short item anticipating the upcoming event, "Transit of Venus", within its feature "Science, Art, Discovery". The American Commission's preparations would assure success. Within the same feature, not related to the transit, was "The Harvest Moon", within which readers interested in astronomy could read about the phenomenon of the harvest moon and what planets might be seen along with it in the days ahead. On 30 November 1882, also in "Science, Art, Discovery", there was a significant piece, "The Transit of Venus", this time describing more completely the purpose and methods of transit observations. The constitution of various expeditions was mentioned, the Americans at Patagonia and New Zealand, and the Germans and Hungarians at Aiken, South Carolina, and Hartford, Connecticut, as well as groups at Costa Rica, Argentina and the Straits of Magellan (e.g. see Fig. 5.26). Included was a reprint of a letter by T.C. Hartshorn encouraging the public to mark their calendars and to observe the transit with their smoked glass, along with the astronomers, as this opportunity would not present itself again for more than 120 years.

An early report after the transit was made in the *Christian Advocate's* feature "Science, Art, Discovery" on 14 December 1882 which was entirely dedicated to its title "The Transit of Venus". A statement of general success was made, followed by specific words provided by Professor George Washington Coakley (1814–1893) regarding observations made at the University of the City of New York. First contact was not observed in the main telescope due to a defect noted in the lens in use that was thereafter replaced. However, a time was noted by Professor R.H. Bull by use of the finder scope. Times were successfully noted for the other contacts with the main telescope. Coakley made an attempt to find a satellite around the planet as it transited the Sun but saw none. Meanwhile, at the College of the City of New York,

Fig. 5.27 Professor John Draper (after *Popular Science Monthly* 1874)

Professor Alfred George Compton (1835–1913) arranged a heliostat apparatus for his students so they could follow the progress of the transit. The heliostat "… threw the Sun's rays into a small telescope, which conveyed them to a screen of white paper, there producing a luminous disk, about six inches in diameter." (p. 794). Also at the College was Professor John William Draper (1811–1882; Fig. 5.27) who took a number of photographs of the event that he shared with the public.

After the transit, on 11 January 1883, the *Advocate* had an item, "Lecture on Astronomy", describing Professor C.A. Young's talk on the recent transit of Venus and the photographs that were taken at Princeton. While introducing Young to the audience, the American clergyman, the Reverend Dr Charles Deems (1820–1893), affirmed the church's belief that "… science was a help to religion, just as religion was a help to science". (p. 29).

The feature "Science and Art", was subtitled "The Transit of Venus" in the issue of 14 December 1882 of the *Christian Union.* It was a summary of the degree of success of the observing parties. Specifically mentioned were those of the U.S. Naval Observatory in Washington and the German astronomers at Trinity College in the United States at Hartford, Connecticut. Parties in Europe were not so fortunate. Snow fell over the Greenwich Observatory and in Paris and Madrid "… bad weather prevailed …" Better weather conditions led to more success at Portsmouth, Penzance, South Wales, South Africa and Jamaica (p. 529).

The Quaker *Friends' Intelligencer* had several articles in anticipation of the 1882 transit. On 6 November 1880 it printed yet another on the experiences of young Jeremiah Horrocks in the article "The Boy Astronomer". On 7 October 1882 "Astronomical Notes" was printed, which consisted of excerpts of intelligence

Fig. 5.28 A photograph of the Great Comet of 1882 (C/1882 R1) which was a conspicuous object in the sky during the last 4 months of 1882 and for a time even was visible in the daytime (courtesy: en.wikipedia.org)

conveyed by Professor Isaac Sharpless (1848–1920) of Haverford College. It included two paragraphs regarding the transit of Venus of 1882. It tells of some officially-sanctioned expeditions, including Professor Newcomb's to the Cape of Good Hope and Professor Lewis Boss' to Chili (*sic*), and described the possibility for involvement by the serious amateur astronomer. Materials were available with instructions for collecting data that would be of value to Washington. On 4 November 1882 in their article "Astronomical Notes" mention was made of the upcoming transit. Three methods were listed to achieve the ultimate goal of measuring the length of the chord which described the path of the transit. The English would emphasize the noting of the times of contact; the Germans would use the heliometer to measure the distance of Venus from the solar limb; and the Americans would rely most heavily on the use of photography to document the advance of the planet across the Sun. Also mentioned in "Astronomical Notes" was the latest on a comet that had been followed for a number of weeks (see Fig. 5.28), the visibility of Jupiter and Saturn at certain times, the upcoming annular eclipse to be seen in the Pacific and the Leonid meteor shower.

In anticipation, on 11 November 1882, the *Friends' Intelligencer* described the rarity and significance of the transit event. It concluded with a paragraph on how an interested individual might observe it, even without a telescope:

> The only essential … is some means of shutting off the excess of the sun's light. A piece of glass smoked in a candle or lamp flame is sufficient. Several pieces of stained glass of different colors afford the advantage that the screen used may be as dark as desired. Another means that can be employed is to use a piece of plate glass as a mirror. Nearly all the light passes through, but a small portion is reflected and forms an image of the sun … Venus will appear as a black circle projected against the sun, and in five hours and twenty-two minutes will pass across from the east to the west. (Child 1882a, p. 618).

On 16 December 1882 it was noted that weather conditions for the observation of the transit were unexpectedly favorable, and "At Friends' Central School opportunity was afforded to all the pupils and visitors to see the transit through the telescope." (Child 1882b, p. 699). Remember, in 1869 at the Friend's Normal School in Tyson, 700 observed the solar eclipse. It seems such astronomical events were significant opportunities for learning in this period in American history. In the same issue of the *Friends' Intelligencer* was another short item listing some specifics of times of contact at several sites (The transit of Venus 1882).

The *Friends' Review: a Religious, Literary and Miscellaneous Journal* published a two-part article on 9 and 16 December 1882, anticipating the results to come from the various observation sites for the transit of Venus (Sharpless 1882a, pp. 276–277; 1882b, pp. 293–294). The significance as a means of determining the distance of the Sun was stated, but with very little explanation. The definitions of the four contacts were given. Instructions to standardize the efforts were described. The Transit of Venus Commission had all its observers of its eight official parties practice with an artificial transit before the event. All telescopes were to be 5 or 6 in. in aperture and should magnify between 150 and 200 times. The exact time of second contact was defined:

> … so long as no light of undulation sweeps across the space which separates the black disk of Venus from the black sky outside, contact has not occurred, no matter if Venus does seem to be entirely within the outline of the sun. The first moment of such undulation is the time to be taken. (Sharpless 1882a, p. 277).

By means of the clockwork of the heliostat the Sun's image would be kept on the photographic plates. Each party would hope to obtain about 200 photographs, noting the times of the exposures. The measurements necessary for the determination of the solar distance would be made from the finished plates. It was noted that there were other events from which the solar distance might be determined but "… it will not do to omit any reasonable method of attaining it." (Sharpless 1882b, p. 294).

The Congregationalist paper *The Independent* on 14 December 1882 published an article, "The Transit", in which it described the varied successes of the observation parties, especially those in the United States.

On 15 June 1882 the *New York Evangelist* printed a paragraph in their feature "Scientific and Useful" stating that there were already some forty expeditions prepared to observe the transit of Venus of that year. This did not include those of the United States, Italy and Austria. Eight French stations and their respective leaders were noted.

One of these was to be in the United States, in northern Florida, under the leadership of Colonel Perrier. Closer to the event, on 16 November 1882, the *New York Evangelist* published an anticipatory article on the transit. It briefly described the limited success of the 1874 observations, and the financial commitment and preparations of the United States for the 1882 observations. The final paragraph was that of encouragement for families to have ready their smoked glass so they, too, could participate in the rare event (The transit of Venus 1882). In a follow-up article on 11 January 1883 there was a short item describing what was on the disk of the Sun on the day of the transit. Sun spots were at a minimum (Scientific and useful 1883).

On 12 April 1882 in Cincinnati's Methodist paper, the *Western Christian Advocate*, there was a review of Simon Newcomb's 4th edition of *Popular Astronomy*, which included an additional section over the previous editions covering the 1882 transit of Venus. The review was favorable. The book was described as understandable to the layman, with all terms explained (*Popular astronomy* 1882).

5.2.1.3 Scientific Periodicals

Scientific American was still a weekly periodical at the time of the 1882 Venus transit. On 1 July 1882 it printed a short article on the 40 expeditions planned at that time. A brief description of the American and French sites was made (Aspects of the planets for July 1882). Later that month, on 29 July 1882, there was a brief scientific explanation of the reason for the rarity of Venus transits. Efforts made by the German and British nations were acknowledged (Aspects of the planets for August 1882). The following week, on 5 August 1882, there were more details on the likely expeditions that had not yet been finalized due to the failure of Congress to agree on the appropriation. Four stations and their leaders for the Southern Hemisphere had been selected, two in South America, one in New Zealand and one in South Africa. There would be four principal stations in the United States, and there would be a number of foreign expeditions on American soil. The Americans would be using photography, with which they had been relatively successful in 1874 (Preparing for the transit of Venus 1882). On 18 November 1882, with the transit less than 1 month away, the magazine described, in more depth, what the now nearly 100 transit expeditions would be attempting to accomplish with their observations. The different methodologies also were described (The transit of Venus 1882).

There were two mentions of the upcoming event in the issue of 25 November 1882 of *Scientific American*. In their regular feature, "Aspects of the Planets for the Month", they anticipated that anyone of average intelligence would be following the transit with their piece of smoked glass. The times for expected contacts were published. A separate article on the transit again gave some of the history. Though all results of the data from thc 1874 transit had not yet been reduced, it was anticipated that 1882 would yield more accurate results (Aspects for the planets for December 1882; Transits of Venus 1882). On 2 December 1882 there was an article by Professor C.A. Young (1882b), then of Princeton University. He discussed the different methodologies that would be employed and the parties that would be involved in the observations.

Fig. 5.29 The Seagrave Observatory housed an 8.25-in. Clark refractor that young Frank Seagrave received as a 16th birthday present from his father (courtesy: www.transitofvenus.nl/history.html)

On 16 December 1882 *Scientific American* published two relevant items. The first, "The Transit of Venus", was an article on the successes of the 1882 expeditions, American and foreign. The second was a more specific report from one of the sites, in "The Transit of Venus as Seen at the Seagrave Observatory". This was a private observatory in Providence, Rhode Island (see Fig. 5.29), where the amateur astronomer, Frank Evans Seagrave (1860–1934), accomplished scientific work and enjoyed "… the grand phenomenon itself." (p. 385).

5.2.2 *The New York Times*

Only 8 years since the last transit, it was apparently not deemed necessary by the *Times* to educate the public on the history and methodologies of such an event by means of lengthy articles, as had been printed for the 1874 transit. However, there

Fig. 5.30 The British transit station at Jimbour House, Queensland, Australia (courtesy: www.transitofvenus.nl/history.html)

was some of this, on a smaller scale, as well as frequent updates on plans and expedition preparations in anticipation of the 1882 transit.

On 14 August 1881 *The New York Times* printed a short item describing the initial efforts in the selection of sites for the American parties. Help from the National Academy of Science was requested (The next transit of Venus 1881). Later that month, on the 20th, the reader would learn that Professor William Harkness read a paper, "The Methods of Determining the Solar Parallax, with Special Reference to the Coming Transit of Venus", the previous day at the meeting of the American Association for the Advancement of Science in Cincinnati (General Telegraph News 1881). On 3 February 1882, an article was reprinted from the *Providence Journal* which related that the upcoming transit would be visible throughout the western hemisphere and would last for 6 h. An "... intelligent observer ... [with] ... keen eyesight ... with the use of smoked glass, might see the tiny dot on the planet with his naked eye." (The coming transit of Venus 1882).

On 31 March 1882 the *Times* printed the speculation by Daniel Kirkwood of Bloomington, Indiana, that the transit might provide an opportunity to watch for a satellite of Venus (General notes 1882). On 3 August 1882 readers would learn that $75,000 was appropriated by the House of Representatives for the upcoming transit expeditions (Speech of President Curtis 1882). Later that month, on the 22nd, there was an article listing all the American parties for transit observations that were subsidized by this appropriation. There were four northern hemisphere sites, all within the boundaries of the country and its territories, and four in the southern hemisphere. The destinations of the expeditions and the members of all the parties were listed. The solar parallax and its significance were explained. There was also brief mention of some British (see Fig. 5.30), French and German parties (Gleanings from the mails 1882).

On 27 November 1882 the *Times* reprinted another item from the *Providence Journal*, a general description of the transit and times it would be visible. Again all intelligent persons were reminded to observe this rare event "... with the aid of a

Fig. 5.31 Jean-Baptiste Dumas (courtesy: en.wikipedia.org)

piece of smoked glass …" (The transit of Venus 1882). On 29 November 1882 there was a request from Professor William R. Brooks (1844–1921) of the Red House Observatory in Phelps, New York, that prayers be made at all churches on Sunday, requesting clear skies for the observation of the transit (Prayers for astronomical science 1882). On 5 December 1882, the day before the transit, there was an article with much information for the general public: the history of transits from the times of Kepler and Horrocks, the goals and methodologies of the observations, and some specifics about the parties. Readers were told how to prepare the smoked glass, and the times that the transit would be visible (Venus crossing the Sun's face 1882).

This transit would find more cooperation among the various nations of the world. *The New York Times* would report on foreign expeditions in foreign countries.

On 30 January 1881, almost 2 years before the Venus transit of 1882, readers of *The New York Times* could learn that the French Academy of Sciences had appointed an International Commission which, under the leadership of Jean-Baptiste André Dumas (1800–1884; Fig. 5.31) would prepare for the expeditions (Scientific gossip 1881). On 12 June 1881 one might further learn that the French Government was sending a scientific expedition to Cape Horn to study terrestrial magnetism, and they would be accompanied by another party which would study the transit of Venus (Scientific gossip 1881). On 6 November 1881, French leadership in international cooperation in the observations of this transit became more apparent. Dumas, the President of the International Commission, would send instructions to all participating astronomers and observatories (Scientific gossip 1881).

Fig. 5.32 Queens University Observatory (far left) was built in 1881 and housed a 6.5-in. Clark refractor (courtesy: www.transitofvenus.nl/history.html)

On 28 November 1881 the *Times* reprinted an article from the Toronto (Canada) *Globe* of 25 November, expressing the opinion that their city could provide a favorable site for transit studies. The Canadians saw this as an opportunity to improve their standing in the astronomical scientific community (Preparing for the transit of Venus 1881). On 14 November 1882 readers would learn that Clement Henry McLeod (1851–1917) from McGill University Observatory and H.V. Payne from the Meteorological Office, Toronto, were going to Winnipeg, Canada, to observe the transit (The transit of Venus 1882). On 6 December 1882 arrangements made at Kingston, Ontario, for observations to be made at Queen's University Observatory (see Fig. 5.32) were published (Little hope of seeing the transit 1882). The next day readers would learn that Canada was mostly cloudy during the time of the transit but "… occasional observations though the rifts in the clouds …" were made (Across the Sun's face 1882, pp. 1–2). On the other hand there was a 2-line article on 29 December 1882: "Toronto, Dec. 28.--Reports from various Canadian stations as to the transit of Venus have been received here. With one exception only they are considered very accurate." (The transit of Venus 1882, p. 2).

On 12 April 1882 the *Times* reported that the French Government would send eight expeditions to study the transit, four to the northern hemisphere and four to the southern (Current foreign topics 1882). On 7 December 1882 it was reported that preparations for viewing the transit in Paris were fruitless due to the heavy cloud cover (Across the Sun's face 1882). On 23 December 1882 one could read that the results from the French party near the Straits of Magellan were awaited "… with great anxiety …" (The late transit of Venus 1882, p. 3). Then on 4 January 1883 it was reported that "The French Commission telegraphed the Académie des Sciences that the results obtained in South America had exceeded all its hopes." (The South American states 1883, p. 2).

On 7 December 1882, in an article previously cited, the *Times* published preliminary results already received from many countries. It was noted that in London clouds and snow made observations at the Greenwich Observatory impossible. The British had more favorable conditions at Cork, Durban and Portsmouth. At Penzance they could see the transit for 2 h. At other English venues clouds interfered with all possible observations. They did, however, have good conditions in Cape Town (Across the Sun's face 1882).

On 17 September 1882 the *Times* noted that there would be four expeditions from foreign nations to the western hemisphere, one at Costa Rica, one at the Straits of Magellan and two in the United States (Scientific gossip 1882). On 23 December 1882 readers would learn that results from the Straits of Magellan were still awaited (The late transit of Venus 1882).

The Belgians were quite successful in South America. One would have read in the *Times* on both 14 December 1882 and 4 January 1883 that they had made 606 observations (The transit of Venus 1882; The South American states 1883).

On 6 December 1882 the *Times* reported that "The Mexican government has supplied instruments to scientific societies throughout the republic for making observations." (Little hope of seeing the transit 1882, p. 5).

On 10 December 1882 on the front page there was a short item received from Havana on the previous day:

> At Manzanillo both the internal contacts of Venus were observed. The external contacts were not seen on account of the interposition of clouds. The ingress of the planet was observed in Porto Rico, but her egress was hidden by clouds. (The transit of Venus 1882, p. 1).

The transit of Venus of 1882 would be the first where the United States, as a sovereign nation, could host scientific expeditions from other countries.

On 19 June 1882 *The New York Times* revealed that the Germans had selected Aiken, South Carolina, as one of their sites for the upcoming transit. Members from their Royal Observatory would arrive in late October (General notes 1882). The next month, from an article dated 9 July one would learn that they also planned to base a second transit party at an unnamed site in this country (Scientific gossip 1882). On 30 August 1882, it was reported that there would actually be four German expeditions to the western hemisphere. The two in the United States would be to the afore-mentioned site in South Carolina, and the other in Connecticut. Each German party would consist of "… two astronomers, a student, and an assistant." (Current foreign topics 1882, p. 1). On 3 November 1882 an article announced the arrival of a German party that would observe from Hartford, Connecticut. The members of the party were identified (Arrival of German astronomers 1882). On 6 December 1882 there was an item on the preparations of the Germans at Hartford the previous day, the last before the transit. Hopes were expressed for good weather:

> If the day is clear three telescopic observations of the contacts at ingress and egress will be made at the station of the astronomers at Trinity College, two by the Germans and one with the college refractor. (Little hope of seeing the transit 1882, p. 5).

Fig. 5.33 The U.S. Naval Observatory (courtesy: tycho.usno.navy.mil/history.html)

Apparently there was some success at this last site, as the Germans participated in the discussion of whether or not Venus had an atmosphere. On 8 December 1882 it was reported: "The German observers at Hartford are quoted as saying affirmatively that there were no indications of an atmosphere." (Article 2 – No title 1882, p. 4).

On 7 December 1882 the *Times* printed an article regarding the parties present in the San Antonio, Texas, area. Besides an American Party, headed by Professor Asaph Hall, there was a Belgian party, headed by Professor Jean-Charles Houzean (*sic*—it should be Houzeau). The first two contacts were missed due to cloudy conditions but the sky cleared and later observations were possible. It was noted that Professor Houzean (*sic*) and his three assistants took no photographs, but obtained 120 measurements, which he planned to compare with the results of the Belgian party in Chile. The Professor took his work very seriously during the transit, allowing no visitors, locking his gate and using police to "... prevent an invasion ...", though he was quite cordial subsequent to the event (Fair success in Texas 1882, p. 2).

The United States also hosted an expedition party from France. On 6 August 1882 the *Times* related: "The Secretary of War has granted permission to a party of French scientists to occupy Fort Marco, at St. Augustine, Florida, for the purpose of making observations of the transit of Venus." (Notes from Washington 1882). On 8 December 1882 it was reported that the French party had clear weather and "... obtained good and complete observations ..." (Watching the transit 1882, p. 5).

All four American Government-subsidized observation sites in the northern hemisphere were within the boundaries of the United States and its territories. Besides these, there were many other observatories and private individuals who took an interest in the event and made what contributions they could to the effort.

The official northern sites for the Americans listed in *The New York Times* on 7 December 1882, included the U.S. Naval Observatory at Washington, D.C. (see Fig. 5.33) under William Harkness; San Antonio, Texas, headed by Professor Asaph

Fig. 5.34 The Sayre Observatory at Bethlehem, Pennsylvania, where the Director, Charles L. Doolittle, observed the transit (courtesy: www.transitofvenus.nl/history.html)

Hall; Fort Selden, New Mexico, headed by Professor Davidson; and Cedar Keys, Florida, headed by Professor Eastman (The government's work 1882). On 6 December 1882 there was an article about the preparations going on at several observatories around the continent. The U.S. Naval Observatory had prepared a similar set-up to that used in the American expeditions for the 1874 transit. A long-frame structure to convey the light to the camera had been built onto the side of the building. The apparatus were listed and it was stated that, with the cooperation of the weather, a successful observation was expected (Little hope of seeing the transit 1882). However, on 7 December readers would learn that the weather did not cooperate. Some readings were taken, some photographs were obtained, but overall the results were disappointing (The government's work 1882). Then the next day Professor Davidson's report on the great success in New Mexico was published (Watching the transit 1882). On 23 December 1882 readers learned that all the Government-subsidized parties employed the same apparatus and arrangements used in 1874. All but the Washington, D.C., parties were quite successful (The late transit of Venus 1882).

Other observatories, colleges and individuals around the country participated in these efforts (e.g. see Figs. 5.34 and 5.35). Professor C.A. Young, active in keeping the public apprised of the progress in observations of this event, participated himself with the facilities at Princeton. When fears were reported by the *Times* on 12 November 1882 that a fire at a small building near the observatory at Princeton would not permit him to take any photographs (The transit of Venus 1882), Young quickly responded on the 14th that all had been restored and his party would be ready (Letters to the editor – Messrs. Harper and Mr. Pym 1882). Indeed on 7

Fig. 5.35 The transit was successfully observed using the 9-in. refractor at the Hobart College Observatory in Geneva, New York. The observer probably was Hamilton Lamphere Smith, the Professor of Astronomy (courtesy: www.transitofvenus.nl/history.html)

December 1882 an article related the success at Princeton. Equipment similar to that employed by the expeditions in 1874 was used, as well as several other telescopes. The Government provided the photographic plates and emulsion. All four contacts were seen. Young also made spectroscopic examinations of Venus' atmosphere (Fine results at Princeton 1882).

On 22 November 1882 the *Times* reported that Harvard University did not expect to take any particular notice of the transit of Venus of 1882 (A large spot on the Sun 1882). However, on 7 December 1882 readers would learn that in fact many observations were made and data were collected (see Fig. 5.36), and that all four contacts were observed (Good work at Harvard 1882).

The *Times* reported on 16 July 1883 that the Litchfield Observatory in New York failed totally due to "... inexorable clouds ..." (Making celestial charts 1883, p. 9).

According to a short item in the *Times* on 4 December 1882, Lafayette College in Pennsylvania (see Fig. 5.37) would make observations as directed by the Naval Department (The transit of Venus 1882). On 11 December 1882 Professor Selden Jennings Coffin (1838–1915) reported that all four contacts were seen. There was also mention of the 'black drop' effect that was apparent and a ring of atmosphere seen around the planet before the third contact (Observations of the transit 1882).

Fig. 5.36 Harvard College Observatory in 1900 (courtesy: en.wikipedia.org)

Fig. 5.37 The Lafayette College Observatory where Professor Coffin successfully observed the 1882 transit of Venus (courtesy: sites.lafayette.edu/specialcollections/tag/selden-jennings-coffin/)

On 8 and 9 December 1882 the *Times* published Professor Langley's observations at Pittsburgh of a peculiar bright spot on the planet when it was halfway onto the disk of the sun. No explanation was proposed. He was partially successful in his observations of the transit (Article 2 – No title 1882; The spot on Venus 1882).

On 2 May 1880 the *Times* reported that the Winchester Observatory at Yale (see Fig. 5.38) ordered a heliometer that would be completed in time for use at the transit of 1882 (Uniformity in time 1880). On 3 and 5 December 1882 the members

Fig. 5.38 Winchester Observatory at Yale University. Both the heliometer, in the left-hand dome, and the 8-in. Reed refractor in the right-hand dome, were used to observe the transit of Venus (courtesy: www.astr.yale.edu/department/history/1856-to-1882)

Fig. 5.39 The Observatory at Vassar College where Maria Mitchell and her party observed the transit (courtesy: pages.vassar.edu/fllaceducation/connections-maria-mitchell-and-her-observatory/)

of their scientific party were identified and their preparations for the transit described (The transit of Venus 1882; The Yale astronomers busy 1882).

On 6 December 1882 the *Times* reported that Vassar College (see Fig. 5.39) was making arrangements in Poughkeepsie, New York, to observe and photograph the upcoming transit (Little hope of seeing the transit 1882). This party was led by Maria Mitchell, whose application to participate in an overseas expedition had been denied due to her sex (Sheehan and Westfall 2004, p. 279).

Fig. 5.40 The Central High School Observatory, Philadelphia (courtesy: hiddencityphila.org/2013/01/wireless-city/)

On 7 December 1882 the *Times* described the results of the efforts at the Central High School Building in Philadelphia (see Fig. 5.40). Contacts were observed but due to hazy conditions photography, spectroscopy and micrometric observations were not attempted (Seen through a hazy sky 1882).

The New York Times listed the four American sites for 1882 Venus transit observations in the southern hemisphere on 7 December of that year to be Santa Cruz, Patagonia, under Lieutenant Samuel W. Very; Cape of Good Hope, under Professor Simon Newcomb; Cordova (Córdoba) near Santiago, Chile, under Professor Boss; and Auckland, New Zealand, under Professor Edwin Smith (The government's work 1882).

On 17 August 1882 the *Times* announced in its regular feature "Notes from Washington" that Lieutenant Samuel W. Very of the Navy would lead the transit party at Santa Cruz, Patagonia. They would leave from New York in a few days in the flag-ship *Brooklyn* (Notes from Washington 1882). A report was made on 4 January 1883 that observations there were marred due to rain (The South American states 1883). On 6 February 1883 readers would learn of the progress of the returning party which had made it as far as Montevideo (Naval intelligence 1883).

Reports on the expedition to Cape Town were initially those of updates on the personnel. In the regular *Times* feature "Army and Navy News" readers learned of the appointment of Lieutenant Thomas Lincoln Casey, Jr., Engineer Corps (1831–1896) to the party on 15 August 1882 (Army and Navy news 1882) and of the appointment of Lieutenant Edward W. Sturdy (1847–1898) as Newcomb's temporary replacement as Superintendent of the Nautical Almanac Office, on 7 September 1882, during the latter's absence (Army and Navy news 1882). On 19 September

1882 the *Times* announced the departure of the expedition group for the Cape of Good Hope on the steamship *Parthia* (Notes from the capitol 1882). Two days later it was related that this, the first of the southern expeditions to leave for its site, would arrive at the Cape Observatory about the first of November (The transit of Venus 1882). On 7 October 1882 it was announced that Professor Newcomb and his party left on the second leg of their journey from Southampton, England, to the Cape of Good Hope on the steamer *Durban* (Current foreign topics 1882). On 8 January 1883 the *Times* related the success of the party, which reported good observations of the internal contacts. They obtained 236 photographs, more than 200 of which were measurable. They had landed in Plymouth, England, on the previous day on their return to the United States (Current foreign topics 1883). In "Army and Navy Matters" on 12 September 1883, it was reported that Simon Newcomb had returned and resumed his duties at the Nautical Almanac Office (National capitol topics 1883).

A second South American expedition went to Valparaiso and ultimately to Córdoba in Chile. It was noted in the *Times* on 21 September 1882 that members at both of the South American venues selected, weather permitting, were to observe the entire transit (The transit of Venus 1882). On 26 October 1882 readers learned that this expedition had departed on October 12 (South American affairs 1882). On 13 December 1882 there was the short item:

> Panama, Dec. 12--Prof. Boss writes from Santiago ... that the American observations of the transit of Venus were completely successful. The weather was splendid, and all the arrangements were carried out. The four contacts were observed, and the photographs and measurements taken were all satisfactory. (The transit of Venus 1882, p. 2).

On 3 February 1883 there was an article subsequent to the return of Professor Lewis, who accompanied Boss, with his party. One learned of the courtesies extended them both by General Maturana of the Army as well as by the President of Chile. The circumstances surrounding the successful transit observations were described (The transit of Venus in Chili 1883).

The remaining American expedition to foreign parts was sent to Auckland, New Zealand. On 18 August 1882 the *Times* published the names of the members of this party which would be under the leadership of Edwin Smith from the United States Coast and Geodetic Survey, as Chief Astronomer. They would sail from San Francisco on 1 September (Notes from Washington 1882). On 3 September 1882, readers learned that Smith would proceed to Japan after completing his transit work to make "... pendulum observations ..." (Notes from Washington 1882). In an article of 21 September 1882 which summarized the expeditions to the southern hemisphere, one learned that only the egress contacts would be visible from New Zealand (The transit of Venus 1882). The summary article of 23 December 1882 told readers that the New Zealand party was successful in observing the last two contacts, and they took more than 200 photographs (The late transit of Venus 1882).

The most complete article found in *The New York Times* dealing with the 1882 Venus transit was printed after the event, on 23 December 1882. In this article of four-plus columns there was a summary of the goals and means of the various expeditions, and the following information summarizing the methods used and the varying degrees of success in observing the contacts. The following summary table, accompanying the key, is taken directly from the article (The late transit of Venus 1882):

Key:

1, 2, 3, 4	numbers denoting contacts observed
P	photographs taken using standardized American methods
P*	photographs taken by different method
h	heliometer measures taken
h*	equivalent measures to heliometer, different means
s	spectroscopic observations
p	photometric observations
m	micrometer measures of planet's diameter

1. Ottawa, Canada--1, 2, 3, 4
2. Kingston, Canada--2, 3, 4
3. Cambridge, Mass.--1, 2, 3, 4, s, p, m; several observers
4. Providence, R.I.--2, P* (23)
5. Amherst, Mass.--3, 4
6. South Hadley, Mass.--3, 4, s
7. Hartford, Conn.--2, 3, 4, h, m; German party
8. New Haven, Conn.--1, 2, 3, 4, P* (150), h, m; several observers
9. Helderburg Mountain, N.Y.--1, 2
10. West Point, N.Y.--1, 2, 3, 4
11. Poughkeepsie, N.Y.--3, 4, P* (9)
12. Brooklyn, N.Y.--1, 2, 3
13. Columbia College, N.Y.--2, 3, 4
14. Western Union Building, New York City--1, 2, 3, 4
15. University City of New York, New York City--1, 2, 3, 4
16. Elizabeth, N.J.--2, 3, 4
17. Princeton, N.J.--1, 2, 3, 4, P (188), s, m; several observers
18. Philadelphia, Penn.--1, 2, 3, 4
19. Easton, Penn.--1, 2, 3, 4
20. Allegheny, Penn.--1, 2, (?), s, m.
21. Pittsburg, Penn.--2, 3
22. Wilmington, Del.--1, 2
23. Baltimore, Md.--2, 3, 4; several observers
24. Annapolis, Md.--2, 3, 4
25. Naval Observatory, Washington, D.C.--1, 2, 3, 4, P (53), m; several observers
26. Coast Survey, Washington, D.C.--2, 3, 4; several observers
27. Signal Service, Washington, D.C.--, 2, 3, 4
28. Charlottesville, Va.--2, 3, 4
29. Aiken, S.C.--3, 4, h, m; German party
30. St. Augustine, Fla.--1, 2, 3, 4, h*, P*(200), m; French party
31. Cedar Keys, Fla.--2, 3, 4, P (180), m; Government Party
32. Chicago, Ill.--1, 2; several observers

33. Madison, Wisc.--1, 2
34. Northfield, Minn.--3, m.
35. Iowa City, Iowa--1, 2
36. Ann Arbor, Mich.--4, m
37. San Antonio, Texas--3, 4, P (200); Government Party
38. San Antonio, Texas--3, 4, h*, m; Belgian party
39. Fort Selden, New Mexico--1, 2, 3, 4, P (216), m; Government Party
40. Lick Observatory, California--2, 4, P (147), m

FOREIGN
Potsdam, Prussia--1, 2, P*, s, m
Jamaica--1, 2, 3, 4
Pueblo, Mexico--1, 2, 3, 4, h*; French party
Chapultenec, Mexico--No contacts, P*(13)
Cape Town, South Africa--1, 2, P (?), American Government Party
Durham, South Africa--1, 2
Tasmania--3, 4, P (?) – American Government Party
Melbourne, Australia--3, 4, P (236[?]); American Government Party
Santiago, Chile--Completely successful, P (?); American Government Party
Santiago, Chile--Completely successful, h*, m; Belgian party

A comparison of this summary of Venus transit expeditions with information gleaned over the previous months reveals that much of the information had been available to the public in previous articles. Interested readers could have followed and compared the relative successes of the various parties around the world and within the boundaries of the United States.

Over the following months there were other articles reflecting a degree of sustained interest in these scientific endeavors.

On the date of the transit itself, 6 December 1882, *The New York Times* printed an instance of a negative judgment on the various expeditions. The writer opined that Venus transits were just excuses for astronomers to request funds to be sent to exotic places. The writer felt the public had been misled at the previous transit of 1874 when it was implied that transits only occurred about once in a century (although this was an entirely unfair statement as numerous instances were found where a full explanation was made of the frequency of transits). He sarcastically remarked:

> No matter where an astronomer might live, the transit was never visible within a thousand miles of his home. The New-York astronomers had to go to Pekin; the Chinese astronomers had to go to Australia; and the Australian astronomers had to go to Europe. (The transit 1882, p. 4).

On 17 December 1882 the *Times* published a compliment to American astronomers from the British popularizer, Richard Proctor, reprinted from the *Gentlemen's Magazine*. Proctor was impressed by the Americans' use of photography and felt the results, once fully interpreted, would be very useful (A compliment to American astronomers 1882).

On 31 December 1882, the *Times* reprinted an item from *Nature* which expressed the sentiment that the recent transits had awakened the intellectual world from "… the slumber of the ages …" (The observations of 2004, 1882, p. 4).

On 18 January 1883 the *Times* published a short item describing a social event at Delmonico's restaurant, where "There was a handsome display of flowers, the most notable of which was a design representing the transit of Venus." (The sheriff's jury 1883, p. 5).

The *Times* printed another negative opinion on the profession of astronomy on 10 February 1883: "An astronomer is a man who is sent at the cost of the nation on scientific picnics in connection with the transits of Venus, and who employs his time in between successive transits in discovering new asteroids." (Wiggins 1883, p. 4).

On 13 June 1883 the *Times* printed a short review of the book by Richard Proctor, *Mysteries of Time and Space*, which included a chapter on the transits of Venus (New publications 1883).

On 27 June 1883 the *Times* printed the obituary of Stephen Alexander from Princeton College. Following a summary of his career as an educator and author there was the statement:

> For several years the aged astronomer had devoted his leisure hours to the study of the heavens from a small observatory in the rear of his residence, and there he observed the recent transit of Venus. (Obituary: Stephen Alexander LL.D. 1883, p. 4).

Over the months following the transit there was not the frequency of references to this particular event that had been seen previously. However, items cited here demonstrate that there was an expectation that readers were aware of much of the basic terminology associated with astronomy, and transits in particular. The education of the public in astronomy might be demonstrated in numerous articles yet to come.

Chapter 6
Discussion

6.1 'Amateurs' and 'Professionals' in Astronomy

Before analyzing the degree of 'popularization of astronomy' one must define the population under consideration and determine what indicators might demonstrate such a popularization.

Americans were just beginning to seek out their own scientific identity at the approach of the middle nineteenth century. Until this time there existed what John Lankford (1997, p. 18) would refer to as the 'preprofessionalism' of the science. It was during this period that American Henry Fitz would first manufacture telescopes that could compete in price and quality with those imported from Europe (Sperling 1989, pp. 54–56). Alvan Clark and Sons and John Brashear (1840–1920; Fig. 6.1) would follow suit (Bell 1986, p. 10). The observatory-building movement began (Williams 2000, p. 87). Local scientific societies, including those specifically for the science of astronomy, were growing in number (Heck 1989, pp. 361, 450; Lankford 1997, p. 14). The national American Association for the Advancement of Science, which started in 1848, did not exclude amateurs (Lankford 1997, p. 34).

The issue of defining 'amateur' versus 'professional' had not yet become problematic. In 1846, one-time sailor Henry Twitchell (1816–1875; see Jensen 2013) presented himself to Ormsby MacKnight Mitchel, the Director of the new Cincinnati Observatory. Twitchell was taken on as an unpaid assistant whose mechanical skills proved invaluable. He would remain at the Observatory for many years (Mitchel 1887, pp. 211–212). When Mitchel accepted the directorship of the Dudley Observatory in Albany, New York, without resigning his position in Cincinnati, Twitchell was left in charge. When Mitchel died while serving as an officer in the Civil War, Twitchell was in fact, if not in title, the Director for about a year (*The Centenary of the Cincinnati Observatory*, p. 42). William Cranch Bond, the first Director of the Harvard College Observatory, did not have a college education. Lewis Morris Rutherfurd (1816–1892; Fig. 6.2), who made significant technical

S. Cottam, W. Orchiston, *Eclipses, Transits, and Comets of the Nineteenth Century*, Astrophysics and Space Science Library 406,
DOI 10.1007/978-3-319-08341-4_6

Fig. 6.1 John Brashear (after Brashear 1924)

Fig. 6.2 Lewis Morris Rutherfurd (after *Popular Science Monthly* 1893: facing 400)

Fig. 6.3 Alexander Dallas Bache (courtesy: en.wikipedia.org)

contributions to astrophotography, was a lawyer by profession (Lankford 1988, p. 482). It seemed that up to this period distinguishing an amateur from a professional astronomer was a nebulous endeavor. This would change with the development of astrophysics. The acquisition and employment of the new scientific and technical knowledge would require the Americans and their British and Continental colleagues to set apart their own professional astronomers and lead to a deterioration of amateur-professional relations on both sides of the Atlantic (see Hetherington 1976; Lankford 1979, 1981; Rothenberg 1981). Thus,

> The amateur-professional nexus in astronomy is … a particularly fruitful area of study in that it focuses on the tensions associated with the evolution of a particular physical science. (Orchiston 1998a, p. 163).

The earliest American scientific professionals had to seek their education elsewhere. The first to receive a Ph.D. degree was Benjamin Apthorp Gould. He did so in Germany (Warner 1979, p. 67). Gould was a member of the 'Lazzaroni' whose membership also included Alexander Dallas Bache (1806–1867; Fig. 6.3) of the U.S. Coast and Geodetic Survey, Joseph Henry of the Smithsonian, and Benjamin Peirce and Louis Agassiz of Harvard (Stephens 1990, p. 31). The 'Lazzaroni' were "… a small network of very influential mid-nineteenth-century American scientists and science administrators … [who] … strove to raise national standards of scientific education and research, shape public opinion, win financial support and thereby lift the level of American science to that of the Europeans." (Hubbell and Smith 1992, p. 262).

Could the amateurs still be significant participants in the science of astronomy and astrophysics? Even as astronomy matured with the incorporation of the methods of astrophysics in the middle nineteenth century, John Lankford listed four functions for amateurs, 'astronomy's enduring resource'. These were: "... recruitment into the profession, instrument design, data collection, and public education." Lankford (1988, p. 483) felt that amateurs could still contribute, but due to scientific advances, the distinction between amateur and professional was no longer so hazy: "Amateurs collect data, but professionals interpret it." Interestingly, Lankford noted that the British amateurs were studying the planets, which at that time in America were considered within the purview of the professional (Lankford 1979, pp. 574–582).

The well-known South African amateur astronomer, the late Dannie Overbeek, would agree with Lankford's assessment of the interchange between the amateur and the professional astronomer. He provides a working definition of amateur astronomers as

> ... those who observe heavenly phenomena and formally report their results to scientific organisations without being paid for these activities. We should also include the group of amateurs who demonstrate or teach astronomy in a formal way. (Overbeek 1999, p. 128).

He goes on to provide examples of the contributions made by amateurs in the observation of variable stars, lunar occultations, the Sun, comets and meteors.

In his studies of Australian, New Zealand and South African amateur astronomers, the New Zealand-born one-time amateur astronomer, Wayne Orchiston, defined an amateur astronomer as "Someone involved in astronomy *for the love of it*, who normally does not earn a primary income from astronomy." (Orchiston 2006, p. 148). The key words in this definition are "*for the love of it*", "normally" and "primary income". In his review paper, "The role of the amateur in early Australian astronomy" (Orchiston 1989), which was written as a companion to "The emergence of professional astronomy in nineteenth century Australia" (Orchiston 1988), Orchiston identified the following ways in which amateur astronomers were able to play a key role in the development of late nineteenth century astronomy:

1) As research astronomers in their own right;
2) As contributors to the research projects of professional astronomers (especially solar eclipse and transit of Mercury and Venus expeditions);
3) As popularizers of astronomy;
4) As manufacturers of telescopes that were suitable for popular or research astronomy;
5) As donors of astronomical instruments to governments and professional observatories;
6) As key participants in the formation and development of early astronomical groups and societies;
7) As sources of professional astronomers—the so-called ATP (amateur-turned-professional) syndrome; and
8) As a political lobby group to governments and professional astronomy (e.g. the Keil Centre which disseminated announcements of newly-discovered comets).

Orchiston, then went on to write a series of papers that examined in detail most of these key elements of nineteenth century amateur astronomy (see Orchiston 1991, 1992, 1993, 1995, 1997a, b, 1998a, b, 1999a, b, 2000a, b, 2001a, b, d, 2002a,

Fig. 6.4 Comet C/1858 L1 Donati on 5 October 1858 (after Weiss 1888)

b, 2003a, 2004a, b, 2006, 2010, 2014; Orchiston and Bembrick 1995; Orchiston and Bhathal 1991; Orchiston and Brewer 1990).

Mirroring what occurred in Britain (see Chapman 1998), amateur astronomers made a major contribution to Australia's nineteenth century research output, due largely to the remarkable achievements of one man, John Tebbutt. Like their professional colleagues, Tebbutt and other Australian amateur astronomers published in the same local and overseas journals, and "Through the sheer weight and quality of their publications, amateurs helped to cement Australia's international astronomical reputation." (Orchiston 1989, p. 21). These remarks also characterize American amateur astronomers, although, comparatively-speaking, their published research output was not so prolific. Australia's John Tebbutt, after all, was an anomaly (e.g. see Orchiston 2001b, 2002b, 2004a).

The second half of the nineteenth century was witness to a remarkable succession of naked eye comets (e.g. see Figs. 6.4 and 6.5), plus transits of Venus and solar eclipses, that graced American skies (see Table 6.1). The general public demanded to learn about these public spectacles. Amateur astronomers responded to this by serving as sources of astronomical (and meteorological) information; running public viewing nights at their observatories; contributing newspaper reports, 'Letters to the Editor' and even regular astronomical columns; writing booklets, books and chapters of books, on astronomy; manufacturing planispheres; delivering public lectures on astronomy; and offering astronomy courses (see Orchiston 1997a). Orchiston (1989, p. 22) maintains that amateur astronomers made a major contribution to the popularization of astronomy in Australia:

Fig. 6.5 Comet C/1861 J1 (Great Comet = Tebbutt) (after Weiss 1888)

> Through their activities an understanding or at least awareness of astronomy reached many more people than the colonial observatories alone could ever have serviced. In this respect they supplemented the official work of the government observatories, but all the while on a voluntary basis, and at considerable personal sacrifice in terms of both time and money. In towns or cities devoid of professional observatories, these individuals ran their own observatories as *de facto* city observatories.

Furthermore, without these amateur astronomers and others who focused on popularizing astronomy through their local newspapers, "… the Australian public would have been far less astronomical aware than was in fact the case." (ibid.). While these comments refer specifically to Australia, they also apply equally to America.

In the late 1970s the University of Calgary Canadian sociologist, Professor Robert Stebbins (b. 1938), pioneered a whole new field of research, the 'sociology of amateurs' (e.g. see Stebbins 1977, 1978, 1979, 1980a, 1992, 2004, 2007). He recognized that amateurs straddle the nebulous boundary between work and leisure, and are 'amators' in that they love astronomy and are happy to invest time, money and effort in it, given the 'rewards' this will bring them (see, also, Williams 1987). Stebbins (1980b, 1981, 1982a, b, 1987) then wrote five papers dealing specifically with the avocation of amateur astronomy, and these form an important contribution to our discipline.

Stebbins (1979, p. 23) describes the 'Professional-Amateur-Public' (P-A-P) system. This is a system of "… functionally interdependent groups …" which he attempted to define. He started with the 'common-sense' description of the professional. He or she gains at least 50 % of his or her livelihood from the pursuit of this interest. The professional spends a significant amount of time in this pursuit.

Table 6.1 Impressive astronomical objects and events visible from the United States, 1850–1910 (adapted from Orchiston 2006, pp. 148–149)

Object or event	Years	Name/date[a]
Naked eye comet		
	1853	C/1853 G1 (Schweizer)
	1853	C/1853 L1 (Klinkerfues)
	1858–1859	C/1858 L1 (Donati)
	1860	C/1860 M1 (Great Comet)
	1861	C/1861 J1 (Great Comet = Tebbutt)
	1862	109P/Swift-Tuttle
	1864	C/1864 N1 (Tempel)
	1874	C/1874 H1 (Coggia)
	1881	C/1881 K1 (Great Comet = Tebbutt)
	1881	C/1881 N1 (Schaeberle)
	1882	C/1882 F1 (Wells)
	1882–1883	C/1882 R1 (Great Southern Comet)
	1887	C/1887 B1 (Great Southern Comet)
	1892	C/1892 E1 (Swift)
	1901	C/1901 G1 (Great Comet)
	1910	C/1010 A1 (Great January Comet)
	1910	1P/Halley (Great Comet)
Transit of Venus		
	1882	
Total Solar Eclipse[b]		
	1851	28 July
	1860	18 July
	1869	7 August
	1878	29 July
	1880	11 January
	1889	1 January
	1900	28 May

[a]Cometary nomenclature is after Marsden and Williams (1996). Note that in Orchiston (1997c) I have made a case for the Great Comet of 1865 to also be named after Hobart's Francis Abbott
[b]The list of solar eclipses is based on the paths of totality shown in Fig. 3.25 in this book

In contrast, amateurs may not derive their livelihood from their activities, nor devote as much time to these. Their chosen form of leisure is "… characterized by a search for satisfaction, taken as an end in itself." (Stebbins 1979, p. 261). However, they are not idle. They are "… doers rather than consumers of what someone else has done." (Stebbins 1979, p. 271). Also significant in the astronomy community is the public. They provide financial and role support, and their interests may influence the selection of subjects of study by the other two segments (Stebbins 1979, p. 22). There is some fluidity and overlap within the P-A-P system. Professionals, for example, may retire and become amateurs. Examples exist demonstrating that amateurs, and even members of the public, have gained the experience and knowledge to become professionals (Stebbins 1979, p. 37; cf. Orchiston 2001a, 2014). There is a synergy whereby any one segment can benefit from the activities and support of the others.

Referring specifically to amateur astronomers, Stebbins uses dedication as a criterion to distinguish 'devotees' from 'dabblers'. America's leading amateur astronomers of the nineteenth century were all devotees, individuals who were happy to make a substantial commitment to the science, in terms of both time and money. Using another dimension, 'knowledge and involvement', Stebbins distinguishes 'active' from 'armchair' amateur astronomers, and he categorizes 'active' amateur astronomers along an 'apprenticeship – journeyman – master' continuum. 'Apprentices' were only beginning their astronomical 'careers', while 'masters' were acknowledged experts who were able to make a useful contribution to astronomy, whatever the field they were in, and communicate effectively with professional astronomers. Many of the American amateur astronomers we have met in this book were 'devotees' who were 'active masters'.

The American astronomer Tom Williams has described recreational participants in astronomy. As part of the public, these might take advantage of a telescope and a clear night to observe "… the aesthetic wonders of the night sky …" (Williams 2000, p. 2). They might be armchair astronomers who read books or attend relevant lectures. Not leaving a mark, these recreational astronomers might be an invisible segment of this scientific community. It is by indirect means that one might determine their presence, perhaps by the number of popular books purchased or by the attendance at relevant public lectures. Williams would also include in this 'recreational' group the telescope-makers, those more interested in making telescopes than in using them (Williams 2000, pp. 2–9; cf. Orchiston and Bembrick 1995). The dedication of some of this last group could contribute significantly to the success of the professionals. The great American telescope-maker John Brashear, who started out as a laborer in the iron mills of Pennsylvania, would provide high-quality instruments for astronomers both in this country and outside of it. He was often called upon to answer a particular request or problem from a professional member of the astronomical community, of which he was to become a significant member (Brashear 1924).

Did the high-profile astronomical events considered here contribute to the popularization of astronomy in America during the second half of the nineteenth century? Did non-professional astronomers develop such an interest that they might ultimately contribute to the science in a number of ways? If so, what was the nature and degree of this popular interest? An attempt will be made here to see how these are reflected in a number of different parameters.

6.2 Articles in *The New York Times*

The New York Times printed many articles over the years under consideration, 1868–1882, dealing with astronomical topics. Their readership encompassed the full spectrum of the literate citizenry of the city, and in order to succeed in the business sense the paper would have to respond to the interests of this population. Figure 6.6 is a graph demonstrating the average number of words printed per page for articles that included the key word 'astronomy' during this period. This word was chosen since the intent is to determine a general interest in astronomy and not just in particular events. It was also deemed the most specific word for this search. Thus, 'eclipse' would bring up a greater number of articles dealing with horse-racing or clipper-ships than with astronomical subjects. One might also find its use as a verb. 'Venus' would bring up such articles as those dealing with the goddess or the statue, the Venus the Milo, or the plant the Venus fly-trap. 'Transit' would often be used as a verb or refer to some sort of transportation. 'Astronomy' was thus chosen to be the most specific word to use in this keyword search although it might not be the most sensitive. Articles about eclipses and transits of Venus might not contain that word.

The ranges of the years chosen have been grouped such that they would contain the years of the significant events, i.e. 1868, 1869, 1874, 1878, 1882 and the preceding and following years. These would then include articles intended to prepare the public for the event, plus those intended to convey what has been learned after the event. The results controlled for the number of issues per year and the number of pages per issue. This last was usually eight pages per issue but over the years there were an increasing number of special editions that contained more pages. Word counts for articles would be taken from those provided by *The New York Times* website when an article dealt with the subject of astronomy in its entirety. When that was not the case, personal word counts of relevant lines or paragraphs were made.

Early in its publication there were already articles on astronomy. A significant drop was evident during the period of 1861–1870, understandably as the eight pages

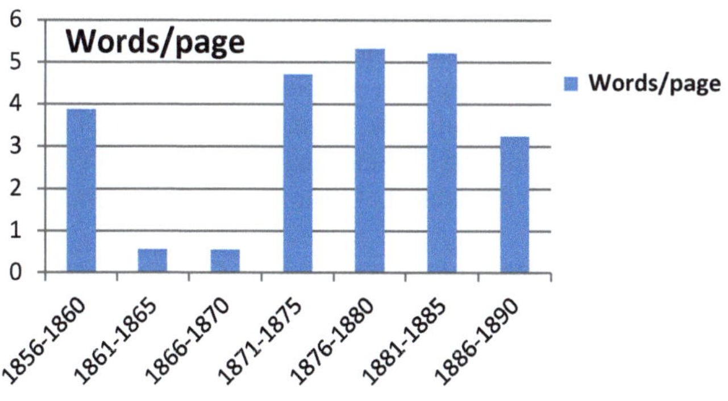

Fig. 6.6 *The New York Times* astronomy articles

would mostly be filled with news of the Civil War, the assassination of President Lincoln and the reports of progress during the ensuing period known as 'Reconstruction'. Unfortunately there was little printed on the scientifically-important eclipse of 1868, the first where spectroscopy was used to study the Sun and its prominences and corona. This event was not visible in the United States, and American scientists did not send out any eclipse expeditions. The public was not interested in the results of a scientific development, i.e. spectroscopy, that they did not yet understand.

In contrast, the eclipse of 1869, seen over much of the United States, was actually covered quite well in the newspaper. There were extensive articles both before the event notifying citizens of locations and times, and afterwards notifying them of successes and failures of scientific observations. However, here the use of the specific keyword 'astronomy' let some significant articles slip through the cracks. In 1869 there were only six articles with the key word 'astronomy', but there were 70 with the word 'eclipse'. Again, many of these had nothing to do with the astronomical event but, nonetheless, there were more relevant articles with the word 'eclipse' than the six picked up by the word 'astronomy'.

The next 5 years demonstrated a dramatic increase in the number of articles on astronomy, reflecting the public's increased interest. The transit of Venus to occur in 1874 was a rare event and the American astronomical community was going to participate in numerous adventurous expeditions. The *Times* would have a series of reports from its very own non-scientific participant, 'Our Own Correspondent', and he would describe to the reader his travels to exotic places, meetings with groups from other countries, and describe in laymen's terms the scientific efforts and results. The *Times* also included a six-column article describing the history and science of the event, educating the public so they could appreciate what was to follow. There would be frequent updates from the various parties around the world of their progress, successes and failures. It was the individual year 1874, of the entire range of years included on the graph, that actually demonstrated the highest word count per page for articles including the keyword 'astronomy', at 11.8 words per page for that year. There were 70 articles including the word 'astronomy', while there were 48 articles (on astronomy) brought up by the phrase 'transit of Venus'.

Over the next several years there seemed to be a continued interest in astronomical topics. Again in 1878 the *Times* did a competent job of reporting to its public on the solar eclipse, both before and after the event, although some of these might have been missed by our keyword search. Selecting just the year 1878 there were 36 articles printed dealing with the subject of 'astronomy' at an average word count of 4.91 per page for the year.

With the last transit of Venus of the century in 1882 we also see a resurgence in the number of relevant articles in *The New York Times,* again not necessarily reflected in our word count based on our keyword search. There were 63 articles that included the word 'astronomy', while there were 62 articles that included the phrase 'transit of Venus'. Many of the transit articles did not contain the word 'astronomy'. The average word count per page using the word 'astronomy' in 1882 was 7.14. It is interesting that across the ocean in Great Britain there seemed to be a decrease in enthusiasm in the transit and the related expeditions. Jessica Ratcliff (2008, p. 165) notes that the British *Times* printed 76 articles on the transit of Venus during the years 1873–1875, but only 21 during the years 1881–1883.

In the years following one can see that there must still have been some continuing interest in astronomical topics. Indeed, by this time there were monthly features informing the public of what celestial objects were in the skies and where and when one should look for them.

Readership of *The New York Times* would include all three of Robert Stebbins' populations: the professional astronomer would want information on the progress of the various expeditions provided within its pages; the interest of the amateur astronomer would be piqued by such events validating the time spent on his hobby; while the literate public would be educated in the science, intrigued by the adventure and perhaps find pride in the accomplishments of his country of citizenship or that of his heritage. *The New York Times* responded to all populations of its general readership in publishing articles on astronomy during these events.

6.3 Popular Lectures

During the peacetime years of the 1800s in the United States lectures became a popular pastime for the public. These could deal with a variety of subjects, including religion, history, literature and the sciences. Astronomy could be a particularly enthralling topic when delivered by such charismatic speakers as O.M. Mitchel or R.A. Proctor, accompanied by the illustrations of the 'magic lantern'. Over the years under consideration here the *Times* would print announcements of upcoming local lectures, or later reports of the contents of these lectures. A number of articles actually alerted the reader to more than a single lecture but rather a series of several lectures. A numerical count of these announcements and reports might be another indicator of popular interest in astronomy over the years. The same time ranges used in the graph for *The New York Times* word counts are used in the following graph (see Fig. 6.7) of relevant lectures announced and reported on over the years. Only popular lectures (i.e. those not of a professional or academic nature) were counted.

Lectures were already a popular form of entertainment, announced and reported on, in the early years of the *Times*. Many of the lectures referred to in these earlier years were by O.M. Mitchel. Mitchel went on the lecture circuit to earn funds to maintain the Cincinnati Observatory that he had established mainly through contributions of an inspired public. He also spoke in New York as a public service to help that city establish an observatory of their own in Central Park (Prof. Mitchel on astronomy 1859). Evidence was printed of the popularity of his lectures:

> Prof. Mitchell (*sic*) had an audience numbering nearly 3,000 to listen to his third lecture on Astronomy, last evening, at the Academy of Music. The subjects discussed were: 'The motions, orbits, and physical characteristics of comets.' (News of the day 1859).

Again there is a decrease in the number of lectures mentioned during the 1860s, presumably due to the greater significance of the Civil War and its aftermath.

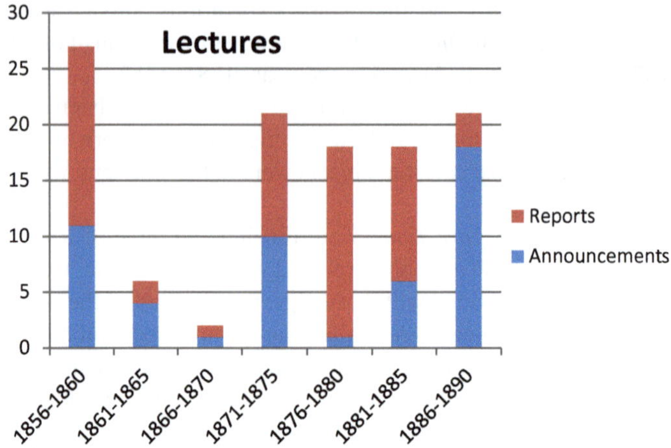

Fig. 6.7 *The New York Times* mentions of popular lectures

Professor Mitchel himself rejoined the military during this war and would die of yellow fever in 1862 while still in service.

Lectures again became items of interest in the news in the 1870s. There were a number of popular lecturers, at this time the most significant arguably being the British astronomer, Richard A. Proctor. Proctor would deliver lecture series covering a variety of astronomical topics such as the Solar System and the stars (The Sun's family of planets 1874; Lecture by Prof. Proctor. 1874; Wonders of the star depths 1874; Birth and growth of the Solar System 1874; The wonders of astronomy 1874; Lectures this week 1874). In March 1874 he would prepare his listeners for "The Coming Transit of Venus" (City and suburban news 1874) and the next month the *Times* made an announcement of a series of four lectures by him intended especially for the ladies. The first in the series had been given the previous day and had indeed been well attended by an audience composed mainly of ladies (The past and future of the world 1874). *The New York Times* summarized a recent successful lecture tour of Proctor in an article of May 1880:

> He delivered 136 lectures from the time he arrived in the United States, last October, to the date of his sailing from San Francisco to Australia, the twelfth of this month. The gross receipts were $50,000, with a clear profit to Proctor of some $15,000 … His short courses here and in Brooklyn were crowded, 11 lectures in both cities, bringing about $8,000. (Article 5, 25 May 1880).

Another popular lecturer was America's own Professor C.A. Young, initially of Dartmouth College and later Princeton University (Prof. C.A. Young dead 1905) who gave a lecture for the public reporting on the previous year's transit of Venus (The transit of Venus 1875). Lankford (1997, p. 32) notes that "Young was in great demand on the lecture circuit." As in the case of the statistical evaluation of word counts, one would find that the individual year with the highest number of mentions of astronomical lectures would be 1874 at 16 (8 announcements of upcoming lectures and 8 reports on those delivered). One might attribute this to the excitement surrounding the transit of Venus and the serendipitous presence of popular lecturers, in particular Richard A. Proctor. The number of relevant lectures seems to indicate that this interest was maintained through the next decade. Frequently also in attendance at these lectures, intended for the general public, were interested professional and amateur astronomers.

6.4 Telescopes

Galileo built his first telescope in 1609. Hussey (1897) noted that since then: "For morc than two hundred years the popular use of the telescope was entirely unknown." Though perhaps still to be considered 'recreational' in their pursuits, per Tom Williams, the purchase of a telescope would indicate a higher degree of commitment to the hobby of astronomy than attendance at a lecture.

The earliest telescopes popular in the United States were refractors, as at that time the mirrors in reflecting telescopes were made of metal that tarnished quickly and required frequent maintenance. On the other hand, refractors were expensive, due to the manufacturer's skill necessary to create the perfect surface of an achromatic lens. Complete telescopes, or even finished lenses, had to be purchased from abroad, adding to the cost. This was to change, as Harvard's 15-in. Merz and Mahler was "… the last great refractor Americans ordered from abroad." (Bell 1986, p. 8).

The earliest American-made telescopes of good quality were those of Henry Fitz in New York City (Bell 1986, p. 9). Fitz had travelled to England and Germany in 1839 to learn how to grind lenses (Sperling 1989), and "… between 1840 and 1855, Fitz alone made nearly four out of five of the astronomical instruments manufactured in the U.S. …" (Bell 1986, p. 10). His telescopes cost a fraction of those from Europe and also expedited availability to the customer as imports could take as long as 3 years (Sperling 1989).

Between 1860 and 1890 Alvan Clark and Sons, and John A. Brashear furnished the best American telescopes and accessories. The Clarks were known for the excellence of their refractors (see Fig. 6.8), and

> All told, the Clarks made 14 telescopes above 16 inches aperture, some 35 telescopes ranging from 10 inches to 16 inches aperture, 128 telescopes in the 6-inch to 9-inch range, and over 400 telescopes of 3 inches to 5 inches aperture. About half of these instruments were installed in academic institutions and used for pedagogical purposes; the other half ended up in private hands. (Warner and Ariail 1995, pp. 30–31).

As evidence of popular appreciation of Clark instruments is a letter from a Mary Hemenway who thanked the Clarks for the "… excellent instrument …" with which she and her children succeeded in seeing Saturn, Jupiter and its moons and resolved a double star (Mary Hemenway to Mr Clarke, 3 July).

In 1877 John Brashear "… invented an improved way of depositing a thin coating of silver on a mirror …" used in reflectors, increasing their popularity, and successfully manufactured quality instruments into the twentieth century (Bell 1986, p. 9). As Bell (1986, p. 12) points out:

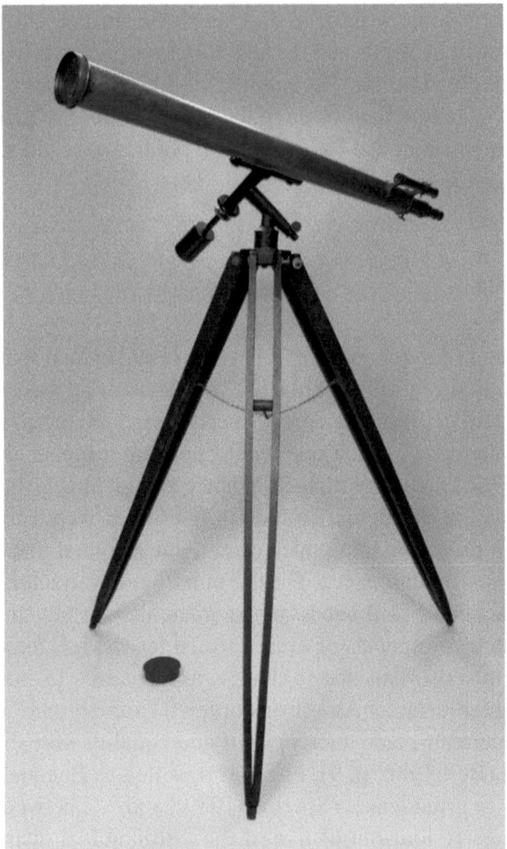

Fig. 6.8 An A-390 Alvan Clark telescope (courtesy: Adler Planetarium and Astronomy Museum, Chicago, Illinois)

> … by 1888 nearly three-quarters of Brashear's smaller instruments were in the hands of private purchasers. This great demand for small telescopes reflected just how genuinely widespread astronomy's fascination was in the general populace.

The popular Clark telescopes were produced during the entire second half of the nineteenth century. In their book *Alvan Clark & Sons: Artists in Optics* Deborah Jean Warner and Robert B. Ariail made a significant contribution by cataloging many of these instruments, including whatever data might be available as to the year of manufacture and the associated customers. The data from the more current second edition of 1995 is used here as the basis for Fig. 6.9 demonstrating telescope sales to individuals over the relevant years. The numbers are chosen that refer to American individuals as customers and not academic institutions or professional observatories. The catalog dates might refer to the dates of manufacture as inscribed on the instrument, dates estimated by the authors based on internal or external evidence, or the earliest date available as determined by the authors. Several other unique

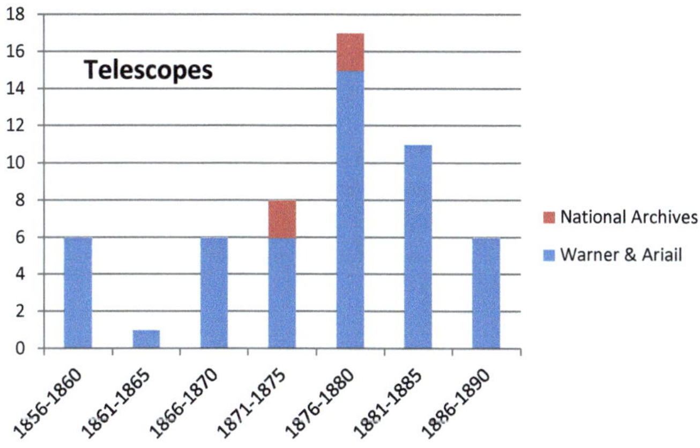

Fig. 6.9 Privately-owned Clark telescopes

Clark instruments referred to in a box of reports sent to the Transit of Venus Commission of 1882, for which the date of manufacture is available, have been included as well (National Archives, Record Group 78, E18, Box 39, Bound mss.). For the sake of consistency the same time frames adopted earlier are used here.

When available, Warner and Ariail include some interesting information on the customers and the use of their telescopes. Among other professions, customers included lawyers, bankers, ministers and physicians. The young teenager and future professional astronomer, George Ellery Hale, obtained a 4-in. Clark refractor in time for the 1882 transit of Venus. Again one may note a decrease in numbers during the years of the Civil War and its aftermath. In fact, three of the telescopes counted for the range 1866–1870, half the number, were actually for the year 1870, at the end of the decade. It might also be noted that the four unique Clark telescopes included from the data obtained at the National Archives, that were used to observe the transit of Venus of 1882, were purchased well before that event. There had been a commitment of some years by these individuals to their recreation.

6.5 Observatories

When Thomas Jefferson founded the University of Virginia in 1819, his plans included an observatory. The structure, ultimately erected in 1828, was poorly considered and likely was little used (Williams 1996). Jefferson's interest in astronomy had not yet taken hold in America. It was about halfway into the century before an astronomy movement there would gain momentum. Regarding the astronomical observatories of nineteenth-century America, Trudy Bell (1986, pp. 3–4) states:

> Of those for which the funding is known (104), nearly half (50) were sponsored by some type of philanthropy – either by individual wealthy patrons (37) or by public subscription (13), which is the collective philanthropy of the common man. In fact, except for the

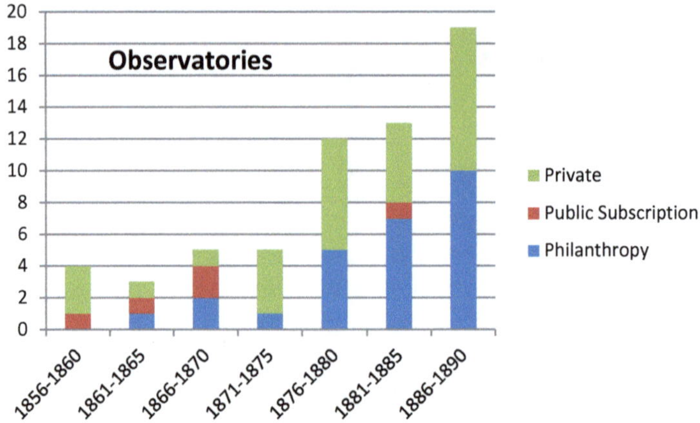

Fig. 6.10 Growth in the number of observatories founded in the United States

> U.S. Naval Observatory, which was financed by the Federal government, *all* the largest observatories built in nineteenth-century America were funded by wealthy philanthropists for whom the institutions were subsequently named.

The Lick Observatory in California was initiated through funds donated by philanthropist James Lick. The seed of his plan in 1873 became a reality by 1888 (Pearson et al. 2011). In 1891, while at the Lick Observatory the famous Irish amateur astronomer, Lord Rosse, commented on the number of people willing to make the two-day stagecoach journey up the mountain in order to make a visit (Bell 1986, p. 4). Observatories were a source of community and national pride. The appreciative public contributed funds and would hope in return to have some degree of access to these 'windows on the heavens'.

The data for Fig. 6.10 are taken from Trudy Bell's chart on the sources of funding for American observatories over the period under consideration here. The observatories included are just the ones established through philanthropic contributions, from either wealthy individuals or public subscription, and those established for private use. The government, schools and universities were also responsible for the existence of a number of observatories, although their contributions were considerably less. Only one school or university and five government-funded observatories were established over the years between 1856 and 1890 (Bell 1986, p. 6).

The growth in the number of observatories seems to be a step behind the interest indicated by the previous indicators considered for this same period. The construction of observatories in America that had started during the 1840s did indeed level during the period of the Civil War and Reconstruction period, but it is later in the century that the numbers seem to increase dramatically. Bell (1986, p. 6) points out that half of the observatories established in America during the 1800s were built during the years 1876–1890, "… the peak of the observatory-building boom in nineteenth-century America." Since observatories were not established overnight, one might consider that the initiation of the process would in many cases have begun several years before and this diagram may indeed parallel the general shape of those presented previously. Remember, the establishment of the Lick Observatory took 15 years. Also to be considered is the fact that one observatory paid for through public subscription may represent hundreds of committed citizens. For example, the initial cost of the Cincinnati Observatory in 1843 was paid for by 300 of its citizens buying shares at the cost of 25 dollars each. Data presented for the latter part of the nineteenth century do indeed reflect an increased public interest in astronomy.

In truth, in the case of observatories, public interest was not always a positive factor for the advancement of astronomy. Ormsby MacKnight Mitchel found the public nights at the Cincinnati Observatory a distraction from his research programs. Bell (1986, p. 4) quotes Henry Rowland's words of 1883:

> The figures show that generosity is a prominent feature of the American people, and that the needs of the country only have to be appreciated to have the funds forthcoming ... What would astronomy have done without the endowments of the observatories? By their means that science has become the most perfect of all branches of physics.

On the other hand Rowland also said:

> Our country has very many excellent observatories and yet little work is done in comparison because no provision has been made for maintaining the work of the observatory, and the wealth, which, if concentrated, might have made one effective observatory which would prove a benefit to astronomical science, when scattered among a half-dozen, merely furnishes telescopes for the people in the surrounding region to view the moon with. (Bell 1986, p. 6).

Considering the figures for the number of private observatories established one must be impressed by the commitment of these individuals to astronomy. They not only committed to the expense of a telescope but to a major structure in which to house it. They were more likely to be participants in the science.

6.6 Publications

The nineteenth century was a period of increased literacy in the United States. Improved lighting and availability of eyeglasses also made reading easier. Periodicals and books were becoming an affordable and profitable means of distributing knowledge of interest to the general public. Trudy Bell (1986, p. 4) points out that, "... by the 1890s ... four major U.S. publications devoted to astronomy had been started in the previous 20 years. Two of them were addressed at least in part to the general public."

O.M. Mitchel's *Sidereal Messenger* was first published in 1846. It was the first astronomical journal in America and one of the few worldwide of a popular nature. It was Mitchel's intent to use the profits from its publications to subsidize the maintenance of the Cincinnati Observatory. Unfortunately it only lasted a little more than 2 years.

In 1849 Benjamin Apthorp Gould began the *Astronomical Journal*. A member of the Lazzaroni, he saw it as a means of validating America's place in the scientific community. Publication ceased in 1861, with the advent of the Civil War, but resumed again in 1883 (Williams 2000, pp. 101–102, 104).

In 1877 the Boston Scientific Society began publication of *The Science Observer*. It was edited by John Ritchie, himself an amateur astronomer. Ostensibly a science journal, it weighed heavily with articles on astronomy. An editorial in the issue of November 1879 noted:

> With the present number ends the second volume of *The Science Observer*, and we trust that the subject matter has been of interest to our readers. As in the previous volume, the articles have been in the main of an astronomical character, and this will probably be the nature of the coming volume. (Editorial, *The Science Observer*, November 1879, p. 92).

In 1882 William Payne of the Carleton College Observatory in Northfield, Minnesota initiated a new journal for the amateur, another *Sidereal Messenger* (Williams 2000, p. 103). Subscription numbers to this journal might be telling, but unfortunately business records have not survived so these data do not exist (Carolyn Sanford and Eric Hillemann, personal communication; Marché II 2005). In 1891 the *Sidereal Messenger* changed its name to *Astronomy and Astrophysics* with the intention of still serving the amateur while allowing publications by professional astrophysicists who had not found a satisfactory venue in which to publish their work in this new field. This did not work well. Within 2 years George Ellery Hale founded *The Astrophysical Journal* and Payne would again publish a journal for amateurs, this time entitled *Popular Astronomy*, which continued until 1951 (Williams 2000, pp. 104–105).

Edward S. Holden, the founder of the Astronomical Society of the Pacific (ASP), thought the Society should have its own publication for its members and to exchange with other groups. Among other things it might include "… observations and papers from amateur members." (Bracher 1989, p. 9). The first issue was published in February 1889, and *Publications of the Astronomical Society of the Pacific* is still in existence today.

An early project of the ASP was the creation of a library for the use of its members. Holden suggested a list of basic books. The types of books available at that time and the nationality of the authors might be one indicator of the maturity of the amateur scientific community. Figure 6.11 is a graphical representation of books

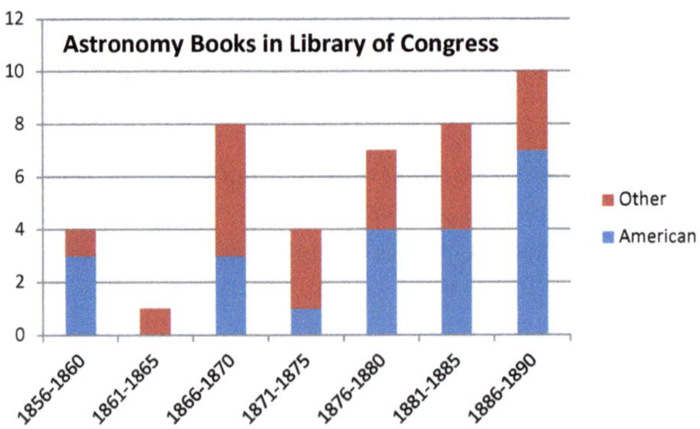

Fig. 6.11 Popular new astronomy books listed in the Library of Congress catalog

listed in the catalog of the Library of Congress. A keyword search was done using 'popular astronomy.' There were 735 hits. These were then selected for the years of publication that have been used for this study, i.e. 1856–1890. There were 42 books in the catalog for those years. An attempt was made at establishing the nationalities of the authors, when these were not obvious, by conducting biographical on-line searches. In case of failure the country of the publisher was assumed. Of course Holden may well have chosen books published before these dates but the limits here are used as an indicator of response to the immediate interest of the public segment by American publishers.

Again one might notice the dearth of data around the years of the Civil War. Richard Proctor's books are counted as 'other' as he was a British citizen, even though he moved to the United States in 1881 and died of yellow fever in New York in 1888. He had four books in the catalog, one each published in the years 1868, 1873, 1875 and 1882. In the second half of the 1870s there are at least as many cataloged books that are American as 'other' and this maintained until the end of the period under consideration.

6.7 Astronomical Societies

The establishment of significant astronomical societies or associations began in the middle 1800s. These could be for amateurs or professionals or both. These could be local or national. The earlier groups tended to be all-inclusive in membership, admitting any individual with an interest in the subject. The earliest scientific groups in the 1840s tended to be local, but "… by the 1860s the scientist had emerged as a legitimate social role and national rather than local scientific societies catered to the needs of the scientific community." (Lankford 1997, p. 14).

An example of an early association would be the Cincinnati Astronomical Society, established in 1842. The first 300 members, none of whom were professional astronomers, each donated 25 dollars for the purpose of building an observatory in their city. Membership would grow with need. Further donations in the form of monies or labor were accepted and the ultimate charter membership would include more than 500 names. The roster of 1870 grew to 865 shareholders, and still demonstrated an open membership, including such names as M.C. Jennings, tailor, and Daniel Robinson, stone cutter. There were also a number of larger entities listed as members, such as the Western Friend's Seamen's Society and St. Paul's Church. Unlike the original roster, which was almost entirely composed of local citizens, the newer one included members from across the country or even the world, among others, Joseph Grinnell of Massachusetts, C.E. Birdsell of Colorado, I.S. Armstrong of France and George W. Brook of England (*Cincinnati Astronomical Society-Stockholders* 1870).

Other local societies established in the late 1800s included the Chicago Astronomical Society, founded in 1862, and the Rittenhouse Astronomical Society, in Philadelphia in 1888. The Chabot Space and Science Center on the west coast of the United States in Oakland, California, was founded as an observatory in 1883, and is now primarily a public education center. According to Trudy Bell (1986, p. 4), "… by the 1890s amateur astronomical societies had sprung up in at least nine cities …" The Boston Scientific Society, ostensibly responding to interests in all sciences, in fact had a substantial emphasis on astronomy. In its constitution is stated:

> Any person may become a member of the Society by proposal and recommendation in writing by any member, nomination of the Council and election by two-thirds vote of the active members present at the regular meeting. (Boston Scientific Society 1886, p. vii).

Its meetings were held on the second and fourth Wednesdays of the month, and a social meeting was every Saturday evening to which all were welcome (Boston Scientific Society 1886, p. 93).

In 1848 the American Association for the Advancement of Science, the AAAS, was established. Its membership was open to all with an interest in any science, including astronomy. However, its mission to "… advance science …" necessitated a core leadership of professionals. Soon after its creation a special section in mathematics and astronomy was created "… in which astronomers communicated through oral presentations as well as informal discussion." (Lankford 1997, p. 22).

The Astronomical Society of the Pacific was founded in San Francisco, California, in 1889. On 7 February 1889 Edward Holden proposed its creation to promote "… cooperation between amateurs and the astronomers at Lick Observatory." (Bracher 1989, p. 7) and he would state: "There should be discussion, questions, remarks, interchange of ideas, contact of active minds." (ibid., p. 9). As the membership grew, noted astronomers would join, such as George Ellery Hale, G.W. Ritchie, W.W. Campbell, David Gill of South Africa and John Tebbutt of Australia. Also significant were the wealthy members who would contribute monies as well as validate the existence of the institution. These included, among others, Andrew Carnegie and Baron Albert von Rothschild (ibid., p. 14). In the first 2 years, under the presidency of Holden, the Society's membership would increase from 40 Californians to 383, from 32 American states and territories and 19 foreign countries (List of members of the Astronomical Society of the Pacific, *Publications of the Astronomical Society of the Pacific* 1892, pp. 1–9).

The first incarnation of an association by the name 'American Astronomical Society' was in 1853, with the subheading 'Cambridge Branch' founded in 1853 by professional astronomers in that area. The subheading was in anticipation of the appearance of other branches around the nation. This did not occur, and the Cambridge group did not last a year (Williams 2000, pp. 95–96). Another association with the same name appeared in Brooklyn Heights, New York, in 1883. There were 18 charter members. This time they were not professional astronomers, although members included Garrett Serviss, a science journalist, and Henry M. Parkhurst, an active and published amateur astronomer. Professionals then were invited to join (Berendzen 1974, p. 35), and among those to accept were C.A. Young, John Brashear, Asaph Hall, Edward Pickering and Maria Mitchell. Simon Newcomb resented the arrogance of this group and refused to participate (Williams 2000, pp. 96–97).

Newcomb would be largely instrumental in the formation of the last-named 'American Astronomical Society' with his suggestion to George Ellery Hale at the dedication of the Yerkes Observatory at Williams Bay, Wisconsin, on 21 October 1897. Subsequent to the Yerkes dedication conference were two more such meetings of astronomers, the first at Harvard and the second again at Yerkes. This last conference would indeed be the first meeting of the Astronomical and Astrophysical Society of America with an initial membership of 113, and Simon Newcomb as President. The cumbersome name was adopted at the time to avoid conflict with the name of the amateur society, the 'American Astronomical Society', or with the initials of the AAAS. However, in 1914, this amateur association no longer being in existence, the professional society adopted the less unwieldy name (Berendzen 1974). Amateurs were not denied membership in this society, but acceptance was not automatic. According to the by-laws anyone nominated by at least two members and who was "... capable of preparing an acceptable paper ..." on a relevant topic might be elected into membership by the Council. The charter membership was 15 % amateur, but over the years this percentage declined (Rothenberg and Williams 1999, pp. 40–52).

What can be gleaned from the limited information available about these early organizations? Lankford saw organizations as one mechanism giving the astronomical community cohesion and giving individual members opportunities to demonstrate their scientific successes and to make contacts that might validate these successes and/or provide resources for further research (Lankford 1997, pp. 4–6). These goals are not the same as those of the amateurs or the public. But where a society or group welcomed both amateur *and* professional astronomers,

> ... individuals from a given region were able to meet together to exchange ideas and news; to increase their skills and knowledge by attending meetings and borrowing from the group's library; to undertake organized programmes of observations or of manufacturing their own telescopes, through the group's specialist sections; and to publish the results of their endeavours in the society's journal or newsletter. On the one hand, such groups had the potential to establish or cement healthy amateur-professional relations, and on the other they provided a structured yet user-friendly way for beginners to be introduced to astronomy. (Orchiston 1998a, p. 163).

It was indicated by Stebbins that the public's primary function was the provision of financial support for astronomy. On 10 December 1880 *The New York Times* printed an article from the *Cincinnati Enquirer* of 5 December listing Cincinnati's millionaires. Early in the city's history there were just three: wine-maker Nicholas Longworth (1783–1863; Fig. 6.12), Judge Jacob Burnet (1770–1853; Fig. 6.13) and James Ferguson. Interestingly, all three were charter members of the Cincinnati Astronomical Society that provided the wherewithal for Mitchel to build their observatory. In fact Nicholas Longworth donated the land and Jacob Burnet was the first President. According to the article of 1880, by that time there were in Cincinnati 74 individual millionaires or entities, such as estates of deceased, in possession of such monies (Cincinnati's rich men 1880). Of those names listed, 38 were found on the Cincinnati Astronomical Society roster of 1870. Of the 848 legible names listed on a roster that should include 865 names according to the cover sheet, these 38 had the means, or would have the means by 1880, to contribute financially to this organization (*Cincinnati Astronomical Society Stockholders*, 7 January 1870). In all cases on the roster where the profession of the member is known, except for

Fig. 6.12 Nicholas Longworth (courtesy: en.wikipedia.org)

Fig. 6.13 Jacob Burnet (courtesy: www.ohiohistorycentral.org/w/File:Burnet_Jacob.jpg

Fig. 6.14 George Ellery Hale as a 20-year old (courtesy: astro.uchicago.edu/Yerkes/history/1892.html)

O.M. Mitchel, the member was not an astronomer. There were a number of doctors, lawyers, clergymen and politicians, public figures interested in the status such an organization would provide to their community. There also were tailors, artists, distillers, entrepreneurs, etc., less public figures but philanthropists important to the success of the Society none-the-less. However, among the membership there was indeed a true degree of interest in the science of astronomy. Many of the members of such organizations might hope to take advantage of such events as public viewing nights, but these might prove to be a distraction to the professional use of an observatory, as was indeed the case for Mitchel in Cincinnati.

What about the amateur, the 'doer'? Can any evidence of an increased commitment to astronomy be found among the membership in indicators other than that of financial philanthropy? As demographic data on membership is incomplete a graphical analysis might be difficult in this case but anecdotal evidence exists demonstrating some serious interest in astronomy among members of these associations. George Ellery Hale, a young member of the Chicago Astronomical Society (see Fig. 6.14), would go on to found the Yerkes, Mount Wilson and Palomar Observatories, as well as invent the spectroheliograph (Dr. George E. Hale 1938). Warner and Ariail (1995, p. 103) note that Hale, as a teenager, was given a second-hand 4-in. Clark refractor in time for use during the 1882 transit of Venus. The 1870 Cincinnati roster includes the name of Frederick Waldo. Waldo was a physician but apparently, as indicated by his membership in the Cincinnati Astronomical Society, he shared with his son Leonard an interest in astronomy. It was Leonard Waldo (1853–1929), who in 1874 would be appointed assistant astronomer to the United States transit of Venus expedition to Hobart, Tasmania (Hall 1883). Leonard also participated in the total solar eclipse expedition to Fort Worth, Texas, in 1878 (Notes, Nature 1878). Andrew P. Henkel, a physician in Cincinnati is listed as a member of the Astronomical Society of the Pacific in their roster of 1901 (List of members of the Astronomical Society of the Pacific, *Publications of the Astronomical*

Fig. 6.15 S.W. Burnham (courtesy: en.wikipedia.org)

Society of the Pacific, 1 February 1901). He was the owner of an 8¼-in. Clark refractor dated 1880 according to Warner and Ariail (1995). This was a large and expensive telescope for an individual, but we do not know if it was used effectively. He donated it to the Cincinnati Astronomical Society in 1925 (Warner and Ariail 1995, p. 112). We cannot know all the individuals who were members of these associations and pursued active roles in astronomy, but it is apparent from these examples that they were not all simply philanthropists.

Graphical demonstration of interest by growth in numbers as done for some of the previous indicators is not possible here due to the lack of data. We know that the membership of the Cincinnati Astronomical Society grew from 300 in 1842 to more than 500 within a short period of time, and up to 865 by 1870. No numbers exist for the rest of the century (personal communication, John Ventre). On the documentation of incorporation of the Chicago Astronomical Society in 1867 there are twelve names listed who with "… their associates and successors forever …" established the organization. Records of numbers of members over the years again are unavailable though certain names associated with particular events might be known. The ubiquitous George Ellery Hale and Sherburne Wesley Burnham (1838–1921; Fig. 6.15), of the Chicago Astronomical Society, participated in the recording of the transit of Venus of 1882, Hale being just a teenager at the time. However, numbers for a particular association might be deceiving, as other local astronomical societies might be created as individuals moved from the heart of large cities to the suburbs. The Chicago Astronomical Society is one of many whose membership varied from 100 to 300, while it may have had a membership of 400–500 at another time (personal correspondence, Eileen Wild). The Astronomical Society of the Pacific

Table 6.2 Membership in the Astronomical Society of the Pacific

Years	Membership
1891	309
1892	383
1893	493
1894	481
1895	433
1896	398
1897	370
1898	349
1899	312
1900	291

started publishing their membership rosters yearly, from 1891. These years are outside the purview of this study but may be of interest as the Society started at the end of the period of interest. There were forty original members of the Astronomical Society of the Pacific in 1889 (Bracher 1989, p. 6). The numbers published for the rest of the century are shown in Table 6.2 (List of members of the Astronomical Society of the Pacific 1891–1900). The last numbers in this table do not seem to indicate an increase in interest in astronomy, but factors other than 'interest' may be the issue in this case. For example, "By 1894 the Society was having financial difficulties, in company with the rest of the nation." (Bracher 1989, p. 15). In that year membership dropped from 481 to 433, and it continued this downward trend for a number of years.

One useful piece of information available is the numbers of visitors to the Chabot Observatory for specific years. This Observatory was originally built through generous gifts from the wealthy hydrological engineer Anthony (Antoine) Chabot (1813–1888). Money was also left to the observatory in his will with the stipulation that the original 8-in. refractor should always be available and free to the students of Oakland and the public. The Observatory opened to the public on 24 November 1883 (Fig. 6.16), the year after the transit of Venus. Visitor numbers are available for the years listed in Table 6.3. Again one sees an increase in public interest with the passage of years reflected in the number of visitors to the Chabot Observatory. The data end in 1891 when the observatory closed briefly for reconstruction. Unfortunately, the only other numbers immediately available were for the much later period of 1975–1976 (Ben Burress, personal communication). As of 2014, the website states that they now are serving 164,000 visitors annually including 52,000 students.

It is apparent that membership in an astronomical association can signify a spectrum of interest and/or participation in the science, ranging from a philanthropic desire to better one's community, to an armchair curiosity about the skies, to the desire to contribute to astronomy either as an amateur or a professional.

Fig. 6.16 The picturesque original Chabot Observatory (courtesy: Chabot Space and Science Center)

Table 6.3 Visitors to the Chabot Observatory

Years	Visitors to Chabot
1884	1,331
1886–1887	1,644
1887–1888	1,738
1888–1889	1,821
1889–1890	2,100
1890–1891	2,165
1975–1976	29,069

6.8 Public Participation in Research

In 1878 the United States Naval Observatory published their 30-page document *Instructions for Observing the Total Solar Eclipse of July 29th, 1878*. It was divided into nine sections, four of which were appropriate for amateur astronomers. Section One described "Observations with the naked eye." Within Section Five were the nine specific types of information to be submitted with reports (*Instructions for Observing the Total Solar Eclipse of July 29th, 1878*). Motivated amateur astronomers had guidelines for studying this astronomical event that would take place in their own country. Besides noting the time of contacts, observers were given instructions on how best to sketch the solar corona. Those with a telescope could search for the suspected planet Vulcan. Templates for sketches (see Fig. 6.17) were provided for sketches at the end of the document (Ruskin 2008, p. 27). This document was reported in such journals as the *New England Farmer* which stated:

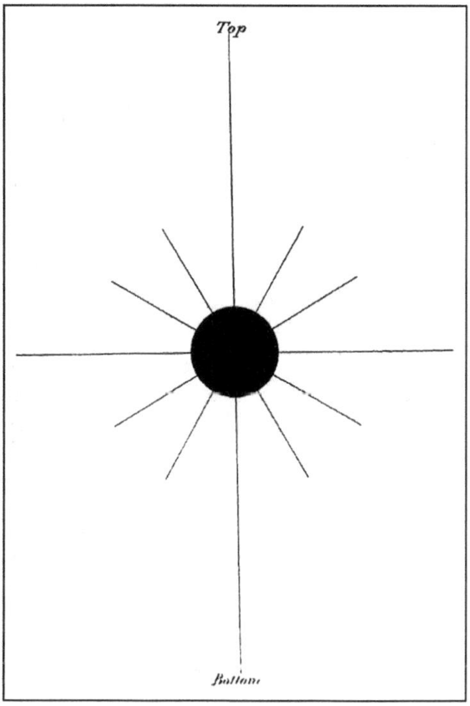

Fig. 6.17 Template for eclipse drawings (after: *Instructions for Observing the Total Solar Eclipse of July 29, 1878*)

> Instructions are given by which unskilled observers, possessed of a spyglass and a watch, may make observations on this eclipse along the line of totality, that are likely to be of use if sent to Washington. (This summer's eclipse of the Sun 1878).

The U.S. Naval Observatory requested submission of the eclipse drawings, and these are now housed in the National Archives in Washington, D.C. (Eddy 1973, p. 342; Ruskin 2008, p. 30).

An early educator at Colorado College, Frank Loud (1852–1927), Head of the Mathematics Department, considered himself an amateur in astronomy. With the help of David Todd, a fellow-Amherst graduate, he borrowed a small telescope. He became involved in the 1878 solar eclipse and subsequently "… became hooked on astronomy …" (Ruskin 2012, p. 116). Local donors would raise the funds for the purchase of a telescope for the College. "In just a few years Loud had transferred himself into one of the most active astronomers in the Rocky Mountain Region." (ibid.).

The previously-mentioned Leonard Waldo from the Harvard College Observatory published his report on the solar eclipse observed at Fort Worth, Texas, on 29 July 1878. Included were individual reports of the four members of the party, as well as his own. These were Mr R.W. Willson from Harvard College, Professor J.K. Rees, from Washington University in St. Louis, Mr W.H. Pulsifer from St. Louis and Mr F.E. Seagrave from Providence, Rhode Island. Finally he also included the contributions of others who had co-operated with his party in response to the published Government instructions. There were six such reports. Two of these listed their

Fig. 6.18 The corona drawn by Professor S.H. Lockett (after: Waldo 1879: Plate III)

results numerically as specified within the Government's instructions. Professor of Mathematics, S.H. Lockett of East Tennessee University, included his drawing of the eclipse (see Fig. 6.18). The one report submitted by J.M. Pearson, John S. Moore and W.J. Finch ended with the comment: "In closing, we would state, as our report will doubtless disclose, that none of us are professional or amateur astronomers." (Waldo 1879, p. 60). Waldo also included the drawings of several individuals for whom there were no further reports. These persons used the template format provided by the Government (see Fig. 6.19). Note their stated professions of builder, farmer, land locator and lumber clerk. These drawings seem to approximate the same orientation and magnitude of the coronal rays as shown by Professor Lockett.

There have been other instances of professional astronomers calling for the participation from the public in sharing data for particular astronomical events. Edward Holden, in order to promote the Lick Observatory, published a pamphlet, *Suggestions for Observing the Total Eclipse of the Sun on January 1, 1889*. Here Holden suggested a number of useful observations an amateur astronomer might make during the eclipse that was to occur in Northern California. One inspired amateur astronomer was a schoolteacher and insurance adjustor, Charles Burckhalter (1849–1923), who would eventually become the Director of the Chabot Observatory. Burckhalter organized an expedition of 65, "… including several ladies who sketched the Sun's corona." (Bracher 1989, p. 6). After learning of the results Holden wrote to Burckhalter:

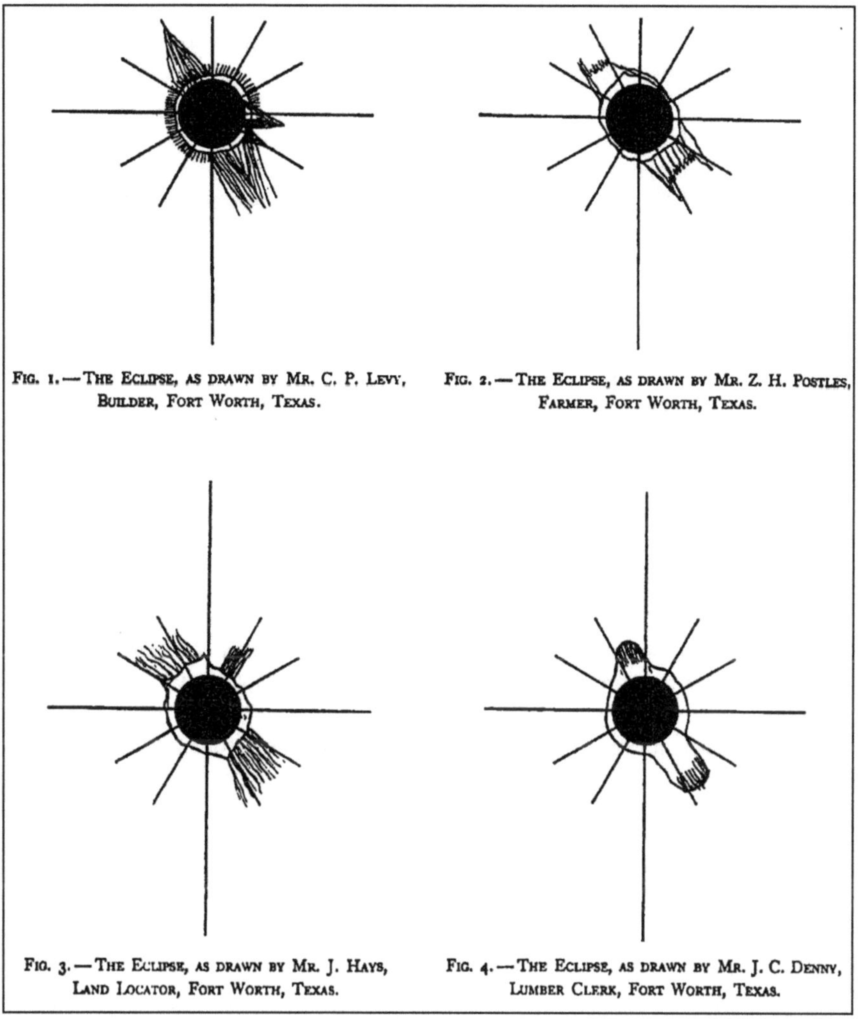

Fig. 6.19 Non-professional drawings of the corona (after: Waldo 1879: Plate IV)

> My dear Mr. Burckhalter:-
>
> I have seen the splendid reports from your parties, in the newspapers of yesterday which arrived here today, and I take the first opportunity to congratulate you and all the members of the expedition on the splendid success you have achieved. Your plan was so capitally conceived that success was sure if the day was fair – but that could not be commanded. Please give my heartiest congratulations to all. (Bracher 1989, p. 6).

It was on the momentum of this enthusiasm that the Astronomical Society of the Pacific was founded (Pearson et al. 2011). In fact, 16 of the 40 original members had participated in the eclipse expedition organized by Burckhalter (Bracher 1989, p. 6).

In 1882 the United States Transit of Venus Commission published instructions for the observation of the upcoming transit. These were to be followed by all the official expeditions to guarantee consistency in the collection of data. It was also intended that they could be "… adapted to the use of amateur observers who desire to be made acquainted with the methods by which they may make observations of value." (United States Transit of Venus Commission 1882). Abbreviated summaries of reports that might be accepted from amateur participants were published in a circular summarized in numerous popular journals, such as the 2 December issue of the *New York Herald*. Ten points to be covered in reports by amateur astronomers to be submitted to the commission were:

1) name of the observing station, including the town, county and state
2) date
3) description of location—street or reference to significant landmark or building
4) description of timepiece with number and maker
5) telescope—aperture, focal length, magnification, maker and number
6) a complete record of the time indicated by the timepiece at the reception of noon signals on 4, 5, 6 and 7 December
7) time of each contact as read on timepiece
8) estimate of uncertainty
9) other remarks?
10) signatures of observers, noting tasks

At the National Archives in Washington, D.C. there is a box containing 93 reports of observations of the transit, submitted by those who were not on official Government expeditions. The first author of this book has personally inspected these and found reports from amateur astronomers as well as professional astronomers. Also in this box is an envelope containing Harkness' summaries of the transit data so submitted.

Of the 93 reports sent to Harkness there were two instances of amateur astronomers submitting their results in a mailing along with a professional astronomer. One was 15-year-old Edward Stockwell, who made his own observations that were submitted with those of his father, John Stockwell (1832–1920), Professor of Mathematics and Astronomy at the Case School of Applied Science. John Hooper, a physician in Baltimore, Maryland, sent his report in the same mailing as Charles Hastings (1848–1932), Professor of Physics at Johns Hopkins University. In both cases it is unclear if Harkness used the amateurs' results. They were not in the envelope of summary reports but these might not be complete. Of the 77 Harkness summaries available, four did not have a corresponding report in the bundle of 93 reports. Thus, the data may be presumed incomplete in both the reports and the summaries. Contributors were here deemed 'professional' astronomers if they worked for an observatory, or in academia as a professor of astronomy, mathematics or physics. Engineers were not necessarily considered 'professional'. This determination was made based on the emphasis of their vocation. In some cases amateur astronomers identified themselves as such. Occupations were on occasion determined by other cited sources. For example, E.E. Barnard (1857–1923; Fig. 6.20) at the time of the transit was a photographer and is herein considered an amateur astronomer, although he would later be associated with the Lick Observatory and become a Professor of Astronomy at the University of Chicago (Edward Emerson Barnard. *Astronomical Journal* 1923). The character of their reports,

Fig. 6.20 E.E. Barnard, later in life when he worked as a professional astronomer (courtesy: en.wikipedia.org)

i.e. incompleteness of information or lack of precision, would be the determining factor when this information was not available. In all cases it was decided that these last were 'amateurs'. Adopting these criteria, there are 62 reports deemed to be from amateur astronomers and 33 from professional astronomers.

Harkness' summary reports were sheets listing the ten points requested in the circular, followed by the appropriate responses. Inclusion of a summary reflected the value of the report submitted. Of the 60 amateur astronomers' reports individually submitted, 47 (i.e. 78 %) had summaries included in Harkness' data. Of the 33 professional astronomers' reports, 26 had summaries (or 79 %). Thus it would seem that amateur astronomers' reports were taken as seriously as those of the professional astronomers. There were also instances of correspondence between Harkness and the respondents, professional and amateur, requesting clarification of points.

How did these amateur astronomers' reports differ from those of the professional astronomers? In many cases they were very similar. It was apparent, however, that some amateur astronomers were not aware of the circular that outlined requested information. A very incomplete report was more likely to be from an amateur astronomer. The amateur astronomers also seemed more anxious to get their results to Harkness and the Commission in a timely manner. Of the dated reports there were ten submitted by amateur astronomers on 6 December, the day of the transit, and only one by a professional astronomer. One amateur astronomer, watchmaker John Herbin, felt obliged to send a telegram with his report on that date. He telegrammed from Alamosa, Colorado: "Transit of Venus passed sun disc one o'clock twelve minutes fifteen seconds Denver time." His enthusiasm is apparent but his information was too incomplete and was not found to be included in one of Harkness' summaries. The two youngest contributors to the Commission were both amateur astronomers. One was Edward Stockwell, the son of professional astronomer John Stockwell, and the other was S.N. Fisk, a student at New York University, class of 1884, who felt "... thankful to be permitted to cooperate with you who are such distinguished and noble workers ..." Fisk's data were included in the Harkness summaries. The accompanying letters submitted by the amateur astronomers would frequently express their enthusiasm and their humility as participants in astronomical research. Note the following examples:

"... my humble efforts ..." (Thomas Bassnett);

"Sir, I have the honor to submit the following report of my observation (as an amateur) of the transit of Venus on the 6th of December last." (David Choat, physician);

"... we fear our observations are too approximate but submit them for what you think them worth." (J.J. Clarke and William Owen);

"I herewith send you my poor relative to the transit of Venus of Dec 6th 1882." (Elwood Garrett, insurance);

"... I can claim credit for good intentions only." (F. Hess, amateur with private observatory);

"It is rather presumptuous to offer these results." (W.L. Marcy, surveyor); and

G.H. Waldin, jewelry store owner, felt "... duty bound ..." to make his report.

William Numsen, packer of oysters, etc., in Baltimore, Maryland, had a private observatory listed in *The Encyclopaedia Britannica* (1889), but he did not want the recipients of his report to "... suppose that I consider my observations of anything like equal to those of the more practiced observer."

J.G. Jackson expressed the significance of the Commission's circular in initiating his endeavors:

> My outfit was only that of an amateur and I had not thought until your circular was published of making any effort towards observations of value, well knowing the number of well equipped observers waiting and eager to do all that science required.

This event proved to be a popular outlet for the non-professional astronomer who wanted to be a 'doer'.

It is apparent from all of the indicators considered in this chapter that astronomy was a popular subject in the second half of the nineteenth century, and there is in some cases a correspondence of this popularity with the particular total solar eclipses and transits of Venus under consideration here.

Chapter 7
Concluding Remarks

In a research paper on popular astronomical journals, Jordan Marché II (2005, pp. 49–50) refers to sociologist Stephen Hilgartner's view on popularization as the process of "… simplifying science for the non-specialist …" This scientific knowledge was often conveyed by non-professionals such as journalists "… for a public that misunderstands much of what it reads." Astronomy fascinated the public, and between 1869 and 1882 there were two transits of Venus, and also two total solar eclipses that were visible from the USA. Our aim in this book was to demonstrate that the education of the public, through such means as newspaper and journal reports of on these rare events, led to an increase in the number of non-professionals interested in astronomy, from the 'invisible' recreational star-gazer to the serious amateurs who actually were able to make a contribution to the science.

The New York Times printed articles for a general population that ranged from readers with minimal interest in astronomy to professional astronomers. A keyword search for 'astronomy' in *Times* articles demonstrated a surge in the number of articles of this kind during the second half of the nineteenth century, subsequent to the Civil War. The highest count was for 1874, the year when the first of the two nineteenth century transits of Venus occurred.

Articles on the transits of Venus were printed that would particularly appeal to the armchair astronomer. In 1874, *The New York Times* even provided its own non-professional correspondent to participate in the official United States transit of Venus expeditions to the Southern Hemisphere. Readers could relate to this individual and might be caught up in the fascination of his travels to exotic places, all in the interest of astronomy. One extensive article, printed on the very day of the transit, described in detail the significance of these transits and the techniques employed in observing and interpreting them, complete with diagrams. Professional astronomers also as well would read *The New York Times,* for they wanted to be updated on the reports submitted by the various transit parties.

The total solar eclipse of 1868 received little mention in the *Times* in spite of its great significance as the first to be studied with the aid of spectroscopy. It was too

S. Cottam, W. Orchiston, *Eclipses, Transits, and Comets of the Nineteenth Century*, Astrophysics and Space Science Library 406,
DOI 10.1007/978-3-319-08341-4_7

remote for the average citizen, as geographically it was not visible in the United States, and the science was too exotic for a non-professional audience. In contrast, the eclipses of 1869 and 1878 attracted considerable public attention as both were visible from the United States. During the eclipse of 1869 international attention was focused on the Americans, as the principal scientists at this event. By the eclipse of 1878, the Americans had established themselves as a force in international astronomy, and this time European astronomers would join them on these scientific expeditions. The non-professional reader of the *Times* would learn of their country's accomplishments and wonder about the answers to such questions as whether or not there was indeed an intra-Mercurial planet. Articles in the *New York Times* covered many aspects of the eclipse expeditions, the science, the adventure and the politics.

Lectures were a popular pastime for the general public during the second half of the nineteenth century. Often supplemented with the visual effects, compliments of the 'magic lantern', they had the potential to both educate and enthrall a non-professional audience. *The New York Times* would announce these lectures to encourage attendance, and afterwards would report on their content. Lectures served as entertainment and even education for the public and as potential fund-raisers for professional astronomers, toward such endeavors as the construction and maintenance of observatories. The graph of mentions of lectures on astronomy (see Fig. 6.7 on p. 266) again shows a surge after the Civil War, and again the greatest number was in the year 1874. Lectures were delivered by such charismatic professional-astronomers as England's Richard Proctor and the United States' own Professor Charles Young for the entertainment and education of the public and amateur astronomers. The lecturer would expect a financial return that might be personal, or it might be used toward scientific endeavors, in either case, a synergistic gain for all.

Personal possession of a telescope might indicate yet a greater commitment to astronomy than the previous parameters. Ostensibly this should indicate that the individual had the desire to personally participate in observation of the heavens. Warner and Ariail (1995) researched the ownership of American-made Alvan Clark telescopes. A graph of privately-owned Clark telescopes sold during the period under consideration peaked during the five-year period of 1876–1880 (see Fig. 6.9 on p. 269). It is known from records in the National Archives in Washington, D.C., that a number of these instruments were used to observe the 1882 transit of Venus in order that these individuals could make a personal contribution to astronomy.

The construction of an observatory to house a telescope would demonstrate an even greater commitment by the public to astronomy. Bell (1986) has provided data indicating that during the nineteenth century many American observatories were built with philanthropic funding, from either wealthy individuals or public subscription. There were also a number of observatories built for private use, by those with more than the casual interest of a recreational star-gazer. Bell (ibid.) pointed out that more than half of the observatories established in America during the 1800s were built between 1876 and 1890. Considering that observatories are not built overnight (for example, the Lick Observatory was initiated in 1873 but was not completed until 1888), one can presume that the seeds of many of these facilities might predate the 1886–1890 peak shown in the graph on page xxx. Public subscribers, inspired by these spectacular astronomical events, might hope to have access to these observatories. Moreover, private individuals, such as the grocery packer, William Numsen, had their own observatories, thereby enabling them to respond to such requests as those made by the Transit of Venus Commission for data collected during observations in 1882.

Several journals devoted to astronomy were initiated in the 1880s. O.M. Mitchel established the first astronomical journal in the United States in 1846, with the intention of raising funds to maintain the Cincinnati Observatory. His *Sidereal Messenger* was of a popular nature, yet it failed after only a little over 2 years. In 1882 William Payne initiated another *Sidereal Messenger* intended for teachers, students and others "... with a love for the elements of this great science ..." (Marché 2005, p. 51). Data on early subscriptions are not available as Payne had ultimate authority over the journal and chose to discard such records (ibid). In 1883 Benjamin Apthorp Gould's *Astronomical Journal*, initiated in 1849 but discontinued during the Civil War, was restored to publication in 1883 (Williams 2000, pp. 101–102,104), and it has continued through to the present day.

Based on the data collected through the Library of Congress catalog, a graph of new books on astronomy published during this period was constructed (see Fig. 6.11 on p. 272). It shows an increase in the number of books published after the Civil War, as well as a steady increase from 1871 to 1890 in those authored by Americans.

In the mid-1800s numerous astronomical societies were established, but very little information exists on an absolute count of their members. The Cincinnati Astronomical Society grew from 300 in 1842 to 865 by 1870. No other numbers exist for that century. The Chicago Astronomical Society was incorporated in 1867 listing twelve names and 'associates'. Again complete numbers over the years are unavailable. The Astronomical Society of the Pacific began in 1889, but did not start to publish their membership lists until 1891. Their numbers increased until 1893 but thereafter decreased steadily until the end of the century, reflecting perhaps the depressing economic conditions experienced at this time by many nations, the United States included.

Some positive data are available for the Chabot Observatory. Its records indicate an increase in public visitors from 1884 to 1891, when it was closed for reconstruction.

In 1878 the United States Naval Observatory published its *Instructions for Observing the Total Solar Eclipse of July 29, 1878*. These guidelines would be referred to by official expeditions, and by individual professional astronomers and amateur astronomers alike, as observations were made of this significant American eclipse. Records were sent to the Government and examples of these, including some by amateur astronomers, may be seen in the report of Leonard Waldo, who led an official expedition to Fort Worth, Texas.

Edward Holden challenged members of the public to observe a total eclipse of the Sun on 1 January 1889. Amateur astronomer, Charles Burkhalter responded and reported on an expedition of 65 members organized by himself. It was this initiative of his that led to the formation of the Astronomical Society of the Pacific.

The United States Transit of Venus Commission invited members of the public, amateur astronomers and professional astronomers, to contribute data on the transit of 1882, in order to establish a better value for the astronomical unit. We know there are 93 reports at the National Archives, 60 of which were provided by amateur astronomers. Doubtless there were other amateur astronomers who observed the event but did not submit reports to the Commission, including a teenage George Ellery Hale, according to Warner and Ariail (1995).

Stebbins (1979) described the Professional-Amateur-Public system as defining functionally-interdependent groups with mutual interests. The professional, according to 'common sense', gains more than 50 % of his/her livelihood from an occupation. The amateur is still a 'doer', but the pursuit is for self-satisfaction. The public provides financial and role support, and may influence the types of study carried out by the amateur and professional astronomers. In the case of astronomy, Williams would add the recreational group that might have this scientific curiosity but leave very little or no trace of this interest.

Do the parameters displayed graphically in Chap. 6 support the premise that there was a significant increase in the number of non-professional astronomers in the USA during the second half of the nineteenth century as a result of the publicity given to the aforementioned transits of Venus and total solar eclipses? The word count for articles on astronomy and mentions of lectures in *The New York Times* are both at their greatest in the year 1874, when the first nineteenth century transit occurred. That newspaper assumed enough interest by its readers to devote much of its content to the subject. The purchase of telescopes by private individuals and the building of observatories through philanthropic contributions seem to peak somewhat later in the century, which indicates a greater commitment to astronomy subsequent to these events. The Library of Congress catalog documents more American publications on astronomy as the century progressed.

More than 20 years ago one of the authors of this book (Orchiston 1989, p. 17) published a graph demonstrating an exponential growth in the number of amateur astronomers in the Sydney-Windsor region of Australia during the second half of the nineteenth century. Unfortunately, absolute numbers for American amateur astronomers are not available for the construction of a similar graph, but the limited numerical and anecdotal data available would tend to suggest an accelerated growth in numbers in this country as well at this time.

Brasch (1916, p. 154) saw astronomy in the American colonies as: "... primarily for the vain art of astrology and for almanac construction." It is apparent that the perception of its significance changed markedly during the nineteenth century, in particular after the appearance of some spectacular astronomical objects and events. These changes led to a better understanding of astronomy by professional scientists, amateur participants, the supporting public and the invisible recreational enthusiasts. All would contribute to, and gain from one another, a better appreciation of the night skies over America.

References

Abbe, C. (1869, August 14). The eclipse – Scientific observations at Sioux Falls City. *Cincinnati Daily Gazette*.

Abbe, C. (1872). Total eclipse of the sun of 1869. *The American Journal of Science and Arts*, 264–267.

Abbe, T. (1955). *Professor Abbe and the isobars: The story of Cleveland Abbe, America's first weatherman*. New York: Vantage Press.

Abney, W. (1874). Dry plate process for solar photography. *Monthly Notices of the Royal Astronomical Society, 34*, 275–278.

Abrahams, P. (1994). Henry Fitz: A preeminent 19th century telescope maker. *Journal of the Antique Telescope Society, 6*, 6–10.

Abstract, by the Secretary, of newspaper accounts of the Comet which have been forwarded to the Society. *Monthly Notices of the Royal Astronomical Society, 5*, 298–302.

Abstract of an article in *Le Cernéen*, a newspaper published in Port Louis, Mauritius, March 14, 1843. *Monthly Notices of the Royal Astronomical Society, 6S*, 8.

Accurate observation. *Scientific American, 19*(19), 297 (4 November 1868).

Across the Sun. *Saturday Evening Post, 61*(5), 7 (20 August 1881).

Across the Sun's face. *The New York Times*, 1 December 1882, pp. 1–2.

Across the Sun's face. *The New York Times*, 7 December 1882, p. 1.

Adams, W. (1865). The faithfulness of God. *New York Evangelist, 36*(44), 2.

Address of President Stokes before the British Science Association. *Appletons' Journal of Literature, Science and Art*, 2(29), 273–277 (16 October 1869).

Advertisement 9 – No title. *American Literary Gazette and Publishers' Circular, 17*(2), 48 (15 May 1871).

Advertisement 21 – No title. *Christian Union, 3*(5), 65 (1 February 1871).

Aid to the Centennial. *The New York Times*, 12 February 1874, p. 4.

Aiken and the transit of Venus. *Story of Aiken – Centennial Edition*, 4–6 April 1935.

Airy, G. B. (1857). On the means which will be available for correcting the measure of the Sun's distance, in the next twenty-five years. *Monthly Notices of the Royal Astronomical Society, 17*(7), 208–221.

Airy, G. B. (1864). On the transit of Venus, 1882, December 6. *Monthly Notices of the Royal Astronomical Society, 24*(8), 173–177.

Airy, G. B. (1868). On the preparatory arrangements which will be necessary for efficient observations of the transits of Venus in 1874 and 1882. *Monthly Notices of the Royal Astronomical Society, 29*(2), 33–43.

S. Cottam, W. Orchiston, *Eclipses, Transits, and Comets of the Nineteenth Century*, Astrophysics and Space Science Library 406,
DOI 10.1007/978-3-319-08341-4

Airy, G. B. (1878). An intra-Mercurial planet. *Nature, 18*, 380–381.

Airy, G. B. (1880). On the preparations to be made for observation of the transit of Venus, 1882, December 6. *Monthly Notices of the Royal Astronomical Society, 40*(7), 381–385.

Allerding, F. (1883). Observations by Mr. Allerding ... *Memoirs of the Royal Astronomical Society*, *47*, 60.

American College and Public School Directory, 1890. St. Louis: C.H. Evans & Co.

American Scientific Association. *The New York Times*, 9 August 1859, p. 1.

American solar eclipse observations. *Scientific American*, *26*(11), 161 (9 March 1872).

The American transit expedition. *Friends' Review: A Religious, Literary and Miscellaneous Journal*, *28*(27), 7 (20 February 1875).

The American transit expedition. *The Friend: A Religious and Literary Journal*, *48*(31), 245–246 (20 March 1875).

Another transit. *The New York Times,* 15 November 1879, p. 4.

Appendix. *American Journal of Science and Arts*, *1*(2), 153–158 (February 1871).

Applebaum, W. (2012). *Venus seen on the Sun: The first observation of a transit of Venus by Jeremiah Horrocks*. Leiden: Brill.

The application of photography to astronomy. *The Methodist Quarterly Review*, *21*, 392–410 (July 1869).

The approaching eclipse of the Sun. *New York Evangelist*, *41*(42), 2 (20 October 1870).

The approaching solar eclipse. *The Galaxy. A Magazine of Entertaining Reading*, *12*(4), 564–565 (October 1871).

The approaching solar eclipse. *Ohio Farmer*, *20*(46), 727 (18 November 1871).

The approaching solar eclipse. *Scientific American*, *25*(17), 261 (21 October 1871).

The approaching total eclipse of the Sun (letter to the editor). *The New York Times*, 28 May 1869, p. 2.

The approaching transit of Venus. *Christian Advocate, 51*(20), 153 (14 May 1874).

Army and Navy news. *The New York Times*, 15 August 1882, p. 2.

Army and Navy news. *The New York Times,* 7 September 1882, p. 2.

Army and Navy news. *The New York Times*, 16 March 1883, p. 2.

Arrival of German astronomers. *The New York Times*, 3 November 1882, p. 8.

Art notes. *The New York Times*, 25 June 1882, p. 4.

Art notes. *The New York Times,* 7 January 1883, p. 4.

Art. V. – The application of photography to astronomy. *The Methodist Quarterly Review*, *21*, 392–410 (July 1869).

Article 1 – No title. *Circular, 6*(20), 160 (2 August 1869).

Article 1 – No title. *Forest and Stream; A Journal of Outdoor Life, Travel, Nature Study, Shooting...*, *2*(2), 18 (19 February 1874).

Article 1 – No title. *New York Evangelist*, *45*(33), 4 (13 August 1874).

Article 1 – No title. *The New York Times,* 2 September 1869, p. 4.

Article 1 – No title. *The New York Times,* 13 December 1875, p. 7.

Article 2 – No title. *Christian Union*, *9*(24), 483 (17 June 1874).

Article 2 – No title. *The New York Times,* 8 December 1882, p. 4.

Article 3 – No title. *Scientific American, 19*(20), 310 (11 November 1868).

Article 4 – No title. *Maine Farmer*, *43*(2), 2 (12 December 1874).

Article 4 – No title. *Scribner's Monthly*, *9*(2), 261–262 (December 1874).

Article 5 – No title. *The New York Times* (25 May 1880).

Article 8 – No title. *Scientific American*, *20*(7), 107 (13 February 1869).

Article 8 – No title. *Zion's Herald*, *51*(11), 85 (1874).

Article 11 – No title. *Scientific American*, *19*(13), 202 (23 September 1868).

Article 17 – No title. *Scientific American, 29*(9), 139 (30 August 1873).

An article extracted from the *Colonial Observer*, of March 8, 1843. Published at Sidney, New South Wales. *Monthly Notices of the Royal Astronomical Society*, *6S*, 8.

Ashe, E. D. (1869). Solar eclipse of August 7, 1869. *Monthly Notices of the Royal Astronomical Society, 30*, 3.

Ashe, E. D. (1870). On his photographs taken during the total solar eclipse, August 7, 1869. *Monthly Notices of the Royal Astronomical Society, 30*, 173–174.

Aspects of the planets for December. *Scientific American, 47*(22), 337 (25 November 1882).

Aspects of the planets for July. *Scientific American, 47*(1), 2–3 (1 July 1882).

Aspects of the planets for August. *Scientific American, 47*(5), 65 (29 July 1882).

Assyrian and Biblical history. *The New York Times*, 11 May 1879, p. 4.

Assyrian records. *Friends' Review; A Religious, Literary and Miscellaneous Journal, 30*(26), 401–402 (10 February 1877).

An astronomer's evil star. *The New York Times*, 30 August 1874, p. 4.

An astronomer's voyage around the world. *The New York Times*, 25 April 1876.

The astronomers' work. *The New York Times*, 5 August 1875, p. 3.

The astronomical event of the century. *The New York Times*, 19 April 1874, p. 6.

The astronomical expedition. *The Galaxy. A Magazine of Entertaining Reading, 11*(1), 137 (January 1871).

Astronomical forecasts. *Friends' Review; A Religious, Literary and Miscellaneous Journal, 32*(9), 134–135 (12 October 1878).

Astronomical lecture. *New-Bedford Mercury*, 26 February 1836.

Astronomical notes. *Friends' Intelligencer, 26*(1331), 542–543 (7 October 1882).

Astronomical notes. *Friends' Intelligencer, 39*(38), 603–604 (4 November 1882).

Astronomical notes. *Scientific American, 31*(23), 356 (5 December 1874).

Astronomical researches. *The New York Times*, 1 March 1879, p. 2.

Astronomical researches rewarded. *The New York Times*, 30 April 1879, p. 1.

The Astrophysical Journal, 25, opposite page 53 (1907).

Aughton, P. (2004). *The transit of Venus: The brief, brilliant life of Jeremiah Horrocks, Father of British Astronomy*. London: Weidenfield & Nicolson.

The August moon. *The New York Times*, 4 August 1878, p. 10.

August skies at night. *Maine Farmer, 46*(38), 4 (17 August 1878).

Austin, G. L. (1878). The coming solar eclipse. *The Youth's Companion, 51*(30), 240.

Balls of the season. *The New York Times*, 28 January 1883, p. 9.

Banneker B. (the Negro astronomer). (1883). *The New York Times*, 2 December 1883.

Barkas, T. P. (1868). The solar eclipse of August 18, 1868 – to the editor of the *Astronomical Register. Astronomical Register, 6*, 135–136.

Barker, G. F. (1879). On the results of the spectroscopic observation of the solar eclipse of July 29th, 1878. *American Journal of Science and Arts, 17*(98), 121–125.

Base-ball. *The New York Times*, 30 July 1878, p. 5.

Baum, R., & Sheehan, W. (2003). *In search of Planet Vulcan. The Ghost of Newton's clockwork universe*. Reading, MA: Perseus Books.

Baxendell, J. (1859). On the corona around the Sun in total eclipse. *Monthly Notices of the Royal Astronomical Society, 29*, 293–294.

Baynes, T. S. (Ed.). (1889). *Encyclopaedia Britannica: A dictionary of arts, sciences and general literature* (Vol. 10, 9th ed., pp. 68–73). New York: J.M. Stoddart.

Beach, F. C. (1889). The effect of electricity on dry plates. *American Journal of Photography, 10*, 130–134.

Becker, B. (2000). Priority, persuasion, and the virtue of perseverance: William Huggins's efforts to photograph the solar corona without an eclipse. *Journal for the History of Astronomy, 31*, 223–242.

Becker, B. (2001). Visionary memories: William Huggins and the origins of astrophysics. *Journal for the History of Astronomy, 32*, 43–57.

Becker, B. (2011). *Unveiling starlight: William and Margaret Huggins and the rise of the new astronomy*. Cambridge: Cambridge University Press.

Belam, J. (1843) A letter from John Belam, Esq., Master of H.M. Sloop Albatross, on the Great Comet of 1843. *Monthly Notices of the Royal Astronomical Society, 5*, 293–294.

Bell, T. (1986). In the shadow of giants: Forgotten nineteenth century American telescope makers and their crucial role in popular astronomy. *Griffith Observer, 50*(9), 3–14.

Berendzen, R. (1974). Origins of the American Astronomical Society. *Physics Today, 27*(12), 32–39.

Bianchi, S., Gasperini, A., Galli, D., Palla, F., Brenni, P., & Giatti, A. (2010). Wilhelm Tempel and his 10.8-cm Steinheil telescope. *Journal of Astronomical History and Heritage, 13*, 43–58.
Birth and growth of the Solar System. *The New York Times*, 23 January 1874.
Bishop, J. E. (1979). The committee of 10. *Sky & Telescope, 57*(3), 212–213.
The boat race. *Zion's Herald, 46*(36), 427 (9 September 1869).
Booker, M. M. (2007). *Among the Stars: The life of Maria Mitchell*. Nantucket: Mill House Press.
Books and pictures at auction. *The New York Times*, 15 January 1875, p. 12.
Books received. *Lippincott's Magazine of Popular Literature and Science, 13*, 136 (January 1874).
Borneo observations of the eclipse of August, 1868, 1869. *The Eclectic Magazine of Foreign Literature, 10*(3), 380.
Boss, B. (1918). *Biographical memoir of Lewis Boss 1846–1912*. Washington: National Academy of Sciences.
Boston Scientific Society. (1886). *The Science Observer V2–5, 1878–1886: A journal for scientists*. Boston: Boston Scientific Society.
Bowers, L. (1879). The stars. *Shaker Manifesto, 9*(10), 231–232.
The boy astronomer. *Friends' Intelligencer, 37*(38), 606–607 (6 November 1880).
The boy astronomer. *New York Evangelist, 45*, 3 (3 December 1874).
Boyle, C. B. (1875, August). System of the Earth. *The American Educational Monthly*, 359–363.
Bracher, K. (1989). *Centennial history of the Astronomical Society of the Pacific* (pp. 4–15, 33–42). San Francisco: Astronomical Society of the Pacific.
Bradford, W. (1869). Scientific and useful: The Science Association at Salem. *New York Evangelist, 40*(36), 3.
Brasch, F. (1916). John Winthrop (1714–1779), America's first astronomer, and the science of his period. *Publications of the Astronomical Society of the Pacific, 28*(165), 153–170.
Brasch, F. (1928). John Winthrop (1714–1779) America's first astronomer and the first disciple of Sir Isaac Newton in the colonies. *Popular Astronomy, 36*, 83–85.
Brashear, J. A. (1924). *A man who loved the stars*. Pittsburgh: University of Pittsburgh Press.
Brashear, R. S. (1995). The Astrophysical Journal: A new journal for a new science. *The Astrophysical Journal, 455*, 403–406.
Brazil. *The New York Times*, 25 January 1875, p. 3.
Brooks, A. J. (1936). Elijah Hinsdale Burritt. *Popular Astronomy, 44*, 293–298.
Brown, A. (1869). The solar eclipse of 1598. *Astronomical Register, 7*, 111–112.
Brown, P., & Jones, J. (1998). Simulation of the formation of the Perseid Meteor stream. *Icarus, 133*, 36–68.
Bruce, R. (1987). *The launching of Modern American Science 1846–1876*. New York: Albert A. Knopf.
Brück, M. (2004). Lord Lindsay's expedition to Mauritius in 1874. In D. Kurtz (Ed.), *Proceedings IAU Colloquium*, 196, pp. 138–145).
A bundle of coincidences. *Richmond Enquirer*, 3 December 1833.
Bureau of Navigation. *The New York Times*, 17 November 1875, p. 5.
Bureau of Steam Engineering. *The New York Times*, 2 December 1878, p. 5.
Burritt, E. (1849). *The geography of the heavens*. New York: Huntington & Savage.
Butterworth, H. (1874, December). The boy astronomer. *St. Nicholas; An Illustrated Magazine for Young Folks, 2*(2), 93.
Caldecott, J. (1843). Observations of the Comet made at the observatory of Trevandrum accompanied by a drawing. *Monthly Notices of the Royal Astronomical Society, 5*, 302–304.
Campbell, J. H. (1879, November 9) An eye-, witness's account of the display in 1833 – from the Columbus (Ga.) *Enquirer*.
Campbell, W. W. (1893). Experiments in photographing the corona. *Publications of the Astronomical Society of the Pacific, 5*, 173.
Campbell, W. W. (1895). Photographing the solar corona without an eclipse. *Publications of the Astronomical Society of the Pacific, 7*, 62.
Campbell, W. W. (1907). The solar corona. *Publications of the Astronomical Society of the Pacific, 19*, 71–80.

Campbell, W. W., & Trumpler, R. (1923). Search for intramercurial bodies. *Publications of the Astronomical Society of the Pacific, 35*, 214–216.

Canada – Arrival of Prince Arthur's horses. *The New York Times*, 21 August 1869, p. 1.

The Cape of Good Hope. *The New York Times*, 23 December 1874, p. 1.

Carrington, R. C. (1858). On the evidence which the observed motion of the solar spots offer for the existence of an atmosphere surrounding the Sun. *Monthly Notices of the Royal Astronomical Society, 18*, 169–177.

Castle, S. N. (1875). Correspondence – Letter from the Sandwich Islands. *New York Evangelist, 46*(15), 2.

catalyst.library.jhu.edu

Centennial of the Declaration of Independence of the United States. *The New England Historical and Genealogical Register and Antiquarian Journal, 27*, 82 (January 1873).

Central eclipse of the Sun in the Island of Madeira, Southern Italy, &. *The New York Times,* 13 January 1867, p. 5.

Chapman, A. (1998). *The Victorian Amateur Astronomer. Independent Astronomical Research in Britain 1820–1920*. Chichester: Wiley.

Chapman, A. (2004). Transit of Venus: Horrocks, Crabtree and the 1639 transit of Venus. *Astronomy & Geophysics, 45*(5), 5–31.

Chase, P. E. (1878). Correspondence – Intra-Mercurial planet. *Friends' Review; A Religious, Literary and Miscellaneous Journal, 32*(5), 74.

Chauvin, M. E. (1993). Astronomy in the Sandwich Islands: The 1874 transit of Venus. *The Hawaiian Journal of History, 27*, 185–225.

Chauvin, M. (2004). *Hokuloa: The British 1874 transit of Venus expedition to Hawaii*. Honolulu: Bishop Museum.

Chicago – The eclipse. *The New York Times,* 11 August 1869, p. 4.

Child, J. M. (1882a, November 11). The transit of Venus. *Friend's Intelligencer, 39*(39), 618.

Child, J. M. (1882b, December 16). The transit of Venus. *Friend's Intelligencer, 39*(44), 699.

Chinese astronomy. *The Galaxy. A Magazine of Entertaining Reading, 18*(1), 127–128 (July 1874).

chroniclingamerica.loc.gov

Cincinnati's rich men. (1880, December 10). The New York Times.

Cincinnati Astronomical Society, Stockholders, 7 January 1870.

Cist, C. (1851). *Sketches and statistics of Cincinnati in 1851*. Cincinnati, OH: W.H. Moore and Co.

City and suburban news. *The New York Times*, 17 March 1874.

Clarke, J. F. (1866). The summer is ended. *The Monthly Religious Magazine, 35*(2), 111–118.

Clerke, A. M. (1881). The great southern comet of 1880. *Littell's Living Age, 150*, 760–766.

Clerke, A. M. (1902). *A popular history of Astronomy during the nineteenth century*. Decorah, IA: Sattre Press.

Close, M. (1843). Notes on its appearance made during a voyage from the Cape of Good Hope. *Monthly Notices of the Royal Astronomical Society, 5*, 293.

Cobb, C. (1896). Some beginnings in science. *Popular Science Monthly, 49*, 768.

Coffin, J. H. (1845). *Solar and lunar eclipses familiarly illustrated and explained, with the method of calculating them according to the theory of Astronomy as taught in New England colleges*. New York: Collins.

Colbert, E. (1878). To the president and directors of the Chicago Astronomical Society. In Chicago Astronomical Society (Ed.), *The Solar Eclipse of July, 1878*. Chicago: Evening Journal Book and Job Printing House.

The Comet of 1843. (1880, March 7). The New York Times.

The Comet of 1843. (1881, August 28). The New York Times.

Comet of 1843. *New-Hampshire Patriot*, 13 April 1843.

The comet of 1843. *New-Hampshire Statesman and State Journal,* 10 March 1843.

The comet of 1843. *Pennsylvania Inquirer and National Gazette*, 29 March 1843.

Comets and commentaries – from the *Boston Post. The Pittsfield Sun*, 16 March 1843.

The great comet of 1843. (1880, March 7). *The New York Times.*

The coming astronomical season. *The New York Times*, 18 September 1875, p. 6.

The coming eclipse. *Every Saturday*, *6*(135), 154–157 (1 August 1868).
The coming eclipse. *New York Times*, 2 August 1868, p. 5.
The coming eclipse of the Sun. *The New York Times*, 20 November 1870, p. 4.
The coming eclipse of the Sun. *The New York Times*, 11 July 1878, p. 1.
The coming solar eclipse. *The New York Times*, 7 July 1869, p. 1.
The coming total eclipse. *The New York Times*, 28 July 1878, p. 10.
The coming total solar eclipse, *Nature*, *17*, 453 (4 April 1878).
The coming transit of Venus. *The American Catholic Quarterly Review*, *7*(28), 712–730 (October 1882).
The coming transit of Venus. *The Catholic World, a Monthly Magazine of General Literature and Science*, *19*(110), 145–162 (May 1874).
The coming transit of Venus. *The New York Times*, 3 December 1874, p. 7.
The coming transit of Venus (from the *Providence Journal). The New York Times*, 3 February 1882, p. 2.
A compliment to American astronomers. *The New York Times*, 17 December 1882, p. 12.
Composition of solar prominences. *Medical News*, *27*(314), 30 (February 1869).
Concerning the weather. *The New York Times*, 6 July 1878, p. 4.
Conteur, 27 August 1922. Cincinnati and her Early Suburbs. *Enquirer*
Cooper, J. F. (1869). The eclipse. *Putnam's Magazine*, *4*(21), 352–359.
Copernicus, K. (1869). The eclipse party. *The Nassau Literary Magazine*, *26*(2), 104–112.
Copy of a letter addressed to the editor of the *New Zealand Colonist* by Captain W.M. Smith, R.A., and extracts from three other letters from New Zealand. *Monthly Notices of the Royal Astronomical Society*, *6S*, 8.
A correction. *The New York Times*, 10 December 1874, p. 4.
Cottam, S., Pearson, J., Orchiston, W., & Stephenson, F. R. (2011). The total solar eclipses of 7 August 1869 and 29 July 1878 and the popularisation of astronomy in the U.S.A. as reflected in the *The New York Times*. In W. Orchiston, T. Nakamura, & R. Strom (Eds.), *Highlighting the history of Astronomy in the Asia-Pacific Region* (pp. 339–375). New York: Springer.
Cottam, S., Orchiston, W., & Stephenson, F. R. (2011). The 1874 transit of Venus and the popularization of astronomy in the U.S.A. as reflected in the *The New York Times*. In W. Orchiston, T. Nakamura, & R. Strom (Eds.), *Highlighting the history of Astronomy in the Asia-Pacific Region* (pp. 225–241). New York: Springer.
Cottam, S., Orchiston, W., & Stephenson, F. R. (2012). The 1882 transit of Venus and the popularization of astronomy as reflected in the *The New York Times. Journal of Astronomical History and Heritage*, *15*(3), 183–199.
Cowper, H. A. (1843). Letter from H. A. Cowper, Esq., H.M. Consul at Pernambuco, Brasil, dated 9th March, 1843. *Monthly Notices of the Royal Astronomical Society*, *5*, 296.
Cozzens, S. E., & Gieryn, T. E (Eds.). (1990). *Theories of science in society* (pp. 1–14, 67–97). Bloomington and Indianapolis: Indiana University Press.
Crookes, W. (1869, December 9). The American eclipse. *Nature 1*, 170.
Curious phenomenon. *The Pittsfield Sun*, 21 November 1833.
Current events. *New York Evangelist*, *39*(44), 8 (29 October 1868).
Current events. *New York Evangelist*, *45*(22), 6 (28 May 1874).
Current events – The eclipse. *New York Evangelist*, *40*(32), 8 (12 August 1869).
Current foreign topics. *The New York Times*, 12 April 1882, p. 1.
Current foreign topics. *The New York Times*, 30 August 1882, p. 1.
Current foreign topics. *The New York Times*, 7 October 1882, p. 1.
Current foreign topics. *The New York Times*, 8 January 1883, p. 1.
Current notes. *The Albion*, *50*(12), 184–185 (23 March 1872).
Curtis, H. D. (1918). Note on the results of the search for an intra-Mercurial planet, eclipse of June 8, 1918. *Publications of the Astronomical Society of the Pacific*, *31*, 234–235.
Daboll, N. (1833, November 27) Remarkable phenomena. *Norwich Courier*.
Danville Quarterly Review, *1*, 614–638 (1861). Cited in Littmann, 1998: 119–120.
Davidson, G. (1875). Transit of Venus – Preliminary account by Professor Davidson of his observations [from the *The New York Times*]. *Friends Intelligencer*, *32*(1), 13–14.

The Davidson transit of Venus Party. *The New York Times*, 21 November 1874, p. 3.
Davis, E. (1921). *History of the The New York Times* (pp. 21–27, 36–38, 46, 62–64, 156, 168–169). New York: Greenwood Press Publishers.
The day of the transit. *Friends' Review; A Religious, Literary and Miscellaneous Journal, 28*(27), 7–8 (20 February 1875).
De Jong, T., & van Soldt, W. H. (1989). The earliest known solar eclipse record redated. *Nature, 338*, 238–240 (16 March 1989).
De la Rue, W. (1858). On lunar photography. *Monthly Notices of the Royal Astronomical Society, 18*, 16–18.
De la Rue, W. (1861). Photographs of the total eclipse. *Monthly Notices of the Royal Astronomical Society, 21*, 177–178.
De la Rue, W. (1862). *On the total Solar Eclipse of July 18th 1860, observed at Rivabellosa, near Miranda de Ebro in Spain*. London: Taylor & Francis.
De la Rue, W., Stewart, B., & Loewy, B. (1867). Researches on solar physics. *American Journal of Science and Arts, 43*(128), 179–192.
De la Rue, W. (1868). On the observation of the transits of Venus by means of photography. *Monthly Notices of the Royal Astronomical Society, 29*, 48–53.
De la Rue, W., Stewart, B., & Mr. Loewy. (1868). Sun-spots and general aspect of the Sun on the day of the total eclipse, August 18th, 1868. *Monthly Notices of the Royal Astronomical Society, 29*, 3–4
De la Rue, W. (1869a). On the Aden and Guntoor photographs of the eclipse of the Sun, August 18, 1868. *Monthly Notices of the Royal Astronomical Society, 29*, 193–194.
De la Rue, W. (1869b). On the solar eclipse of August 18th, 1868. *Monthly Notices of the Royal Astronomical Society, 29*, 73–82.
De la Rue, W. (1869c). The great solar eclipse of 1868. *Astronomical Register, 7*, 45.
De la Rue, W. (1870). On some attempts to render the luminous prominences of the Sun visible without the use of a telescope. *Monthly Notices of the Royal Astronomical Society, 30*, 22–24.
Débarbat, S. (2004). Astronomers Français et passage de Vénus sur le Soleil. *L'Astronomie, 118* (Mai), 268–273.
Departure of Prof. Davidson for Nagasaki. *The New York Times*, 30 August 1874, p. 5.
Devens, R. M. (1878). *Our first century: Being a popular descriptive portraiture of the great and memorable events of perpetual interest in the history of our country* (pp. 196–203, 329–336, 424–430). Springfield, MA: C.A. Nichols & Co.
DeVorkin, D. H. (Ed.). (1999). *The American Astronomical Society's first century* (pp. ix–xii). College Park, MD: American Institute of Physics.
Dick, S. J. (1998). Observation and interpretation of the Leonid meteors over the last millennium. *Journal of Astronomical History and Heritage, 1*, 1–20.
Dick, S. J. (1995). The American transit of Venus expeditions of 1882. *Bulletin of the American Astronomical Society, 27*, 1331.
Dick, S. J. (2003). *Sky and Ocean joined: The U.S. Naval Observatory, 1830–2000* (pp. 263–273). New York: Cambridge University Press.
Dick, S. J. (2004). The American transit of Venus expeditions of 1874 and 1882. In D. W. Kurtz (Ed.). Transits of Venus: New views of the solar system and galaxy. In *Proceedings IAU Colloquium No. 196, 2004* (pp. 100–110).
Dick, S. J., Orchiston, W., & Love, T. (1998). Simon Newcomb, William Harkness and the nineteenth-century American transit of Venus expeditions. *Journal for the History of Astronomy, 29*, 221–255.
The discovery of Vulcan. *The New York Times*, 8 August 1878, p. 5.
Discussing scientific study; Papers before the National Academy. *The New York Times*, 16 November 1881, p. 2.
The distance of the Sun (from the *Times*). *Littell's Living Age, 159*, 319–320 (3 November 1883).
Donati, G. B. (1868). On the approaching solar eclipse, and the present condition of practical astronomy in Italy. *Astronomical Register, 6*, 170–173.
Dr. George E. Hale, astronomer, dead. *The New York Times*, 22 February 1938, p. 21.
Dr. Isaac Sharpless dies. *The New York Times*, 17 January 1920.

Draper, H. (1878). The solar eclipse of July 29th, 1878. *American Journal of Science and Arts, 16*(93), 227–230.

The Draper eclipse expedition. *The New York Times,* 18 July 1878, p. 4.

Draper's eclipse expedition. *The New York Times*, 26 July 1878, p. 5.

Duerbeck, H. W. (2004). The German transit of Venus expeditions of 1874 and 1882: Organization, methods, stations, results. *Journal of Astronomical History and Heritage, 7*(1), 8–17.

Duerbeck, H. W. (2007). Die Photographen des Venusdurchgangs von 1874. *Acta Historica Astronomiae, 33*, 358–397.

E.C.H (1843). Great Comet of 1843. *American Journal of Science and Arts*, *21*, 175.

Eastman, J. R. (1882). Correspondence to Stephen Clegg Rowan on 6 December 1882, National Archives.

Eclipse expedition from New-York. *The New York Times*, 14 July 1878.

Eclipse of the Sun. *Ohio Farmer*, *17*(16), 247 (18 April 1868).

The eclipse to-day. *The New York Times*, 7 August 1869, p. 4.

The eclipse. *Cincinnati Daily Gazette*, 14 August 1869.

The eclipse. *New York Observer and Chronicle*, *56*(31), 246 (1 August 1878).

The eclipse – A difficulty explained and a question answered. *Cincinnati Commercial*, 11 August 1869.

The eclipse – Accounts from the scientific parties. *Cincinnati Commercial*, 10 August 1869.

The eclipse – Darkness upon the face of the Earth. *Cincinnati Commercial*, 8 August 1869.

The eclipse – How it was seen in New-York and elsewhere. *The New York Times*, 8 August 1869, p. 1.

The eclipse – Its meteorological effects. *The New York Times*, 11 August 1869, p. 8.

The eclipse – Observations at Behring Straits. *The New York Times*, 3 October 1869, p. 1.

The eclipse – Observations at Springfield, Illinois. *The New York Times*, 12 August 1869, p. 4.

The eclipse – Scientific observations at Sioux Falls City. *Cincinnati Daily Gazette*, 14 August 1869.

The eclipse and music in Central Park today. *The New York Times*, 7 August 1869, p. 2.

The eclipse at Aden. *New York Observer and Chronicle, 46*(39), 310 (24 September 1868).

The eclipse in Alaska. *The New York Times,* 5 September 1869, p. 1.

The eclipse in Siberia. *Overland Monthly and Out West Magazine*, *6*(6), 519–524 (June 1871).

The eclipse in Siberia. *The New York Times*, 4 October 1869, p. 1.

The eclipse of 1806. *The New York Times*, 2 August 1869, p. 2.

The eclipse of 1806. *The New York Times*, 29 July 1878, p. 2.

The eclipse of December, 1870. *Scientific American*, *26*(11), 161 (9 March 1872).

The eclipse of the Moon. *The New York Times*, 7 January 1871, p. 5.

The eclipse of the Sun – Successful observations in the West. *The New York Times,* 30 July 1878, p. 5.

The eclipse of the Sun (from Special Correspondent of Daily News). *Nature, 18*, 430–433 (22 August 1878).

The eclipse of the Sun [from the *Chicago Tribune*]. *Friends' Review; A Religious, Literary and Miscellaneous Journal*, *31*(51), 810–812 (3 August 1878).

The eclipse of the Sun. *Astronomical Register*, *6*, 193 (1868).

The eclipse of the Sun. *Every Saturday: A Journal of Choice Reading*, *2*(58), 108–110 (4 February 1871).

The eclipse of the Sun. From Mr. De la Rue to the Editor of the Athenaeum, 1868. *Astronomical Register*, *6*, 215–219.

The eclipse of the Sun. *Nature, 18*(457), 353 (1878).

The eclipse of the Sun. *Nature*, *18*, 425–426 (15 August 1878).

The eclipse of the Sun. *Nature*, *18*, 430–433 (22 August 1878).

The eclipse of the Sun. *The Albion, 46*(16), 192 (18 April 1868).

The eclipse of the Sun. *The New York Times*, 21 July 1878, p. 7.

The great solar eclipse of (1868) Data from graphic discussion of the Guntoor photographs Data from graphic discussion of the Guntoor photographs, *Astronomical Register*, 7, 45.

Eddy, J. A. (1973). The great eclipse of 1878. *Sky and Telescope, 45*(6), 340–346.

Edison, T. A. (1879). On the use of the tasimeter for measuring the heat of the stars and the Sun's corona. *American Journal of Science and Arts, 17*(97), 52–54.
Edison's tasimeter. *The New York Times*, 8 December 1879, p. 5.
Editorial article 1 – No title. *The New York Times*, 30 July 1878, p. 4.
Editorial article 5 – No title. *The New York Times*, 8 December 1874, p. 6.
Editorial article 5 – No title. *The New York Times*, 7 December 1882, p. 4.
Editorial article 7 – No title. *The New York Times*, 4 December 1879, p. 4.
Editorial notes, 1883. *Sidereal Messenger*, 1, 267
Editor's table – The great solar eclipse. *The Ladies' Repository: A Monthly Periodical, Devoted to Literature, Art and Religion*, *29*, 158–160 (August 1869).
Educational. *Zion's Herald*, *52*(19), 147 (13 May 1875).
Educational. *Zion's Herald*, *55*(20), 159 (16 May 1878).
Edward Emerson Barnard. *Astronomical Journal*, *35*, 25–26 (19 May 1923)
Eleven passengers dead. *The New York Times*, 26 June 1892.
Ellinwood, F. F. (1874). On the Pacific. *New York Evangelist, 45*(40), 1.
Emery, E., & Emery, M. (1984). *The Press and America* (5th ed., pp. 151–153, 326–333). New Jersey: Prentice-Hall, Inc., Englewood Cliffs
The English and American transit campaigns compared. *Scientific American, 32*(12), 181 (20 March 1875).
Espenak, F. (2003). Transits of Venus – Six millennium catalog: 2000 BCE to 4000 CE. NASA/Goddard Space Flight Center Eclipse Home Page. Fn. 8.
European news – The solar eclipse. *The New York Times*, 20 September 1868, p. 1.
The eventful year of our Lord 1868. *Scientific American*, *20*(1), 9 (1 January 1869).
The examiners of the Sun. *The New York Times*, 8 August 1878, p. 5.
Fair success in Texas. *The New York Times*, 7 December 1882, p. 2.
The Far East. *The New York Times,* 23 January 1875, p. 2.
Fernie, D. (2002). *Setting sail for the universe: Astronomers and their discoveries*. New Brunswick, NJ: Rutgers University Press.
Fine results at Princeton. *The New York Times*, 7 December 1882, p. 2.
Flammarion, C., (re-edited by G.C. Flammarion). (1955). *Astronomie Populaire* (pp. 356–357). Paris: Draeger Frères.
Flammarion, C. (1875). Le passage de Vénus: résultats des expéditions françaises. *La Nature, 3*, 356–358.
Foreign. *New York Observer and Chronicle, 46*(38), 302 (17 September 1868).
Foreign items. *The New York Times,* 13 November 1870, p. 3.
Foreign gleanings. *The Independent, 20*(1024), 2 (16 July 1868).
Foreign news by mail – Science in Europe – The solar eclipse. *The New York Times*, 27 September 1868, p. 1.
Foreign notes. *The New York Times*, 11 April 1874, p. 8.
Forster, T. (1843a). Letter from T. Forster, Esq., dated Bruges, March 22, 1843, on the Great Comet. *Monthly Notices of the Royal Astronomical Society*, *5*, 269.
Forster, T. (1843b). Second letter from T. Forster, Esq., dated Bruges, March 28, 1843, on the Great Comet. *Monthly Notices of the Royal Astronomical Society*, *5*, 269.
Forster, T. (1843c). Letter from T. Forster, Esq., dated Bruges, April 22, 1843. *Monthly Notices of the Royal Astronomical Society*, *5*, 296.
Forty-Second Congress. *The New York Times*, 7 March 1872, p. 5.
Frederick Scott Archer. *British Journal of Photography, 22*, 102–104 (26 February 1875).
French preparations for the transit of Venus. *Nature*, 190, July 9, 1874.
From our exchanges: The Sun. *Zion's Herald*, *54*(35), 1 (30 August 1877).
From the *Cincinnati Daily Gazette* – Meteors. *Baltimore Patriot & Mercantile Advertiser*, 25 August 1834.
From the *New Haven Herald. New-Bedford Mercury*, 22 November 1833.
From the *New Haven Herald* – Meteoric shower of Nov. 1837. *Rhode Island Republican*, 22 November 1837.

From the *New Haven Herald* – The meteors. *Baltimore Patriots,* 22 November 1834.
Gardiner, J. H. (1878). Satellites of Jupiter and Saturn. *Astronomical Register, 16*, 105.
Gardiner, J. H. (1880). Report of Astronomical Society's meeting – Mars and Betelgeux. *English Mechanic and World of Science, 774*, 482.
General European news – The eclipse. *The New York Times*, 26 December 1870, p. 5.
General notes. *The New York Times*, 31 March 1882, p. 4.
General notes. *The New York Times*, 19 June 1882, p. 4.
General Telegraph News. *The New York Times*, 20 August 1881, p. 5.
van Gent, R. (1993). De Nederlandse Venus-expedities van 1874 en 1882. *Zenit, 20*, 332–337.
Genuth, S. S. (1990). Blazing stars, open minds, and loosened purse strings: Astronomical research and its early Cambridge audience. *Journal for the History of Astronomy, 21*, 9–19.
Gerry, S. L. (1891). Old masters of Boston. *New England Magazine, 3*(6).
German observations of the eclipse of the Sun. (1868, October 18). *Daily Alta California.*
General Telegraph News. *The New York Times*, 20 August 1881, p. 5.
Gingerich, O. (1999). Benjamin Apthorp Gould and the founding of the *Astronomical Journal. The Astronomical Journal, 117*, 1–5.
Gleanings. *Western Christian Advocate, 41*(7), 50–51(18 February 1874).
Gleanings from the mails. *The New York Times*, 22 August 1882, p. 3.
Goldfarb, S. (1969). Science and democracy, a history of the Cincinnati Observatory. *Ohio History, 78*, 172–178.
Golub, L., & Pasachoff, J. M. (1997). *The Solar Corona*. Cambridge: Cambridge University Press.
Good work at Harvard. *The New York Times*, 7 December 1882, p. 2.
Gossip about the eclipse [from *The Methodist*]. *Friends' Review; A Religious, Literary and Miscellaneous Journal*, *25*(83), 519–520 (6 April 1872).
The government's work. *The New York Times,* 7 December 1882, p. 2.
Grease. *The New York Times*, 3 September 1881, p. 4.
The great eclipse [from *The Evening Post*]. *Littell's Living Age* 1272, 183–184 (17 October 1868).
The great eclipse. *The Friend: A Religious and Literary Journal*, *42*(49), 388–390 (31 July 1869).
The great eclipse. *The Round Table 210*, 68–69 (30 January 1869).
The great eclipse of August 17, 1868. *The Eclectic Magazine of Foreign Literature, 8*(3), 1152–1155 (September 1868).
Great eclipse of the Sun in 1868. *New York Observer and Chronicle, 46*(1), 6 (2 January 1868).
Great shower of meteors expected. *Maine Farmer, 34*(47), 2 (1 November 1866).
The great solar eclipse. *Flag of our Union, 24*(33), 515 (14 August 1869).
The great solar eclipse. *Flag of our Union, 24*(33), 515 (14 August 1869).
The great solar eclipse. *The Ladies' Repository; A Monthly Periodical, Devoted to Literature, Art and Religion*, pp. 252–256 (October 1869).
The great solar eclipse. *New York Observer and Chronicle*, *47*(28), 222 (15 July 1869).
The great solar eclipse. *Scientific American, 21*(8), 117 (21 August 1869).
The great solar eclipse – City scenes. *The New York Times*, 9 August 1869, p. 8.
The great solar eclipse – Where it begins and ends. *The New York Times,* 5 August 1869, p. 5.
Great solar eclipses. *The Eclectic Magazine of Foreign Literature, 8*(4), 1181–1191 (October 1868).
Great solar eclipses. *The Friend; A Religious and Literary Journal*, *42*(9), 65–66 (24 October 1868)
Great solar eclipses. *The Friend; A Religious and Literary Journal*, *42*(10), 73–74 (31 October 1868).
Great solar eclipses. *The Friend; A Religious and Literary Journal*, *42*(11), 83–83 (7 November 1868).
Great shower of meteors expected. *Maine Farmer*, *34*(47), 2 (1 November 1866).
Greene, J. C. (1958). Science and the public in the age of Jefferson. *Isis, 49*, 13–25.
Guillemin, A. (1877). *The World of Comets*. London: Sampson Low, Marston, Searle and Rivington.
Gummere, S. J. (1868a). The solar eclipse of 1869. *Friends' Intelligencer, 25*(16), 252–254.
Gummere, S. J. (1868b). The solar eclipse of 1869. *Friends' Review; A Religious, Literary and Miscellaneous Journal, 21*(42), 646–648, 650.

Gummere, S. J. (1869a). The solar eclipse of eighth month 7th, 1869. *Friends' Review; A Religious, Literary and Miscellaneous Journal*, *22*(45), 705–707 (3 July 1869).
Gummere, S. J. (1869b). The solar eclipse of eighth month 7th, 1869. *Friends' Review; A Religious, Literary and Miscellaneous Journal*, *22*(46), 726 (10 July 1869).
Gummere, S. J. (1869c). Correspondence – Description of the eclipse education in North Carolina *Friends' Review; A Religious, Literary and Miscellaneous Journal*, *23*(1), 1–3.
Haig, C. T. (1868–1869). Spectroscopic observations of the eclipse of the Sun, August 18, 1868. *Proceedings of the Royal Society*, *17*, 74–80.
Hainaut, O. R., Meech, K. J., Boehnhardt, H., & West, R. M. (1998). Early recovery of Comet 55P/Tempel-Tuttle. *Astronomy and Astrophysics*, *333*, 746–752.
Hale, G. (1913). The work of Sir William Huggins. *Monthly Notices of the Royal Astronomical Society*, *37*, 145–153.
Hall, A. (1871). Art. V. – On the application of photography to the determination of astronomical data. *American Journal of Science and Arts*, *2*(7), 25–30.
Hall, A. (1879). *Reports on telescopic observations of the transit of Mercury, May 5–6, 1878* (pp. 58–59). Washington, DC: Government Printing Office.
Hall, A., (1882). Correspondence to Stephen Clegg Rowan on 8 December 1882, National Archives.
Hall, J., compiler. (1883). *The genealogy and biography of the Waldos of America from 1650 to 1883*. Danielsonville, Connecticut: Press of Scofield and Hamilton.
Hamilton College. *New York Evangelist*, *40*(30), 4 (29 July 1869).
Harkness, W. (1891). Solar parallax from the transit of Venus photography. *Publications of the Astronomical Society of the Pacific*, *3*(14), 46.
Harper's Encyclopaedia of United States History (Vol. IV). New York: Harper & Brothers (1905).
Hartshorn, T. C. (1882, November 16). The transit of Venus. *New York Evangelist*, *53*(46), 6.
Haywood, J. (1886). Electric phenomena in our solar system. *Sidereal Messenger*, *5*, 81–85.
Heck, A. (1989). *International directory of astronomical societies*. Strasbourg, France: Observatoire Astronomique.
Hemenway, M. [undated]. Letter to Mr Clarke [sic]. Adler Planetarium and Astronomy Museum, Folder 18, Box 1, Series 1.
Henderson, Professor. (1843). Elements of the great comet of 1843, with an ephemeris, from the observations of March 20, 22 and 25, reported in Professor Schumacher's Circular. *Monthly Notices of the Royal Astronomical Society*, *5*, 266.
Henderson, Professor. (1843). Letter from Lieutenant G.B.G. Downes, Royal Engineers, to Professor Henderson. *Monthly Notices of the Royal Astronomical Society*, *5*, 266–267.
Henderson, Professor. (1843). Letter from Professor Henderson to the Secretary, on the Great Comet of 1843. *Monthly Notices of the Royal Astronomical Society*, *5*, 267–269.
Hentschel, K. (1999a). Photographic mapping of the solar spectrum 1864–1900, part I. *Journal for the History of Astronomy*, *30*, 93–113.
Hentschel, K. (1999b). Photographic mapping of the solar spectrum 1864–1900, part II. *Journal for the History of Astronomy*, *30*, 201–219.
Herdendorf, C. (1986). Captain James Cook and the transits of Mercury and Venus.
Herschel, J. F. W. (1842). *Treatise on Astronomy (American edition)*. Philadelphia: Lea & Blanchard.
Herschel, J. (1868–1869). Account of the solar eclipse of 1868, as seen at Jamkhandi in the Bombay Presidency. *Proceedings of the Royal Society*, *17*, 103–125.
Hetherington, N. S. (1976). Amateur versus professional – The British Astronomical Association and the controversy over canals on Mars. *Journal of the British Astronomical Association*, *86*, 303–308.
Hicks, I. R. (1894). *Rev. Irl R. Hicks Almanac*. St. Louis, Missouri, 126.
The highest inhabited point. *The New York Times*, 7 March 1879, p. 3.
Hilgard, J. E. (1874). The transit of Venus. *The International Review*, *1*(2), 160–172.
Himes, C. F. (1874). Art. II. – Preparation of photographic dry-plates by daylight, by desensitizing and re-sensitizing the silver compounds. *American Journal of Science and Arts*, *8*(43), 16–18.

Hind, J. R. (1869). Past and future solar eclipses. *Astronomical Register, 7*, 87–88.

Hind, J. R. (1870). The approaching total eclipse of the Sun. *The Albion, 48*(51), 806.

Hindle, B. (1956). *The pursuit of science in revolutionary America, 1735–1789*. New York: W.W. Norton & Company, Inc.

The Hindu view of the late eclipse. *The Eclectic Magazine of Foreign Literature*, 9(1), 22–27 (January 1869).

Hingley, P. D. (2005). The priest and the stuffed penguin: Father Stephen Perry, S.J. and the transit of Venus expeditions to Kerguelen Island, 1874, and Madagascar, 1882. *Journal of the British Astronomical Association, 115*(3), 150–158. 168–170.

The Historical and Philosophical Society of Ohio and the University of Cincinnati. (1944). *The Centenary of the Cincinnati Observatory*. Cincinnati, OH: Court Index Press, Inc.

Historical Sun-darkenings. *The New York Times*, 3 August 1879, p. 4.

History of photography in America. *The Phrenological Journal and Science of Health, 54*(4), 250–258 (April 1872).

Hitchcock, E. (1834). Art. XIII – On the meteors of November 13, 1833. *American Journal of Science and Arts, 25*, 354–363.

Hoffleit, D. (1992). *Astronomy at Yale, 1701–1968*. New Haven: The Connecticut Academy of Art and Sciences.

Hogg, H. S. (1951). Le Gentil and the transits of Venus of 1761 and 1769. *Journal of the Royal Astronomical Society of Canada*, 45, 37–44, 89–92, 127–134, 173–178.

Holden, E. S. (1875). Historical note on the observation of the corona and red prominences of the Sun. *American Journal of Science and the Arts, 10*(56), 81–83.

Holden, E. S. (1897). The beginnings of American astronomy. *Science, 5*, 929–935.

Home news. *The Albion, 48*(28), 403–404 (10 July 1869).

Home news. *The Albion, 57*(41), 607 (9 October 1869)

Hometown memories – A pictorial history of Shelby County, Kentucky. 1994. Marceline, MO: Heritage House Publishing.

Hopkins, Captain. (1843). Notes on the Comet, accompanied by a pencil sketch, by Captain Hopkins, commanding the East India Company's Ship, Seringapatam, on a voyage from the Cape of Good Hope. *Monthly Notices of the Royal Astronomical Society, 5*, 295.

Horner, M. (1883). Observations of the transit of Venus, 1882, December 6, made at Mells, ten miles south of Bath. *Monthly Notices of the Royal Astronomical Society, 43*(5), 276–277.

Hough, G. W. (1878). Report by Prof. Hough – To the Cincinnati Astronomical Society. In Chicago Astronomical Society (Ed.), *The solar eclipse of July, 1878*. Chicago: Evening Journal Book and Job Printing House.

How to observe the eclipse and save your eyes. *Scientific American, 21*(7), 107 (14 August 1869).

Howlett, F. (1883). Notes on the transit of Venus. *Monthly Notices of the Royal Astronomical Society, 43*(5), 278–279.

http://etcweb.princeton.edu/CampusWWW/Companion/nassau_lit.html.

Hubbard, J. S. (1849). On the orbit of the Great Comet of 1843. *Astronomical Journal, 1*, 10–13.

Hubbell, J. G., & Smith, R. W. (1992). Neptune in America: Negotiating a discovery. *Journal for the History of Astronomy, 23*, 261–285.

Huggins, W. (1856). Description of an observatory erected at Upper Tulse Hill. *Monthly Notices of the Royal Astronomical Society, 16*, 175–176.

Huggins, W. (1869). On a possible method of viewing the red flames without an eclipse. *Monthly Notices of the Royal Astronomical Society, 29*, 4.

Huggins, W. (1885a). Photographing the corona. *The Observatory, 8*, 376–377.

Huggins, W. (1885b). The solar corona. *The Observatory, 8*, 153–159.

Huggins, W. (1900). A suggested explanation of the solar corona. *Astrophysical Journal, 12*, 279–280.

Hughes, D. W. (1982). The history of meteors and meteor showers. *Vistas in Astronomy, 26*, 325–345.

Hughes, D. W. (1995a). The Perseid meteor shower. *Earth, Moon, and Planets, 68*, 31–70.

Hughes, D. W. (1995b). The world's most famous meteor shower picture. *Earth, Moon, and Planets, 68*, 311–322.

Hughes, D. W., & Stott, C. (1995). The planisphere: A brief historical review. *Journal of the British Astronomical Association, 105*(1), 35–39.

Humboldt, A., & Bonpland, A. (1907). *Personal narrative of travels to the Equinoctial regions of America during the years 1799–1804* (trans: Ross, T.) (pp. 351–360). London: George Bell & Sons.

Hussey, W. J. (1897). Astronomy and astronomers in their relations to the public. *Publications of the Astronomical Society of the Pacific, 9*(55), 53–64.

The Illustrated London News, 7 January 1871, p. 20.

Illustrations. *Every Saturday, 2*(59), 126–135 (11 February 1871).

Important discovery during the late eclipse. *Every Saturday, 1*(9), 233–235.

Inkster, I. (1978). Robert Goodacre's astronomy lectures (1823–1825), and the structure of scientific culture in Philadelphia. *Annals of Science, 35*, 353–363.

Inkster, I. (1980). Robert Goodacre (1777–1835) and popular astronomy. *Journal of the British Astronomical Association, 90*, 245–252.

Inkster, I. (1982). Advocates and audiences – Aspects of popular astronomy in England, 1750–1850. *Journal of the British Astronomical Association, 92*, 119–123.

Instructions for Observing the Total Solar Eclipse of July 29th, 1878, Issued by the U.S. Naval Observatory. Washington: Government Printing Office. 1878. p. 30 (Review). *Astronomical Register, 16*, 278 (1878).

Interplanetary Signalling. *The Galaxy. A Magazine of Entertaining Reading, 17*(1), 126–127 (January 1874).

Is the weather affected by solar eclipses? *Scientific American, 21*(10), 154 (4 September 1869).

Israel, P. (1998). *Edison: A lifetime of invention* (pp. 160–162). New York, NY: Wiley.

Italian Affairs. *The New York Times*, 11 January 1875, p. 10.

Items – The coming eclipse. *Friends' Intelligencer, 25*(11), 176 (16 May 1868).

Items – The solar eclipse. *Friends' Intelligencer, 25*(29), 464 (19 September 1868).

Items – Photographing the solar eclipse [from the *New York Evening Post*]. *Friends' Intelligencer, 25*(30), 480 (26 September 1868).

Items – The solar eclipse [from the *New York Evening Post*]. *Friends' Intelligencer, 26*(10), 160 (8 May 1869).

Items. *Friends' Intelligencer, 31*(17), 272 (20 June 1874).

Items. *Friends' Intelligencer, 31*(21), 336 (18 July 1874).

Items. *Friends' Intelligencer, 32*(17), 272 (19 June 1875).

Items., *Friends' Intelligencer, 35*(24), 383–384 (3 August 1878).

Jacob, W. S. (1843). An account of the Comet as seen on board the ship Childe Harold on her voyage from Bombay to London. *Monthly Notices of the Royal Astronomical Society, 5*, 295–296.

Janiczek, P. M. (1983). Remarks on the transit of Venus expedition of 1874. In S. J. Dick & L. E. Doggett (Eds.), *Sky and Ocean joined* (pp. 52–73). Washington: U.S. Naval Observatory.

Janssen, J. (1869a). The eclipse of the Sun of August 18, 1868. *The Catholic World, A Monthly Magazine of General Literature and Science, 8*(45), 697–703.

Janssen, J. (1869b). Observations spectrale prises pendant l'éclipse du août 1868, et methode d'observation des protubérances en dehors des éclipses. *Comptes Rendus de l'Académie des Sciences, 68*, 367–376.

Janssen, J. (1869c). The total solar eclipse of August, 1868, Pt. I. *Astronomical Register, 7*, 107–110.

Janssen, J. (1869d). The total solar eclipse of August, 1868, Pt. II. *Astronomical Register, 7*, 131–133.

Jarrell, R. A. (1990). Ashe, Edward David. In *Dictionary of Canadian Biography* (Vol. 12). Toronto: University of Toronto/Université Laval.

Jenniskens, P., Betlem, H., de Lignie, M., ter Kuile, C., Van Vliet, M. C. A., Van't Leven, J., Koop, M., Morales, E., & Rice, T. (1998). On the unusual activity of the Perseid meteor shower

(1989–96) and the dust trail of comet 109P/Swift-Tuttle. *Monthly Notices of the Royal Astronomical Society, 301*, 941–954.

Jensen, W. B. (1989). Robert Bowne Warder 1848–1905. *CINTACS Newsletter, 27*(1), 10–15.

Jensen, W. B. (2013). Notes from the Oesper Collections. The Twitchell Acidometer. University of Cincinnati, *Museum Notes*, May/June, 1–3.

Jewitt, D. (1996). Debris from Comet P/Swift-Tuttle. *Astronomical Journal, 111*, 1713.

Jewitt, D. (2004). From cradle to grave: The rise and demise of the comets. In M. C. Festou, H. E. Keller, & H. A. Weaver (Eds.), *Comets II* (pp. 659–676). Tuscon: University of Arizona.

Joe Jefferson goes a-fishing. *The New York Times*, 8 August 1878, p. 3.

Johnson, R. (1868). Correspondence. *Friends' Review – A Religious, Literary and Miscellaneous Journal, 21*(42), 650–651.

Jones, H. M. & I. B. Cohen with the assistance of Everett Mendelsohn. (1963). *A treasury of scientific prose: A nineteenth-century anthology* (pp. 7–11, 79–121). Boston, MA: Boston, Little and Brown.

Jones, R. I. G. (2005). Sir Robert Ball: Famous Victorian astronomer and lecturer par *excellence. The Antiquarian Astronomer, 2*, 27–36.

Jorda, L., Lecacheux, J., Colas, F., Frappa, E., & Laques, P. (1998). Comet 55P/Tempel-Tuttle. *Minor Planet Circular, 6818.*

Joseph Winlock. *American Academy of Arts and Sciences, Boston. Proceedings, 11*, 339–350 (May 1875–May 1876).

Kalb, A. J. (1869). The eclipse – A remarkable one. *Christian Advocate, 44*(31), 242.

Kanas, N. (2009). *Star maps. History, artistry, and cartography*. Berlin: Springer.

Kapoor, R. (2014). Indian astronomy and the transit of Venus. 2: The 1874 event. *Journal of Astronomical History and Heritage, 17*(1), 113–135.

Keay, C. S. L. (1980a). Anomalous sounds from the entry of meteor fireballs. *Science, 210*, 11–15.

Keay, C. S. L. (1980b). Audible sounds excited by aurorae and meteor fireballs. *Journal of the Royal Astronomical Society of Canada, 74*, 253–260.

Keene. (1833). *New Hampshire Sentinel*, 21 November 1833 (Thursday).

Kendall, E. O. (1843a). Letter from Professor Kendall, containing observations of the Comet made at Philadelphia. *Monthly Notices of the Royal Astronomical Society, 5*, 304–308.

Kendall, E. O. (1843b). Schreiben des Herrn Professors *Kendall* au den Herausgeber. *Astronomische Nachrichten, 20*, 387.

Kendall, P. M. (compiler). (1896). *Maria Mitchell, life, letters and journals* (pp. 165–178, 224–237). Boston, MA: Lee and Shepard Publishers

King, F. T. (1869). Correspondence – Education in North Carolina. *Friends' Review; A Religious, Literary and Miscellaneous Journal, 23*(1), 3–4.

Knapp, P. (2004). From the archives – Trinity and the transit of Venus, 1882. *Trinity Reporter*, Spring, 13–15.

Knight, G. H. (1869). Observations of the eclipse as seen at Westport, KY. *Scientific American, 21*(11), 165.

Kobre, S. (1969). *Development of American Journalism* (pp. 255–261, 400–405). Dubuque, Iowa: Wm. C. Brown Company Publishers.

Kohlstedt, S. G. (1990). Parlors, primers and public schooling: Education for science in nineteenth-century America. *Isis, 81*, 424–445.

Koorts, W. P. (2003). The 1882 transit of Venus and the Huguenot Seminary for Girls. *Monthly Notices of the Astronomical Society of South Africa, 62*(3&4), 76–87.

Koorts, W. P. (2004). The 1882 transit of Venus: The British expeditions to South Africa. *Monthly Notices of the Astronomical Society of South Africa, 63*(3&4), 34–57.

Koorts, W. P. (2006). Transit of Venus observations and relics in South Africa (based on a presentation at the African Astronomical History Symposium, 2005 November 8 & 9). *Monthly Notices of the Astronomical Society of South Africa, 65*(1&2), 8–22.

Kopper, M. (2004). Austria's scientific contribution to the observation of the 1874 transit of Venus. *Journal of Astronomical Data, 10*(7), 175–210.

Kragh, H. (2009). The solar element: A reconsideration of helium's early history. *Annals of Science, 66*, 158–175.
Kronk, G. W. (2003). *Cometography: A catalogue of Comets. Volume 2, 1800–1899*. Cambridge: Cambridge University Press.
Kwas, M. (1999). The spectacular 1833 Leonid meteor storm: The view from Arkansas. *The Arkansas Historical Quarterly, 58*(3), 314–324.
Langley, S. P. (1897). Biographical sketch of William Crawford Winlock. *Astronomische Nachrichten, 142*, 271.
The language of light. *Appleton's Journal of Literature, Science and Art, 7*(161), 465–468 (27 April 1872).
Lankford, J. (1979). Amateur versus professional: The transAtlantic debate over the measurement of Jovian longitude. *Journal of the British Astronomical Association, 89*, 574–582.
Lankford, J. (1981). Amateurs versus professionals: The controversy over telescope size in late Victorian science. *Isis, 72*, 11–28.
Lankford, J. (1987). Photography and the 19th-century transits of Venus. *Technology and Culture, 28*, 648–657.
Lankford, J. (1988). Astronomy's enduring resource. *Sky & Telescope*, 482–483.
Lankford, J. (1997). *American Astronomy*. Chicago: University of Chicago Press.
A large spot on the Sun. *The New York Times*, 22 November 1882, p. 5.
The last transit of Venus; Progress toward completion of the records. *The New York Times*, 22 December 1877, p. 1.
The late aurora. *Friends' Intelligencer, 28*(40), 635–636 (2 December 1871).
The late eclipse. [from the *Cornhill Magazine*]. *Friends' Review – A Religious, Literary and Miscellaneous Journal, 24*(32), 507–509 (1 April 1871).
The late eclipse. [from the *Cornhill* Magazine]. Littell's *Living Age*, (1398), 729–736 (18 March 1871).
The late eclipse – An old Negro and his chickens. *The New York Times*, 26 August 1869, p. 5.
The late Emperor of China. *The New York Times*, 5 March 1875, p. 4.
Late observations in India. *The Albion, 46*(40), 478–479 (3 October 1868).
The late solar eclipse. [from the *Cornhill Magazine*] *The Eclectic Magazine of Foreign Literature, 13*(4), 412–420 (April 1871).
The late solar eclipse *Friends' Intelligencer, 35*(25), 397–399 (10 August 1878).
The late transit of Venus. *The New York Times*, 23 December 1882, p. 3.
The latest advance in solar science. *The New York Times*, 6 December 1868, p. 5.
Latest news by cable. *The New York Times*, 2 December 1874, p. 1.
Latest news by cable. *The New York Times*, 11 December 1874, p. 1.
Latest news by cable. *The New York Times*, 9 August 1875, p. 5.
The latest novelty in photography. *Scientific American, 18*(26), 407–408 (27 June 1868).
Lauga, J. (2004). Les missions Françaises pour le passage de Vénus de 1874. *L'Astronomie*, 118 (Mai), 296–305.
Launay, F. (2012). *The Astronomer Jules Janssen: A Globetrotter of celestial physics*. New York: Springer.
Launay, F., & Hingley, P. (2005). Jules Janssen's 'Revolver Photographique' and its British derivative, the 'Janssen Slide'. *Journal for the History of Astronomy, 36*, 57–79.
Le Gentil de la Galaisière, G. J. (1781). *Voyage dans les Mers de L'Inde*, En Suisse, Chez les Libraires Associés. (Elibron Classics Replica Edition, 2006).
Lecture on astronomy. *Christian Advocate, 58*(2), 29 (11 January 1883).
Leach, J. H. (1872a). To the editor of *Scientific American, 26*(12), 179.
Leach, J. H. (1872b). The discovery and history of the chromosphere. *Scientific American, 26*(15), 227.
Leach, J. H. (1872c). The discovery and history of the chromosphere. *Scientific American, 26*(16), 243.
Leavitt, D. (1833, December 16). Remarks on the phenomena and nature of meteors, *New Hampshire Patriot*.
Lecture by Prof. Proctor. *The New York Times*, 20 January 1874.
Lectures. *The New York Times*, 3 December 1874, p. 7.

Lectures. *The New York Times*, 25 April 1876.
Lectures this week. *The New York Times*, 30 March 1874.
Ledger, E. (1878). Elementary arithmetic and the duration of totality in a solar eclipse. *The Observatory, 16*, 105–110.
Lee, A. M. C. (1937). *The Daily Newspaper in America* (p. 272). New York: The Macmillan Company.
The lessons of the eclipse. *Appletons' Journal of Literature, Science and Art*, *2*(23), 89–90 (4 September 1869).
A letter from Prof. Watson. *The New York Times*, 25 February 1875, p. 2.
Letters to the editor – The approaching total eclipse of the Sun. *The New York Times*, 28 May 1869, p. 2.
Letters to the editor – Messrs. Harper and Mr. Pym. *The New York Times*, 14 November 1882, p. 4.
Letters to the editor – The professorship of mathematics at West Point. *New York Times*, 17 April 1878, p. 2.
Letters to the editor – The tall tower and the transit. *The New York Times,* 3 May 1878, p. 7.
Lewis, I. M. (1946). The total solar eclipse of June 16, 1806. *Astronomical Journal, 52*, 70–72.
List of members of the Astronomical Society of the Pacific, 1891. *Publications of the Astronomical Society of the Pacific*, *3*(13), 1–8.
List of members of the Astronomical Society of the Pacific, 1892. *Publications of the Astronomical Society of the Pacific*, *4*(20), 1–10.
List of members of the Astronomical Society of the Pacific, 1893. *Publications of the Astronomical Society of the Pacific*, *5*(27), 1–15.
List of members of the Astronomical Society of the Pacific, 1894. *Publications of the Astronomical Society of the Pacific*, *6*(33), 1–16.
List of members of the Astronomical Society of the Pacific, 1895. *Publications of the Astronomical Society of the Pacific*, *7*(39), 1–13.
List of members of the Astronomical Society of the Pacific, 1896. *Publications of the Astronomical Society of the Pacific*, *8*(46), 1–12.
List of members of the Astronomical Society of the Pacific, 1897. *Publications of the Astronomical Society of the Pacific*, *9*(54), 1–13.
List of members of the Astronomical Society of the Pacific, 1898. *Publications of the Astronomical Society of the Pacific*, *10*(60), 1–11.
List of members of the Astronomical Society of the Pacific, 1899. *Publications of the Astronomical Society of the Pacific*, *11*(66), 1–10.
List of members of the Astronomical Society of the Pacific, 1900. *Publications of the Astronomical Society of the Pacific*, *12*(72), 1–12.
List of members of the Astronomical Society of the Pacific, 1901. *Publications of the Astronomical Society of the Pacific*, *13*(77), 1–9.
Literary notes. *The New York Times*, 2 April 1877, p. 3.
The little folks' carnival. *The New York Times*, 3 February 1883, p. 5.
Little hope of seeing the transit. *The New York Times*, 6 December 1882, p. 5.
Littmann, M. (1998). *The heavens on fire: The Great Leonid Meteor Storms*. Cambridge, UK: Cambridge University Press.
Littrow, M. C. L. (1843). Letter from M.C.L. Littrow, Director of the Observatory of Vienna, on the Great Comet. *Monthly Notices of the Royal Astronomical Society, 5*, 271–272.
Local miscellany – Arrival of the *Swatara. The New York Times*, 1 June 1875, p. 8.
Locke, J. (1834, August 25). From the *Cincinnati Daily Gazette* – Meteors. *Baltimore Patriot & Mercantile Advertiser*.
Lockyer, J. N. (1869). The recent total eclipse of the Sun. *Nature, 1*, 14–15.
Lockyer, J. N. (1872, February 24). The solar eclipse [from *Nature*]. *Littell' s Living Age*, Issue 1446, 483–487.
Lockyer, J. N. (1874). *Contributions to solar physics*. London: Macmillan and Co.
Lockyer, J. N. (1878a). The coming total solar eclipse [from *Nature*]. *Littell's Living Age, 137*(1772), 557–563.

Lockyer, J. N. (1878b). The coming total solar eclipse. *Nature, 17*, 481–483.
Lockyer, J. N. (1878c). The coming total solar eclipse, pt. II. *Nature, 17*, 501–503.
Lockyer, J. N. (1878d). The eclipse of the Sun. *Nature, 18*, 353.
Lomb, N. (2011). *Transit of Venus 1631 to the present*. Sydney: New South Publishing.
Loomis, E. (1866). Miscellaneous readings – Astronomical theory of shooting stars. *Farmers' Cabinet, 65*(19).
Loss of gravity. *The New York Times*, 23 April 1881, p. 4.
Lunar assistance. *The Eclectic Magazine of Foreign Literature, 8*(5), 1398–1399 (November 1868).
Lunar assistance. *Scientific American, 19*(22), 338–339 (25 November 1868).
Lyot, B. (1939). A study of the solar corona and prominences without eclipses. *Monthly Notices of the Royal Astronomical Society, 99*, 580–594.
MacPherson, H. (1905). *Astronomers of today*. London: Gall & Inglis.
MacPherson, H. (1908). Pierre Jules Cesar Janssen. *Popular Astronomy, 16*, 72–74.
Mailly E. (1860). *Précis de L'Histoire de l'Astronomie aux États-Unis D'Amérique*, Bruxelles: Imprimeur de l'Académie de Belgique.
Making celestial charts. *The New York Times*, 16 July 1883, p. 9.
Maley, P. D. (2005). 1882 U.S. Naval Observatory Expedition Commemorated. *StarScan, 21*(1), 4.
Maor, E. (2004). *Venus in transit*. Princeton, NJ: Princeton University Press.
Marché, J. D., II. (2005). 'Popular' journals and community in American astronomy, 1882–1951. *Journal of Astronomical History and Heritage, 8*(1), 49–64.
Mars and his moons. *The New York Times,* 4 September 1877, p. 4.
Marsden, B. G. (1996). Searches for planets and comets. In T. W. Rettig & J. M. Hahn (Eds.), *Completing the inventory of the solar system* (Vol. 107). ASP conference series.
Marsden, B., & Williams, G. V. (1996). *Catalogue of cometary Orbits 1996.* Cambridge, MA: Minor Planet Center.
Mason, J. W. (1995). The Leonid meteors and Comet 55P/Tempel-Tuttle. *Journal of the British Astronomical Association, 105*, 219–235.
Mason, J. W. (2004). Observing the transit of Venus. *Journal of the British Astronomical Association, 114*(3), 119–120.
Massachusetts – Observers of the eclipse. *The New York Times*, 23 July 1869, p. 5.
Matthews, W. E., & Cleveland, J. (2008). *Portrait of the Past – A pictorial history of Shelby County, Kentucky, 1865–1980.* Morley, MO: Acclaim Press.
Maunder, E. W. (1870). Bright lines of oxygen in the solar spectrum. *The Observatory, 3*(28), 118–120.
Maunder, E. W. (1886a). Note on some recently published spectroscopic observations. *Monthly Notices of the Royal Astronomical Society, 46*(5), 282–289.
Maunder, E. W. (1886b). Note on M. Trouvelot's paper. *Monthly Notices of the Royal Astronomical Society, 46*, 334.
Mauritius Expedition, 1874 Division 1, Chapter 1, The Heliometer, Dun Echt Observatory Publications, 2, 1–10.
McMormmach, R. (1966). Ormsby MacKnight Mitchel's *Sidereal Messenger*, 1846–1848. *Proceedings of the American Philosophical Society, 110*(1), 35–47.
McNarry, L. R. (1964). Out of old books. *Journal of the Royal Astronomical Society of Canada, 58*, 279–282.
Meadows, A. J. (1972). *Science and controversy. A biography of Sir Norman Lockyer*. Cambridge, MA: MIT Press.
The Mediterranean solar eclipse. *The Galaxy. A Magazine of Entertaining Reading, 12*(2), 179–194 (August 1871).
The meeting of the National Academy of Science. *Scientific American, 30*(19), 293 (9 May 1874).
Memoranda. *Scribner's Monthly, 9*(5), 642 (March 1875).
Menzel, D. H. (1934). The solar reversing layer – A question of nomenclature. *The Observatory, 58*, 84–85.
Meteoric shower. *The Pittsfield Sun*, 17 May 1838.
Meteoric shower of November, 1837. *New-Bedford Mercury*, 24 November 1837.

Meteoric showers. *Advocate*, 13 March 1925.
The meteoric theory. *The New York Times*, 4 September 1879.
The meteors. *Baltimore Patriot & Mercantile Advertiser*, 22 November 1834.
The meteors of 1833. *The New York Times*, 14 November 1884.
Minor topics. *The New York Times,* 19 August 1867, p. 4.
Minor topics. *The New York Times*, 5 January 1871, p. 4.
Minutes – Reformed Church in America, 1881. New York, Board of Publication of the Reformed Church in America, 42.
Miscellaneous city news – National Academy of Science. *The New York Times*, 9 November 1878, p. 8.
Miscellaneous news. *Zion's Herald, 51*(30), 237.
Miscellaneous readings – Astronomical theory of shooting stars. *Farmer's Cabinet*, 29 November 1866.
Misch, A., & Sheehan, W. (2004a). A movie of the 1882 transit of Venus assembled from plates taken at Lick Observatory by David P. Todd. *The Journal of Astronomical Data, 10*(4), 1–9.
Misch, A., & Sheehan, W. (2004b). A remarkable series of plates of the 1882 transit of Venus. *Journal of Astronomical Data, 10*, 211–218.
Mitchel, F. A. (1887). *Ormsby MacKnight Mitchel: Astronomer and general – A biographical narrative by His Son F.A. Mitchel.* Cambridge: The Riverside Press.
Mitchel, O. M. (1867). *The planetary and stellar worlds.* New York: Oakley & Mason.
Mitchel, O. M. (Ed.). (1846–1848). *Sidereal messenger* (Vol. I–III) (July 1846–October 1848).
Mitchell, M. (1869a). The total eclipse of 1869. *Friends' Intelligencer, 26*(37), 587–589.
Mitchell, M. (1869b). The total eclipse of 1869. *Friends' Intelligencer, 26*(38), 603–605.
Mitchell, M. (1877). The planet Jupiter – In two parts – Part II. *The Youth's Companion, 50*(30), 239–240.
Mitchell, M. (1878). Astronomy popularized. *Christian Union, 17*(16), 328.
Mitchell, P. M. (1896). *Maria Mitchell: Life, letters, and journals.* Boston, MA: Lee & Shepard Publishers.
Modern astronomy. *Littell's Living Age, 125* (1611), 195–196 (24 April 1975)
Monday's eclipse of the Sun. *The New York Times*, 31 July 1878, p. 5.
Montgomery, J. (1982). Nathaniel Bowditch, American mathematical astronomer. *Bulletin of the American Astronomical Society, 14*, 917.
Montojo, M., (1843). Notes on the Comet as seen by M. Montojo, at San Fernando. *Monthly Notices of the Royal Astronomical Society*, *5*, 295.
Monument to the late Richard Anthony Proctor. *Publications of the Astronomical Society of the Pacific, 5*(32), 222 (November 1893)
Morris, C. (1869). Spectrum analysis. *Lippincott's Magazine of Literature, Science and Education, 3*, 540–546.
Morrison, L. V., & Stephenson, F. R. (2004). Historical values of the Earth's clock error ΔT and the calculation of eclipses. *Journal for the History of Astronomy, 35*, 327–335.
Mott, F. L. (1930). *A history of American magazines 1741–1850* (Vol. 1, pp. 89, 130–131, 302–304, 317–319, 372, 442, 444–445, 488, 529–535, 562–565). New York: D. Appleton and Company.
Mott, F. L. (1938a). *A history of American magazines 1850–1865* (Vol. 2, pp. 14, 18, 43, 63, 66–67, 74, 78-79, 84, 89, 99, 129, 140, 158–159, 175–176, 207, 262–274, 301–305, 316–324, 364–379, 419–431, 466–468). Cambridge, MA: Harvard University Press.
Mott, F. L. (1938b). *A history of American magazines 1865–1885* (Vol. 3, pp. 32–33, 35, 64, 68, 69, 74, 84, 152, 165, 210, 256, 259–261, 319–324, 329–330, 336–339, 357–381, 388–390, 396–409, 417–439, 457–480, 500–505, 510–516, 518–532, 544–551). Cambridge, MA: Harvard University Press.
Mott, F. L. (1941). *American journalism: A history of newspapers in the United States through 250 years* (pp. 428–429). New York: The Macmillan Company.
Mott, F. L. (1957). *A history of American magazines 1885–1905* (Vol. 4, pp. 671–693). Cambridge, MA: Harvard University Press.
De Mourão, R. R. F. (2004). *Transit of Venus: 1639–2012.* Brazil (in Portuguese).

Mr. R.A. Proctor and the transit of Venus. *The New York Times,* 8 February 1875, p. 3.
Mr. Seward in Alaska. *The New York Times*, 14 September 1869, p. 5.
Murray, D. (1874). The transit of Venus. *Scribner's Monthly, 9*(2), 183–190.
Musto, D. F. (1967). A survey of the American observatory movement. In A. Beer (Ed.), *Vistas in Astronomy* (pp. 87–92). Oxford: Pergamon.
N.O. Bulletin, *Mississippi Free Trader and Natchez Daily Gazette,* 24 March 1843.
Nasmyth, J. (1843). Letter from James Nasmyth, Esq., to Sir J.F.W. Herschel, on an early observation of the train of the Great Comet. *Monthly Notices of the Royal Astronomical Society, 5*, 270–271.
Nath, B. B. (2013). *The story of helium and the birth of astrophysics.* New York: Springer.
National Archives. [Responses to circular of instruction issued November 28, 1882] Group 78, E18, Box 39, bound mss.
National capitol topics; Army and Navy matters. *The New York Times*, 12 September 1883, p. 2.
The national observatory. *Scientific American, 39*(26), 401 (28 December 1878).
Nature of the elements. *The New York Times*, 7 January 1879, p. 3.
Naval intelligence. *The New York Times,* 28 November 1882, p. 2.
Naval intelligence. *The New York Times,* 6 February 1883, p. 2.
The nebular hypothesis. *The New York Times*, 11 August 1878, p. 5.
The new meridian line. *The New York Times*, 19 May 1869, p. 1.
New publications. *The New York Times,* 14 March 1874, p. 8.
New publications. *The New York Times,* 23 September 1874, p. 2.
New publications. *The New York Times*, 22 October 1874, p. 2.
New publications. *The New York Times*, 30 December 1877, p. 10.
New publications. *The New York Times*, 13 June 1883, p. 3.
The new solar theory. *The New York Times*, 14 January 1879, p. 2.
The new solar theory. *The New York Times,* 15 January 1879, p. 2 [duplicate of article of previous day].
New way of raising beans. *Massachusetts Ploughman and New England Journal of Agriculture, 28*(45), 3 (7 August 1869).
New York State agricultural fair. *New York Observer and Chronicle, 46*(37), 294 (10 September 1868).
The New York Times – The Complete Front Pages, 1851–2008. New York: Black Dog & Leventhal Publishers.
Newcomb, S. (1867). Investigation of the distance of the Sun, and the elements which depend on it, from the observations of Mars, made during the opposition of 1862, and from other sources. In *Washington observations, 1865*. Washington, DC: Government Printing Office.
Newcomb, S. (1880). *Observations of the transit of Venus, December 8–9, 1874 made and reduced under the direction of the commission created by Congress. Part I.* Washington: Government Printing Office.
Newcomb, S. (1895). *The elements of the four inner planets and the fundamental constants of Astronomy*. Washington: Government Printing Office.
News of interest – Chinese astronomy. *Saturday Evening Post*, 53(50), 4 (11 July 1874).
News of the day. *The New York Times*, 22 January 1859.
The next transit of Venus. *The New York Times*, 14 August 1881, p. 2.
Noble, W. (1888). Richard A. Proctor. *The Observatory, 11*, 366–368.
A nocturnal mystery. *The New York Times*, 8 September 1880, p. 4.
Norton, W. A. (1871). Art II – On the corona seen in total eclipses of the Sun – Results of observations at the total eclipse of August 7, 1869. *American Journal of Science and Arts, 1*(1), 5–15.
Notes. *Nature, 18*, 423–424 (15 August 1878).
Notes and queries. *New York Observer and Chronicle, 54*(11), 85 (16 March 1876).
Notes from the capitol. *The New York Times*, 19 September 1882, p. 5.
Notes from Washington, *The New York Times*, 2 March 1882, p. 1.
Notes from Washington. *The New York Times*, 6 August 1882, p. 7.
Notes from Washington. *The New York Times*, 17 August 1882, p. 1.
Notes from Washington. *The New York Times*, 18 August 1882, p. 1.

Notes from Washington. *The New York Times*, 3 September 1882, p. 1.

Notes in Science and Agriculture – The recent solar eclipse. *The Phrenological Journal and Science of Health*, *67*(5), 261.

The November meteors. *Afro-American Advocate*, 11 November 1892, p. 4.

Nunis, D. B. (Ed.). (1982). *The 1769 transit of Venus: The Baja Californian observations of Jean-Baptiste Chappe d"Auteroche, Vicente de Doz an Joaquin Velázquez Cárdenas de León*. Los Angeles: Natural History Museum of Los Angeles County.

Obituary: Stephen Alexander LL.D., *The New York Times*, 27 June 1883, p. 4.

Obituary: Henry James Anderson. *The New York Times*, 14 December 1875, p. 5.

Obituary: Henry Draper. *Scientific American*, *47*(23), 352 (2 December 1882).

Obituary: The Earl of Crawford. *The New York Times*, 16 December 1880, p. 2.

Obituary: Capt. Carlile P. Patterson. *The New York Times*, 16 August 1881, p. 5.

Obituary: Richard A. Proctor. (1888). *The Observatory*, *11*, 366–368.

Obituary: Richard Anthony Proctor. (1889). *Monthly Notices of the Royal Astronomical Society*, *49*, 164–167.

Obituary: William Henry Wesley. (1922). *Monthly Notices of the Royal Astronomical Society*, *83*, 255–259.

Obituary: William Henry Wesley. (1922). *The Observatory*, 45, 341, 354–355.

Obituary notices – Henry Draper, *Medical and Surgical Reporter*, 1882: 644.

Observations in the Hawaiian Islands. *The New York Times*, 30 December 1874, p. 4.

Observations made during the recent annular eclipse of the Sun. *Monthly Notices of the Royal Astronomical Society*, *33*, 246 (1873).

The observations of 2004. *The New York Times,* 31 December 1882, p. 4.

Observations of the bright lines in the spectrum of the solar atmosphere. *Monthly Notices of the Royal Astronomical Society*, *33*, 248–249 (1873).

Observations of the solar eclipse of 18th, August, 1868, received from the Hydrographic Department of the Admiralty; And from the Peninsular and Oriental Company. *Monthly Notices of the Royal Astronomical Society*, *29*, 1.

Observations on the total eclipse of the Sun of 1869. *The American Journal of Science and Arts [Third Series]*, 1872, 264–267.

Observations of the transit. *The New York Times*, 11 December 1882, p. 1.

The observations of 2004. *The New York Times*, 31 December 1882, p. 4.

An observatory on Etna. *Scientific American*, *36*(15), 227 (14 April 1877).

Observed changes in a solar prominence. *Scientific American*, *26*(12), 179 (16 March 1872).

The occultation season. *The New York Times*, 19 November 1875, p. 4.

The old year – Chronology of 1868. *The New York Times*, 1 January 1869, p. 2.

Olmsted, D. (1834a). Art. XIV – Observations on the meteors of November 13th, 1833. *American Journal of Science and Arts*, *25*, 363–411.

Olmsted, D. (1834b). Art. XXV – Observations on the meteors of November 13th, 1833. *American Journal of Science and Arts*, *26*, 132–174.

On a possible method of viewing the red flames without an eclipse (extract from letter from Sir John Herschel) *Monthly Notices of the Royal Astronomical Society 29*, 5–6 (1868).

The opposition of Mars. *The New York Times*, 16 July 1877, p. 4.

Orchiston, W. (1988). The emergence of professional astronomy in nineteenth-century Australia. *Search, 19*, 76–81.

Orchiston, W. (1989). The role of the amateur in early Australian astronomy. *Search, 20*(1), 16–23.

Orchiston, W., & Brewer, A. (1990). David Ross and the development of amateur astronomy in Victoria. *Journal of the British Astronomical Association, 100*, 173–181.

Orchiston, W. (1991). Amateur-professional co-operation in astronomy education: A nineteenth century Australian model. *Proceedings of the Astronomical Society of Australia, 9*(1), 186–188.

Orchiston, W., & Bhathal, R. (1991). Illuminating incidents in Antipodean astronomy: Section A of the Royal Society of New South Wales. *Australian Journal of Astronomy, 4*, 49–67.

Orchiston, W. (1992). The contribution of Francis Abbott to Tasmanian and Australian astronomy. *Vistas in Astronomy, 35*, 315–344.

Orchiston, W. (1993). John Grigg, and the genesis of cometary astronomy in New Zealand. *Journal of the British Astronomical Association, 103*, 67–76.
Orchiston, W., & Buchanan, A. (1993). Illuminating incidents in Antipodean astronomy: Campbell Town, and the 1874 transit of Venus. *Australian Journal of Astronomy, 5*(1), 11–31.
Orchiston, W. (1995). John Grigg, and the development of astrophotography in New Zealand. *Australian Journal of Astronomy, 6*, 1–14.
Orchiston, W., & Bembrick, C. (1995). The role of the large reflecting telescope in amateur astronomy: An Australian case study. *Australian Journal of Astronomy, 6*, 53–74.
Orchiston, W. (1997a). The role of the amateur in popularizing astronomy: An Australian case study. *Australian Journal of Astronomy, 7*(2), 33–66.
Orchiston, W. (1997b). The role of the large refracting telescope in Australian amateur astronomy: An historical perspective. *Australian Journal of Astronomy, 7*, 89–114.
Orchiston, W. (1997c). The "tyranny of distance" and Antipodean cometary astronomy. *Australian Journal of Astronomy, 7*, 115–126.
Orchiston, W. (1998a). Amateur-professional collaboration in Australian science: The earliest astronomical groups and societies. *Historical Records of Australian Science, 12*, 163–182.
Orchiston, W. (1998b). Illuminating incidents in Antipodean astronomy: John Tebbutt and the Great Comet of 1861. *Irish Astronomical Journal, 25*, 167–178.
Orchiston, W. (1998c). Mission impossible: William Scott and the first Sydney observatory directorship. *Journal of Astronomical History and Heritage, 1*(1), 21–43.
Orchiston, W. (1999a). Comets and communication: Amateur-professional tension in Australian astronomy. *Publications of the Astronomical Society of Australia, 16*, 212–221.
Orchiston, W. (1999b). Of comets and variable stars: The Afro-Australian astronomical activities of J.F. Skjellerup. *Journal of the British Astronomical Association, 109*, 328–338.
Orchiston, W. (2000a). Amateur telescope making in Australia: An historical perspective. *Amateur Telescope Making Journal, 15*, 10–26.
Orchiston, W. (2000b). John Tebbutt of Windsor, New South Wales: A pioneer southern hemisphere variable star observer. *Irish Astronomical Journal, 27*, 47–54.
Orchiston, W., Love, T., & Dick, S. J. (2000c). Refining the astronomical unit: Queenstown and the 1874 transit of Venus. *Journal of Astronomical History and Heritage, 3*(1), 23–44.
Orchiston, W. (2001a). From amateur astronomer to observatory director: The curious case of R.T.A. Innes. *Publications of the Astronomical Society of Australia, 18*, 317–328.
Orchiston, W. (2001b). 'Sentinel of the Southern Heavens': The Windsor Observatory of John Tebbutt. *Journal of the Antique Telescope Society, 21*, 11–23.
Orchiston, W. (2001c). Wellington astronomers of the 1840's. *Southern Stars, 40*(1), 5–7.
Orchiston, W. (2001d). William John Macdonnell and the development of astronomy in New South Wales. *Journal of the British Astronomical Association, 111*, 13–25.
Orchiston, W. (2002a). Joseph Ward: Pioneer New Zealand telescope-maker. *Southern Stars, 41*, 13–21.
Orchiston, W. (2002b). Tebbutt versus Russell: Passion, power and politics in nineteenth century Australian astronomy. In S. M. R. Ansari (Ed.), *History of Oriental Astronomy* (pp. 169–201). Dordrecht: Kluwer.
Orchiston, W. (2003a). Amateur telescope making in Australia. An historical perspective. In W. J. Cook (Ed.), *The best of amateur telescope making journal* (Vol. 2, pp. 208–239). Richmond: Willman-Bell.
Orchiston, W. (2003b). Australia's earliest planispheres. *Journal of the British Astronomical Association, 113*, 329–332.
Orchiston, W. (2004a). James Cook's 1769 transit of Venus expedition to Tahiti. In D. W. Kurtz (Ed.), *Proceedings IAU Colloquium No. 196, 2004* (pp. 1–15).
Orchiston, W. (2004b). John Tebbutt and observational astronomy at Windsor Observatory. *Journal of the British Astronomical Association, 114*, 141–154.
Orchiston, W. (2004c). The nineteenth century transits of Venus: An Australian and New Zealand overview. *Journal of Astronomical Data, 10*, 219–308.

Orchiston, W., & Buchanan, A. (2004). 'The Grange', Tasmania: Survival of a unique suite of 1874 transit of Venus relics. *Journal of Astronomical History and Heritage, 7*(1), 34–43.

Orchiston, W. (2006). Amateurs in the Antipodes: The common denominators of nineteenth and early twentieth century South African, Australian and New Zealand astronomy. *Monthly Notices of the Astronomical Society of South Africa, 65*, 148–159; 194–208.

Orchiston, W., Chen, K-. Y., Lee, E. -H., & Ahn, Y. -S. (2006). British observations of the 1868 total solar eclipse from Guntoor, India. In K. -Y. Chen, W. Orchiston, B. Soonthornthun, & R. Strom (Eds.), *Proceedings of the Fifth International Conference on Oriental Astronomy* (pp. 23–34). Chiang Mai: Chiang Mai University.

Orchiston, W. (2010). The changing role of the 'Catts Telescope': The life and times of a nineteenth century 20-inch Grubb reflector. *Journal of Astronomical History and Heritage, 13*, 235–254.

Orchiston, W., & Pearson, J. (2011). The Lick Observatory solar eclipse expedition to Jeur (India) in 1898. In *Mapping the Oriental Sky: Proceedings of the Seventh International Conference on Oriental Astronomy* (pp. 197–206). From: ICOA-7 Mapping the Oriental Sky: Seventh International Conference on Oriental Astronomy, 6–10 September 2010, Tokyo, Japan

Orchiston, W. (2014). The amateur-turned-professional syndrome: Two Australian case studies. In W. Orchiston, D. A. Green, & R. Strom (Eds.), *New insights from recent studies in historical astronomy: Following in the footsteps of F. Richard Stephenson.* New York: Springer.

Orchiston, W., & Soonthornthum, B. (2014). French observations of the 18 August 1868 total solar eclipse from Wah-Koa, Thailand. In T. Nakamura & W. Orchiston (Eds.), *The development of astronomy and emergence of astrophysics in Asia*. New York: Springer.

Osterbrock, D. E. (1999). AAS meetings before there was an AAS; The pre-history of the Society. *The American Astronomical Society's first century* (pp. 3–19). Washington, DC: American Institute of Physics.

Other suns than ours; Lecture by Prof. Richard A. Proctor at Association Hall, *New York Times*, 7 April 1874.

Our astronomical column, *Nature, 18*, 426 (15 August 1868).

Overbeek, M. D. (1999, October). Roberts to the CCD – A review of Southern Africa amateur astronomy during the 20th century. In *Proceedings of the Annual General Meeting 1999, Astronomical Society of Southern Africa*, pp. 127–138.

The Pacific Coast – Arrival and reception of Mr. Seward. *The New York Times*, 28 August 1869, p. 5.

The Pacific Coast – Arrival of the Davidson expedition at Sitka. *The New York Times*, 31 July 1869, p. 1.

The Pacific Coast – The astronomical expedition to Siberia unsuccessful. *New York Times,* 23 September 1869, p. 5.

The Pacific Coast – Discovery of iron in Alaska. *The New York Times*, 29 August 1869, p. 1.

The Pacific Coast – Return of an eclipse-observation party. *The New York Times*, 8 November 1869, p. 1.

The Pacific Slope; Sailing of the Colorado for Yokohama. *The New York Times,* 1 June 1874, p. 1.

Paine, R. T. (1869a). Solar eclipse of August 7th, 1869. *Monthly Notices of the Royal Astronomical Society, 30*, 1–3.

Paine, R. T. (1869b). On the solar eclipse of August, 1869. *Monthly Notices of the Royal Astronomical Society, 29*, 285–293.

Pang, A. S.-K. (1994). Victorian observing practices, printing, technology, and representations of the solar corona, (1): The 1860s and 1870s. *Journal for the History of Astronomy, 25*, 249–269.

Pang, A. S.-K. (1995). Victorian observing practices, printing, technology, and representations of the solar corona, (2): The 1860s and 1870s. *Journal for the History of Astronomy, 26*, 63–73.

Pang, A. S.-K. (2002). *Empire of the Sun. Victorian solar eclipse expeditions*. Stanford: Stanford University Press.

Pannekoek, A. (1961). *A history of astronomy*. New York: Dover Publications, Inc.

Pardee, D., & Swerdlow, N. (1993). Not the earliest solar eclipse. *Nature, 343*, 406.

Parnell and Gladstone. *The New York Times*, 30 October 1881, p. 3.

Pasachoff, J. M. (1998). Williams College's Hopkins observatory: The oldest extant observatory in the United States. *Journal of Astronomical History and Heritage, 1*(1), 61–78.

Pasachoff, J. M. (1999). Halley as an eclipse pioneer: His maps and observations of the total solar eclipses of 1715 and 1724. *Journal of Astronomical History and Heritage, 2*, 39–54.

Pasachoff, J. M. (2001). Solar-eclipse science: Still going strong. *Sky & Telescope*, 40–47 (February).

Pasachoff, J. M., Schneider, G., & Golub, L. (2005). The black-drop effect explained. In D. W. Kurtz & G. E. Bromage (Eds.), *Transits of Venus: New views of the solar system and galaxy, IAU Colloquium No. 196* (U.K., 2004), pp. 242–253.

Passage de Vénus du 6 décembre 1882 : rapports préliminaires, 1883. Académie des Sciences, 7, 47, 50–56.

The past and future of the world; A lecture for ladies by Prof. A. Proctor at Association Hall. *The New York Times*, 3 April 1874.

Past and future solar eclipses, *Astronomical Register*, *7*, 87–88 (1869).

Payne, W. W. (1904). Charles Hall Rockwell. *Popular Astronomy, 12*, 262–265.

Payne, W. W. (1911). Robert White McFarland. *Popular Astronomy, 19*, 195–200.

Pearson, J., & Orchiston, W. (2008). The 40-foot solar eclipse camera of the Lick observatory. *Journal of Astronomical History and Heritage, 11*(1), 25–37.

Pearson, J., Orchiston, W., & Malville, J. (2011). Some highlights of the Lick observatory solar eclipse expeditions. In W. Orchiston, T. Nakamura, & R. Strom (Eds.), *Highlighting the history of astronomy in the Asia-Pacific region* (pp. 243–337). New York: Springer.

Peirce, B. (1843). Letter in *Boston Courier* of April 1, 1843. *Monthly Notices of the Royal Astronomical Society, 5*, 298.

Peltier, L. (1965). *Starlight nights: The adventures of a Stargazer*. New York: Harper & Row.

Penrose, F. C. (1878a). The total eclipse of the Sun, July 29, 1878. *Monthly Notices of the Royal Astronomical Society, 39*, 48–51.

Penrose, F. C. (1878b). The eclipse of the Sun, 1878, July 29. *The Observatory, 2*, 187–191.

Peregrinations of a Bullet. *The New York Times*, 15 December 1874, p. 8.

Peters, C. H. F. (1869, July 28). The solar eclipse – The Litchfield expedition. *The New York Times*, 5.

Peters, C. H. F. (1875, February 25). The transit of Venus. *New York Evangelist*, *46*(8), 7.

Peters, C. H. F. (1881). Section VII. Observations made at Queenstown … In S. Newcomb (Ed.), *Observations of the transit of Venus, December 8–9, 1874 … Part II*. Washington: Government Printing Office (unpublished typescript, U.S. Naval Observatory Library).

Phelps, A. (1867). Work of the spirit. *Friends' Review – A Religious, Literary and Miscellaneous Journal, 21*(12), 179–180.

Phillimore, R. H. (1968). *Historical records of the survey of India eighteenth century* (Vol. 5). Dehra Dun: The Surveyor General of India.

Photographs of the eclipse [from the *New York Evening Post*]. *Friends' Intelligencer*, *24*(25), 397–398 (21 August 1869).

The Physical Laboratories of the University of Manchester. Manchester, University of Manchester (1906).

Pickering, E. C. (1871). The eclipse of 1870. *Old and New, 3*(5), 634–638.

Pickering, E. C. (May 1882–May 1883). Observations of the transit of Venus, December 5 and 6, 1882, made at the Harvard College Observatory. In *American Academy of Arts and Sciences, Boston, Proceedings*, pp. 15–40.

Pickering, W. H. (1878). Total eclipse of the Sun, July 29, 1878. *Monthly Notices of the Royal Astronomical Society, 39*, 137–139.

Pigatto, L., & Zanini, V. (2001). Spectroscopic observations of the 1874 transit of Venus; The Italian party at Muddapur, eastern India. *Journal of Astronomical History and Heritage, 4*(1), 43–58.

The planet Venus. *Massachusetts Ploughman and New England Journal of Agriculture* 42(6), 4 (11 November 1882).

The planet Vulcan. *The New York Times*, 4 October 1876, p. 2.

The planet Vulcan: Discovered by Professor Watson. *The New York Observer and Chronicle, 56*(35), 278 (29 August 1878).

Plotkin, H. (1977). Henry Draper, the discovery of oxygen in the Sun, and the dilemma of interpreting the solar spectrum. *Journal for the History of Astronomy, 8*, 44–49.

Pogson, N. R. (1868). *Report of the Government Astronomer upon the Proceedings of the Observatory, in Connection with the Total Eclipse of the Sun on August 18, 1868, as Observed in Masulipatam, Vunpurthy, Madras, and Other Stations in Southern India*. Madras: Madras Observatory.

Polariscope observations of the corona during the total solar eclipse of December, 1871. *Monthly Notices of the Royal Astronomical Society*, *33*, 245 (1873).

Pollock, R. (1843). An account of the Comet as seen on board the ship Malabar on her passage from the Cape of Good Hope. *Monthly Notices of the Royal Astronomical Society, 5*, 296.

Pope Hennessey, J. (1868). Account of observations of the total eclipse of the Sun, made August 18th, 1868, along the coast of Borneo. In *Proceedings of the Royal Society of London*, *17*, 81–90.

Popular astronomy. *Western Christian Advocate 49*(15), 119 (12 April 1882).

Popular Science Monthly, 3 (1872–1873).

Popular Science Monthly, 4, January (1874).

Popular Science Monthly, 7 (1875).

Popular Science Monthly, 19 (1880–1881).

Popular Science Monthly, 32 (1887–1888).

Popular Science Monthly, January 1893.

Popular Science Monthly, 50 (1896)

Popular Science Monthly, 63, 86 (1903).

Popular Science Monthly, 81 (1912).

Portolano, M. (2000). John Quincy Adams's rhetorical crusade for astronomy. *Isis, 91*, 480–503.

Powell, C. S. (1988). J. Homer Lane and the internal structure of the Sun. *Journal for the History of Astronomy*, *19*, 183–197.

A powerful report. *The New York Times*, 19 August 1880, p. 4.

Practical reform. *Farmer's Cabinet*, 31 March 1843.

Prayers for astronomical science. *The New York Times*, 29 November 1882, p. 2.

Preparations for observing the transit of Venus, *Monthly Notices of the Royal Astronomical Society, 34*, 186–190 (1873).

Preparations to photograph the great eclipse. *The New York Times*, 26 April 1868, p. 10.

Preparing for the transit of Venus. *Scientific American*, *47*(6), 81 (5 August 1882).

Preparing for the transit of Venus (from the Toronto Canada *Globe*), *The New York Times*, 28 November 1881, p. 2.

Prince, C. L. (1883). Note on the transit of Venus, 1882, December 6. *Monthly Notices of the Royal Astronomical Society*, *43*(5), 278.

Princeton College – The approaching commencement. *The New York Times*, 30 May 1878, p. 5.

Princeton College. *The New York Times,* 24 December 1878, p. 8.

The Princeton commencement. *The New York Times*, 16 June 1878, p. 1.

Pritchard, A. (1897). *Charles Pritchard ... memoirs of his life*. London: Seeley and Company.

Pritchard, C. (1869). Perceiving without seeing – A romance in astronomy. *The Eclectic Magazine of Foreign Literature, 9*(5), 530–539 (May 1869).

Problems of solar eclipses. *The New York Times*, 3 December 1879, p. 2.

Proceedings of the Royal Society, 86A, facing page i (1910).

Proctor, R. A. (1869). Strange solar discoveries. *The Eclectic Magazine of Foreign Literature*, *9*(3), 280–284 (March 1869).

Proctor, R. A. (1870a). Notes on the solar corona and the zodiacal light; With suggestions respecting observations to be made on the total solar eclipse of December 24th, 1870. *Monthly Notices of the Royal Astronomical Society, 30*, 138–150.

Proctor, R. A. (1870b). Further remarks on the corona. *Monthly Notices of the Royal Astronomical Society, 30*, 221–223.

Proctor, R. A. (1871a). *Light science for leisure hours*. New York: D. Appleton and Company.

Proctor, R. A. (1871b). Note on Oudemann's theory of the coronal radiations. *Monthly Notices of the Royal Astronomical Society, 31*, 71–73.

Proctor, R. A. (1871c). On the nomenclature of matter exterior to the Sun's globe. *Monthly Notices of the Royal Astronomical Society, 31*, 73–74.

Proctor, R. A. (1871d). Note on the corona. *Monthly Notices of the Royal Astronomical Society, 31*, 153–154.
Proctor, R. A. (1871e). Theoretical considerations respecting the corona. *Monthly Notices of the Royal Astronomical Society, 31*, 184–194.
Proctor, R. A. (1871f). Theoretical considerations respecting the corona: Part II. *Monthly Notices of the Royal Astronomical Society, 31*, 254–262.
Proctor, R. A. (1871g). On the shallowness of the real solar atmosphere. *Monthly Notices of the Royal Astronomical Society, 31*, 196–197.
Proctor, R. A. (1871h). On the motion of matter projected from the Sun: With special reference to the outburst witnessed by Prof. Young of America. *Monthly Notices of the Royal Astronomical Society, 32*, 42–53.
Proctor, R. A. (1871i). The Sun's atmosphere at length discovered. *The Eclectic Magazine of Foreign Literature, 14*(1), 59–67.
Proctor, R. A. (1872a). The late solar eclipse. *The Eclectic Magazine of Foreign Literature, 16*(2), 227–235 (August 1872).
Proctor, R. A. (1872b). The December eclipse [from the *Cornhill Magazine*]. *Littell's Living Age*, (1440), 88–97.
Proctor, R. A. (1873). Note on the expulsion theory of the solar corona, comets, &c., with special reference to Prof. Norton's original enunciation of that theory. *Monthly Notices of the Royal Astronomical Society, 33*, 536–539.
Proctor, R. A. (1874). The Sun a bubble [from the *Cornhill Magazine*]. *The Eclectic Magazine of Foreign Literature, 20*(6), 689–699.
Proctor, R. A. (1875). The success of the transit expeditions. *The Eclectic Magazine of Foreign Literature, 22*(1), 80–92.
Proctor, R. A. (1876). Astronomy in America [from *Popular Science Review*]. *The Eclectic Magazine of Foreign Literature, 24*(6), 679–687.
Proctor, R. A. (1877). Suns in flames [from *Belgravia Magazine*]. *The Eclectic Magazine of Foreign Literature, 25*(5), 545–557.
Proctor, R. A. (1882a). The end of the world. *Potter's American Monthly, 18*(126), 632–638.
Proctor, R. A. (1882b). *Transits of Venus: A popular account of past and coming transits* (4th ed.). London: Longmans, Green, and Co.
Prof. C.A. Young Dead. *The New York Times*, 5 January 1908.
Prof. Draper's discovery. *The New York Times*, 27 January 1879, p. 8.
Prof. J.R. Eastman dies. *The New York Times*, 27 September 1913.
Professor Abbe's Eclipse Observation. *Cincinnati Commercial*, 14 August 1869.
Professor Appel's lectures. *Reformed Church Messenger, 40*(7), 8–9 (18 February 1874).
Professor Henry Draper. *The Chautauquan, 3*(4), 234–235 (January 1883).
Prof. C.A. Young dead. *The New York Times*, 5 January 1905.
Prof. Mitchel on astronomy. *The New York Times*, 15 January 1859.
Prof. Mitchell [sic] on astronomy. *The New York Times*, 27 January 1859.
Professor Abbe's eclipse observations. *Cincinnati Commercial*, 4 August 1869.
Professor Proctor's lecture on the transit of Venus, *The New York Times*, 2 April 1874, p. 5.
Professor R.A. Proctor. *Scientific American, 30*(8), 115 (21 February 1874).
Professor Winlock. *Scientific American, 33*(10), 149 (4 September 1875).
Progress and prospects of solar chemistry. *Scientific American, 32*(10), 144 (6 March 1875).
Progress of astronomy. *The New York Times*, 11 November 1879, p. 2.
Pryor and Co.'s Madison City Directory, 1877. Madison, WI: Pryor & Co. Publisher, pp. 69–71.
Pugilistic Society excited. *The New York Times*, 7 December 1882, p. 8.
Radau, M. (1874). The transit of Venus (translation). *Appleton's Journal of Literature, Science and Art, 12*(281), 170–174.
Ranyard, A. C. (1878). Does the gaseous part of the corona vary with sun spots? *Astronomical Register, 16*, 281–282.
Ratcliff, J. (2008). *The transit of Venus enterprise in Victorian Britain*. London: Pickering & Chatto.
Rayet, G. (1869). The total solar eclipse of August, 1868. Report of M. Rayet. *Astronomical Register, 7*, 133–134.

Read and run. (26 September 1874) *Massachusetts Ploughman and New England Journal of Agriculture, 33*(52), 2.
Recent advances in science. *Scientific American, 31*(9), 129 (29 August 1874).
The recent eclipse. *The New York Times*, 27 December 1870, p. 4.
The recent phenomena in the heavens. *Independent Inquirer*, 23 November 1833.
Recent progress in solar astronomy. *The New York Times*, 22 February 1880, p. 6.
The recent solar eclipse. *The New York Times*, 4 August 1878, p. 1.
Recent solar researches [from *The Spectator*]. *Littell's Living Age, 1366*, 375–376 (6 August 1870).
Recent solar researches. *Maine Farmer, 38*(36), 4 (13 August 1870).
Recent solar researches. *The New York Times*, 18 July 1870, p. 6.
Reception to Dr. Peters at Utica, N.Y. *The New York Times*, 1 January 1875, p. 1.
Reddy, V., Snedegar, K., & Balasubramanian, R. K. (2007). Scaling the magnitude: The fall and rise of N.R. Pogson. *Journal of the British Astronomical Association, 117*, 237–243.
Reed, E. W. (n.d.) Hannah Mary Bouvier Peterson 1811 – 1870. In *American women in science before the civil war*. www.catherinereed.com/book/peterson.html.
Reich, J. (Ed.). (1900). *Appleton's cyclopaedia of American biography* (Vol. 4). New York: D. Appleton and Co.
Report of Mr. Janssen. *Astronomical Register, 7*, 107–110 (1869).
Report of Mr. Janssen, part II. *Astronomical Register, 7*, 131–133 (1869).
Report of Mr. Rayet. *Astronomical Register, 7*, 133–134 (1869).
Resignation of Dr. Hastings. *The New York Times*, 4 December 1883.
Review 1 – No title. *Littell's Living Age, 1182*, 207–218 (26 January 1867).
Review 3 – No title. *Massachusetts Teacher, 27*(1), 39–41 (January 1874).
Richard Anthony Proctor [by the editor]. *The Eclectic Magazine of Foreign Literature, 19*(3), 371–373 (March 1874).
Robert Treat Paine. *Proceedings of the American Academy of Arts and Sciences, 21*, 532–535 (1885).
Robinson, S. W. (1864). The arch truss girder again – More upon the new system. *Journal of the Franklin Institute, 77*, 368.
Rogers, S. (edited by his son Elder John I. Rogers). (1880). *Autobiography of Elder Samuel Rogers*. Cincinnati, OH: Standard Publishing Company.
Rossiter, M. W. (1971). Benjamin Silliman and the Lowell Institute: The popularization of science in nineteenth-century America. *New England Quarterly, 44*, 602–626.
Rothenberg, M. (1981). Organization and control: Professionals and amateurs in American astronomy, 1899–1918. *Social Studies of Science, 11*, 305–325.
Rothenberg, M. (1990). Patronage of Harvard College observatory, 1839–1851. In *Two Astronomical Anniversaries: HCO & SAO* (pp. 37–47). Cambridge, MA: Harvard-Smithsonian Center for Astrophysics.
Rothenberg, M. (1999). Observers, publications, and surveys: Astronomy in the United States in 1849. *The Astronomical Journal, 117*, 6–8.
Rothenberg, M., & Williams, T. (1999). Amateurs and the Society during the formative years. *The American Astronomical Society's First Century* (pp. 40–52). Washington, DC: American Institute of Physics.
Rufus, W. C. (1924). Astronomical observatories in the United States prior to 1848. *The Scientific Monthly, 19*, 120–139.
Ruskin, S. (2008). Spring. "Among the Favored Mortals of Earth" – The press, state pride, and the eclipse of 1878. *Colorado Heritage*.
Ruskin, S. (2012). Advancing astronomy on the American Frontier: The career of Frank Herbert Loud. *Journal of Astronomical History and Heritage, 15*(2), 115–124.
Sabine, Lieut.-Col., (1843). Observations of the Comet by Mr. S.C. Walker and Professor Kendall, at the observatory of the High School at Philadelphia. *Monthly Notices of the Royal Astronomical Society, 5*, 294–295.

Sands, B. F. (1869). Supplementary notes on the observations for magnetism and position, made in the U.S. Naval Observatory expedition to Siberia to observe the solar eclipse of August 7th, 1869. *Astronomische Nachrichten, 75*, 323–330.
Sands, B. F. (1870). *Reports on observations of the total eclipse of the Sun, August 7, 1869.* Washington: Government Printing Office.
Saum, L. (1999). The Proctor interlude in St. Joseph and in American astronomy, romance and tragedy. *American Studies International, 37*(1), 34–51.
Sayings and doings. *Harper's Bazaar, 2*(38), 599 (18 September 1869).
Sayings and doings. *Harper's Bazaar, 4*(8), 119 (25 February 1871).
Sayings and doings. *Harper's Bazaar, 7*(6), 95 (7 February 1874).
Sayings and doings. *Harper's Bazaar, 7*(15), 239 (11 April 1874).
Sayings and doings. *Harper's Bazaar, 7*(25), 399 (20 June 1874).
Sayings and doings. *Harper's Bazaar, 7*(28), 477 (11 July 1874).
Sayings and doings. *Harper's Bazaar, 11*(31), 495 (3 August 1878a).
Sayings and doings. *Harper's Bazaar, 11*(32), 511 (10 August 1878b).
Sayings and doings. *Harper's Bazaar, 11*(35), 559 (31 August 1878c).
Schaeberle, J. M. (1890a). A mechanical theory of the solar corona. *Publications of the Astronomical Society of the Pacific, 2*, 68–70.
Schaeberle, J. M. (1890b). A mechanical theory of the solar corona. *Monthly Notices of the Royal Astronomical Society, 50*, 372–373.
Schaeberle, J. M. (1890c). The solar corona. *Publications of the Astronomical Society of the Pacific, 2*, 260.
Schaeberle, J. M. (1891). Some physical phenomena involved in the mechanical theory of the corona. *Publications of the Astronomical Society of the Pacific, 3*, 70–73.
Schaefer, B. E. (2001). The transit of Venus and the notorious black drop effect. *Journal of History for Astronomy, 32*, 325–335.
Schimpff, R. D. (1885). The Biela meteors. *Science, 6*(149), 519–520.
Schultz, C. (1964). Glimpses into Cincinnati's Past. *Ohio History, 73*, 157–179.
Schumacher, Professor. (1843). Circular letter from Professor Schumacher, dated March 26, 1843, on the Great Comet. *Monthly Notices of the Royal Astronomical Society, 5*, 272–273.
Schumacher, Professor. (1843). Second circular letter from Professor Schumacher, dated March 31, 1843, on the Great Comet. *Monthly Notices of the Royal Astronomical Society, 5*, 273–275.
Schuster, A. (1878a). Some remarks on the total solar eclipse of July 29, 1878. *Monthly Notices of the Royal Astronomical Society, 39*, 44–47.
Schuster, A. (1878b). The Sun's corona during the eclipse of 1878. *The Observatory, 2*, 262–266.
Schuster, A. (1879). On the polarisation of the solar corona. *Monthly Notices of the Royal Astronomical Society, 40*, 35–36.
Science. *Appleton's Journal of Literature, Science and Art, 9*(250), 30–31, (3 January 1874).
Science. *Appleton's Journal of Literature, Science and Art, 11*(255), 190–191, (7 February 1874)
Science. *Christian Advocate, 53*(5), 75 (31 January 1878).
Science. *Christian Advocate, 53*(25), 395 (20 June 1878).
Science. *The Eclectic Magazine of Foreign Literature, 4*(5), 638–639 (November 1866).
Science. *The Eclectic Magazine of Foreign Literature, 7*(4), 518 (April 1868).
Science. *The Eclectic Magazine of Foreign Literature, 7*(5), 647 (May 1868).
Science. *The Eclectic Magazine of Foreign Literature, 8*(5), 1416 (November 1868).
Science. *The Eclectic Magazine of Foreign Literature, 9*(4), 507–508 (April 1869).
Science. *The Independent, 24*(1215), 2 (14 March 1872).
Science. *The Independent, 26*(1331), 6 (4 June 1874).
Science. *The Independent, 27*(1395), 8 (26 August 1875).
Science. *The Independent, 30*(1537), 7 (16 May 1878).
Science. *The Independent, 31*(1588), 6 (8 May 1879).
Science – The great eclipse. *The Eclectic Magazine of Foreign Literature, 7*(5), 647 (1868).
Science – The solar eclipse. *The Independent, 30*(1550), 8 (15 August 1878).

Science and art. *Christian Union, 18*(10), 194 (4 September 1878).
Science and art. *Christian Union, 26*(24), 529 (14 December 1882).
Science and invention. *Appleton's Journal of Literature, Science and Art, 11*(264), 478–479 (11 April 1874).
Science and invention. *Appleton's Journal of Literature, Science and Art, 12*(276), 30–31 (4 July 1874).
Science and invention. *Appleton's Journal of Literature, Science and Art, 12*(278), 94–95 (18 July 1874).
Science and the arts. *The New York Teacher and American Educational Monthly,* 496 (December 1868).
The Science Observer V2–5, 1878–86: A Journal for Scientists, 1886. Boston: Boston Scientific Society, 92 (November 1879).
Science, art, discovery. *Christian Advocate*, *57*(39), 10 (28 September 1882).
Science, art, discovery. *Christian Advocate*, *57*(48), 762 (30 November 1882).
Science, art, discovery. *Christian Advocate*, *57*(50), 794 (14 December 1882).
Science in America. *Scientific American, 35*(17), 261 (21 October 1876).
Science on the platform. *The New York Times*, 8 November 1878, p. 2.
Science, statistics, explorations, etc. *Putnam's Magazine. Original Papers on Literature, Science, Art and National Interests, 3*(14), 252–254 (February 1869).
Scientific. *Circular*, *5*(7), 56 (4 May 1868).
Scientific. *The Independent, 22*(1144), 4 (3 November 1870).
Scientific. *The Independent, 23*(1155), 2 (19 January 1871).
Scientific and practical information – Recent astronomical discoveries. *Scientific American, 32*(21), 324 (22 May 1875).
Scientific and sanitary – Astronomical instruments. *Christian Union*, *3*(7), 109 (15 February 1871).
Scientific and sanitary – Astronomy in 1873. *Christian Union, 9*(1), A20 (7 January 1874).
Scientific and sanitary – Constitution of the Sun *Christian Union*, *8*(21), 424 (19 November 1873).
Scientific and sanitary – A handsome courtesy. *Christian Union*, *2*(25), 396 (24 December 1870).
Scientific and sanitary – The transit in Tasmania. *Christian Union*, *11*(11), 236 (13 March 1875).
Scientific and useful. *New York Evangelist*, *39*(32), 3 (6 August 1868).
Scientific and useful. *New York Evangelist, 46*(7), 7 (1875).
Scientific and useful. *New York Evangelist, 53*(24), 7 (15 June 1882).
Scientific and useful. *New York Evangelist, 54*(2), 7 (1883).
Scientific and useful – The eclipse and intra-Mercurial planets [from the *Journal of Commerce*]. *New York Evangelist*, *40*(26), 7 (1 July 1869).
Scientific and useful – Eclipses of the Sun and Moon. *New York Evangelist*, *48*(4), 7 (25 January 1877).
Scientific and useful – English astronomers versus American astronomers. *New York Evangelist, 42*(9), 3 (2 March 1871).
Scientific and useful – Expedition of observation. *New York Evangelist*, *42*(6), 7 (9 February 1871).
Scientific and useful – The Science Association at Salem. *New York Evangelist, 39*(37), 3 (10 September 1868).
Scientific and useful – Total solar eclipse in August. *New York Evangelist*, *40*(29), 7 (22 July 1869).
Scientific and useful – Total solar eclipse in August. *Reformed Church Messenger*, *34*(46), 7 (14 July 1869).
Scientific gossip. *The New York Times*, 15 July 1878, p. 3.
Scientific gossip. *The New York Times*, 15 September 1878, p. 10.
Scientific gossip. *The New York Times*, 30 January 1881, p. 4.
Scientific gossip. *The New York Times*, 12 June 1881, p. 4.
Scientific gossip. *The New York Times*, 6 November 1881, p. 6.
Scientific gossip. *The New York Times*, 5 March 1882, p. 4.
Scientific gossip. *The New York Times*, 26 March 1882, p. 6.
Scientific gossip. *The New York Times*, 9 July 1882, p. 4.

Scientific gossip. *The New York Times*, 3 September 1882, p. 4.
Scientific gossip. *The New York Times*, 17 September 1882, p. 5.
Scientific gossip. *The New York Times,* 19 November 1882, p. 6.
Scientific gossip. *The New York Times*, 10 December 1882, p. 6.
Scientific gossip. *The New York Times,* 7 January 1883, p. 4.
Scientific intelligence. *American Journal of Science and Arts, 7*(38), 139–166 (February 1874).
Scientific notes. *Appleton's Journal of Literature, Science and Art, 3*(49), 277–278 (5 March 1870).
Scientific notes. *Appleton's Journal of Literature, Science and Art, 6*(119), 51–52 (8 July 1871).
Scientific notes. *Appleton's Journal of Literature, Science and Art, 8*(172), 51–52 (13 July 1872).
Scientific notes. *Appleton's Journal of Literature, Science and Art, 9*(207), 349–350 (8 March 1873).
Scientific notes. *Friends' Review; A Religious, Literary and Miscellaneous Journal, 31*(29), 463 (2 March 1878).
Scientific notes. *Friends' Review; A Religious, Literary and Miscellaneous Journal, 32*(10), 159 (19 October 1878).
Scientific notes. *Friends' Review; A Religious, Literary and Miscellaneous Journal, 32*(27), 431 (15 February 1879).
Scientific notes. *Overland Monthly and Out West Magazine, 12*(2), 190–192 (February 1874).
Scientists at Saratoga. *The New York Times*, 28 August 1879, p. 1.
Seamon, W. H. (Ed.). (1888). *Albemarle.* Charlottesville, VA: Wm. H Prout.
Seargent, D. (2009). *The greatest Comets in history – Broom stars and celestial scimitars.* New York City: Springer.
Secchi, Father. (1872a). The solar prominences. *Monthly Notices of the Royal Astronomical Society, 32*(5), 226–230.
Secchi, Father. (1872b). Observations of the solar prominences, from Jan. 1 to April 29. *Monthly Notices of the Royal Astronomical Society, 32*, 318–320.
See, T. J. J. (1920). Historical notice of John Nelson Stockwell of Cleveland. *Popular Astronomy, 28*, 565–584.
See, T. J. J. (1895). The services of Nathaniel Bowditch to American Astronomy. *Popular Astronomy, 2*(9), 385–394.
Seen through a hazy sky. *The New York Times*, 7 December 1882, p. 2.
Sekanina, Z., & Chodas, P. W. (2008). A new orbit determination for the bright Sungrazing comet of 1843. *The Astrophysical Journal, 687*, 1415–1422.
Sener, S. M. (1895a). The Lancaster barracks, where the British and Hessian soldiers were detained during the Revolution. Reprinted from *Harrisburg Daily Telegraph.*
Sener, S. M. (1895b). Old-time heroes of the war of the Revolution, and of 1812–14. Reprinted from *Harrisburg Daily Telegraph.*
Shankland, P. D., & Orchiston, W. (2002). Nineteenth century astronomy at the U.S. Naval Academy. *Journal of Astronomical History and Heritage, 5*(2), 165–179.
Sharpless, I. (1882a). The transit of Venus. *Friends' Review; A Religious, Literary and Miscellaneous Journal, 36*(18), 276–277.
Sharpless, I. (1882b). The transit of Venus. *Friends' Review; A Religious, Literary and Miscellaneous Journal, 36*(19), 293–294.
Shearman, T. (1913). Norman Robert Pogson. *Popular Astronomy*, 21, 479–484.
Sheehan, W., & Misch, A. (2004). Ménage à trois: David Peck Todd, Mabel Loomis Todd, Austin Dickinson, and the 1882 transit of Venus. *Journal for the History of Astronomy, 35*(2), 123–134.
Sheehan, W., & Westfall, J. (2004). *The transits of Venus.* Amherst, NY: Prometheus Books.
The sheriff's jury. *The New York Times*, 18 January 1883, p. 5.
Shower of stars. *Broad Axe*, 16 January 1926.
Shy, J. R. (2002). Early astronomy in America: The role of the College of William and Mary. *Journal of Astronomical History and Heritage, 5*, 41–64.
Sicard, M. (1998). Passage de Vénus: le revolver photographique de Jules Janssen. *Études Photographiques, 4*, 44–63.

Singular phenomenon in the solar sierra. *Monthly Notices of the Royal Astronomical Society, 33*, 246–247 (1873).

Smiley, P. T. (1874, November). The transit of Venus. *St. Nicholas, an Illustrated Magazine for Young Folks* 2(1), 8–10.

Smith, W., Lukens, J., Rittenhouse, D., Sellers, J., Smith, D., & Maskeline, N. (1769, June 3). Account of the transit of Venus over the Sun's disk, as observed at Norriton, in the county of Philadelphia and province of Pennsylvania. *Philosophical Transactions*, 59, 289–326.

Smyth, P. (1843). Observations made at the Royal Observatory, Cape of Good Hope, by Piazzi Smyth, Esq. *Monthly Notices of the Royal Astronomical Society, 5*, 296–297.

Solar discovery. *New York Evangelist*, *42*(4), 1 (26 January 1871).

Solar eclipse. *Astronomical Register*, *6*, 243 (1868).

The solar eclipse. *The Albion*, *47*(33), 481 (14 August 1869).

The solar eclipse. *Christian Advocate*, *44*(33), 257 (19 August 1869).

The solar eclipse. *Christian Union*, *18*(6), 107–108 (7 August 1878).

The solar eclipse. *Maine Farmer*, *37*(36), 2 (14 August 1869).

The solar eclipse. *The New England Farmer, and Horticultural Register*, *57*(31), 2 (3 August 1878).

The solar eclipse. *The New York Times*, 1 August 1878, p. 1.

The solar eclipse. *The New York Times,* 2 August 1878, p. 2.

The solar eclipse. *Scientific American*, *21*(9), 137 (28 August 1869).

The solar eclipse – Arrangements by Commodore Sands. *The New York Times*, 6 July 1869, p. 3.

The solar eclipse – The Litchfield expedition. *The New York Times,* 28 July 1869, p. 5.

The solar eclipse – Successful observations at various points [from the Boston Daily Advertiser]. *Massachusetts Ploughman and New England Journal of Agriculture*, *37*(44), 2 (3 August 1878).

The solar eclipse, December 1871. *Scientific American*, *25*(6), 88 (3 February 1872).

The solar eclipse in August. *The New York Times*, 23 May 1869, p. 3.

The solar eclipse of 1880. *Scientific American*, *40*(6), 90 (8 February 1879).

The solar eclipse of 22nd December, 1870. *Scientific American*, *24*(1), 7–8 (1 January 1871).

The solar eclipse of April 16. *Scientific American*, 31(2), 15 (11 July 1874).

The solar eclipse of Aug. 17 1868. *The Albion, 46*(32) 376 (8 August 1868).

The solar eclipse of March 25, 1876. *Scientific American*, *34*(11), 163 (11 March 1876).

Solar mysteries. *The New York Times*, 30 July 1878, p. 4.

The solar prominences. *Monthly Notices of the Royal Astronomical Society*, *33*, 247–248 (1873).

South American affairs. *The New York Times*, 26 October 1882, p. 2.

The South American states, *The New York Times*, 4 January 1883, p. 2.

Speech of President Curtis. *The New York Times*, 3 August 1882, p. 3.

Spectrum analysis. *Lippincott's Magazine of Literature, Science and Education*, *3*, 540–546 (May 1869).

Sperling, N. (1989). Fair play for Fitz: Henry Fitz introduces the all-American telescope. *Rittenhouse, 3*(2), 49–57.

The spot on Venus. *The New York Times*, 9 December 1882, p. 1.

Stars of the night. *The New York Times*, 21 August 1871, p. 5.

Staubermann, K. (2004). The oldest collection of astronomical plates: Recreating photographic practice of the 1880s. *Journal for the History of Astronomy, 35*, 447–455.

Stebbins, R. A. (1977). The amateur: Two sociological definitions. *Pacific Sociological Review, 20*, 582–606.

Stebbins, R. A. (1978). Towards amateur sociology: A profession for the profession. *The American Sociologist, 13*, 239–252.

Stebbins, R. A. (1979). *Amateurs – On the margin between work and leisure*. Beverly Hills, CA: Sage Publications, Inc.

Stebbins, R. A. (1980a). "Amateur" and "hobbyist" as concepts for the study of leisure problems. *Social Problems, 27*, 413–417.

Stebbins, R. A. (1980b). Avocational science: The amateur routine in archaeology and astronomy. *International Journal of Comparative Sociology, 21*, 34–48.

Stebbins, R. A. (1981). Science amators? Rewards and cost in amateur astronomy and archaeology. *Journal of Leisure Research, 13*, 289–304.
Stebbins, R. A. (1982a). Amateur and professional astronomers: A study of their interrelationships. *Urban Life, 10*, 433–454.
Stebbins, R. A. (1982b). Looking downwards: Sociological images of the vocation and avocation of astronomy. *Southern Stars, 29*, 177–188.
Stebbins, R. A. (1987). Amateurs and their place in professional science. In D. S. Hayes, R. M. Genet, & D. R. Genet (Eds.), *New generation small telescopes* (pp. 217–225). Mesa: Fairborn Press.
Stebbins, R. A. (1992). *Amateurs, professionals, and serious leisure*. Montreal: McGill-Queens University Press.
Stebbins, R. A. (2004). *Between work and leisure: The common ground of two separate worlds*. New Brunswick: Transaction Publications.
Stebbins, R. A. (2007). *Serious leisure: A perspective for our time*. Piscataway: Transaction Publishers.
Steele, J. D. (1870). Storms in the Sun. *Christian Union, 1*(2), 26.
Stephens, C. E. (1990). Astronomy as public utility: The Bond years at the Harvard College Observatory. *Journal for the History of Astronomy, 21*, 21–33.
Stephenson, B., Bolt, M., & Friedman, A. F. (2000). *The universe unveiled: Instruments and images through history*. Chicago: Adler Planetarium and Astronomy Museum.
Stephenson, F. R. (2008). How reliable are archaic records of large solar eclipses? *Journal of the History for Astronomy, 39*, 229–250.
Sterken, C., Duerbeck, H., Cuypers, J., & Langenaken, H. (2004). Jean-Charles Houzeau and the 1882 Belgian transit of Venus expeditions. *Journal of Astronomical Data, 10*(7), 309–330.
Sterken, C. (2009). Venus 1882 and Jean-Charles Houzeau. *Astronomische Nachrichten*, AN 330, No. 6:582–585.
Stone, E. J. (1868). A rediscussion of the observations of the transit of Venus, 1769. *Monthly Notices of the Royal Astronomical Society, 28*(6), 255–266.
The storm – Have we a prophet among us? *Zion's Herald, 46*(41), 487 (14 October 1869).
The 'Strange Light' again. *Hudson River Chronicle*, 11 March 1843.
Struve, W. O. (1873). List of stations selected for observation of the transit of Venus by Russian astronomers. *Monthly Notices of the Royal Astronomical Society, 33*(7), 415–417.
The study of the Sun. *The New York Times*, 30 August 1879, p. 2.
Summary of events. *The Friend; A Religious and Literary Journal, 51*(38), 304 (4 May 1878).
Summary of events. *The Friend, A Religious and Literary Journal, 51*(48), 384 (13 July 1878).
The Sun a bubble. *Every Saturday: A Journal of Choice Reading, 2*(17), 458–463 (24 October 1874).
The Sun's corona [from the *Cornhill Magazine*]. *Friends' Review; A Religious, Literary and Miscellaneous Journal, 24*(18), 294–296 (24 December 1870).
The Sun's corona [from the *Cornhill Magazine*]. *Friends' Review; A Religious, Literary and Miscellaneous Journal, 24*(20), 307–309 (7 January 1871).
The Sun's corona. [from the *Cornhill* Magazine]. The *Eclectic Magazine of Foreign Literature, 13*(1), 52–59 (January 1871).
The Sun's crown. *The Eclectic Magazine of Foreign Literature, 11*(2), 223–224 (February 1870).
The Sun's distance. *Littell's Living Age, 3*, 766–768 (22 December 1877).
The Sun's family of planets; Lecture by Prof. Proctor at Association Hall. *The New York Times*, 16 January 1874.
The Sun's long streamers [from the *Cornhill Magazine*]. *The Eclectic Magazine of Foreign Literature, 29*(2), 185–191 (February 1879).
The Sundry Civil Appropriation Bill. *The New York Times*, 21 June 1874, p. 2.
The *Swatara*, *The New York Times*, 31 May 1874, p. 7.
The *Swatara*; Arrival of the steamer from Brazil. *The New York Times*, 31 December 1875, p. 1.
The *Swatara*; Destinations of the Observatory Corps. *The New York Times*, 28 May 1874, p. 2.
Swift, L. (1878, September 19). Discovery of Vulcan (letter to the editor). *Nature, 18*, 539.

Table talk. *The Round Table, 178*, 396–397 (20 June 1868a).

Table talk. *The Round Table, 184*, 75–77 (1 August 1868b).

Table talk. *The Round Table, 190*, 177–180 (12 September 1868c).

Tacchini, M. (1873). Singular phenomenon on the solar sierra. *Monthly Notices of the Royal Astronomical Society, 33*, 246–247.

A talk about eclipses. *The Youth's Companion, 41*(47), 188 (19 November 1868).

Taylor, R. N. (1872, February 24). The solar eclipse. *Littell's Living Age,* 1446, 490.

Taylor, W. C. (1869). The eclipse. *Cincinnati Daily Enquirer,* 9 August 1869.

Tebbutt, J. (1868). Observations of the solar eclipse, August 18th, 1868. *Monthly Notices of the Royal Astronomical Society, 29*, 2.

Tebbutt, J. (1883). Note on Professor Newcomb's remarks on the Windsor observations of the transit of Venus in 1874. *Monthly Notices of the Royal Astronomical Society, 43*(5), 279–280.

Telegrams – India – The total eclipse of the Sun. *The New York Times*, 25 August 1868, p. 4.

Tennant, J. F. (1867a). On the solar eclipse of 1868, August 17. *Monthly Notices of the Royal Astronomical Society, 27*, 79–80.

Tennant, J. F. (1867b). On the eclipse of August 1868. *Monthly Notices of the Royal Astronomical Society, 27*, 173–178.

Tennant, J. F. (1868). The total solar eclipse of August 17–18, 1868. *Monthly Notices of the Royal Astronomical Society, 28*, 245–246.

Tennant, J. F. (1869a). On the solar eclipse of 1871. *Monthly Notices of the Royal Astronomical Society, 29*, 284–285.

Tennant, J. F. (1869b). Report on the total eclipse of the Sun, August 17–18, 1868. *Memoirs of the Royal Astronomical Society, 37*, 1–41.

Tennant, J. F. (1872a). Letter from Col. Tennant to Dr. Huggins. *Monthly Notices of the Royal Astronomical Society, 32*(3), 70–72.

Tennant, J. F. (1872b). Report on observations made by the government of India on the total eclipse of the Sun on December 11–12, 1871. *Monthly Notices of the Royal Astronomical Society, 32*, 253–255.

Tennant, J. F. (1896). Superstition respecting eclipses. *The Observatory, 19*, 276–277.

Terrestrial magnetism – The eclipse. *Friends' Review; A Religious, Literary and Miscellaneous Journal, 24*(33), 526 (8 April 1871).

Thanks to Captain Chandler. *The New York Times,* 3 June 1876, p. 5.

Their wrathful deity. *The New York Times,* 17 August 1878, p. 2.

This summer's eclipse of the Sun. *The New England Farmer and Horticultural Register, 57*(24), 4 (15 June 1878).

Thompson, J. P. (1874). America as seen from Europe. *Every Saturday: A Journal of Choice Reading, 1*(8), 201–205.

The *Times* catalogue of new books and new editions. *The New York Times*, 17 April 1883, p. 5.

To the editor of the *Pennsylvania Inquirer. Pennsylvania Inquirer and National Gazette,* 18 April 1843.

Today's eclipse. *The New York Times*, 22 December 1870, p. 4.

Todd, D. P. (1881a). On the use of the electric telegraph during total solar eclipses. *The Observatory, 4*, 198–202.

Todd, D. P. (1881b). The solar parallax as derived from the American photographs of the transit of Venus. *The Observatory, 51*, 202–205.

Todd, D. P. (1883). On the observations of the transit of Venus, 1882, December 5–6, made at the Lick Observatory, Mount Hamilton, California. *Monthly Notices of the Royal Astronomical Society, 43*(5), 273–276.

Todd, M. L. (1894). *Total eclipses of the Sun*. Boston: Robert Brothers.

Tolles, F. B. (1956). Philadelphia's first scientist. *Isis, 47*(1), 20–30.

The Tombs police report. *The New York Times*, 26 December 1874.

Topics of the season. *The New York Times*, 19 July 1878, p. 5.

Total eclipse in 1869. *Scientific American, 19*(17), 266 (1868).

The total eclipse in America [from *The Spectator*]. *Littell's Living Age* 1321, 807–809 (25 September 1869).

The total eclipse of 1869. *Hours at Home; A Popular Monthly of Instruction and Recreation*, *9*(6), 555–560 (October 1869).
The total eclipse of 1870. *The Friend; A Religious and Literary Journal*, *44*(29), 226–227 (11 March 1871).
The total eclipse of 1878. *Friends' Review; A Religious, Literary and Miscellaneous Journal*, *32*(7), 103 (28 September 1878).
The total eclipse of August seventh. *The Catholic World, a Monthly Magazine of General Literature and Science*, *10*(55), 106–114 (October 1869).
The total eclipse of the Sun. *Scientific American*, *39*(4), 56–57 (27 July 1878).
The total solar eclipse [from *The London Daily News*]. *Friends' Review; A Religious, Literary and Miscellaneous Journal*, *28*(38), 603–604 (8 May 1875).
The total solar eclipse. *The New England Farmer and Horticultural Register*, *57*(28), 2 (13 July 1878).
Total solar eclipse in August. *Western Christian Advocate*, *36*(31), 242 (4 August 1869).
The total solar eclipse of 1868. *Astronomical Register*, *7*, 186–187 (1869).
The total solar eclipse of July 29th, 1878. *The American Catholic Quarterly Review*, *3*(12), 635–648.
The total solar eclipse of September 17–18, 1876. *Scientific American*, *34*(19), 194 (25 March 1876).
The transit. *The Independent*, *34*(1776), 17 (14 December 1882).
The transit. *The New York Times*, 6 December 1882, p. 4.
The transit expeditions. *The Galaxy. A Magazine of Entertaining Reading*, *20*(2), 271 (August 1875).
The transit of Mercury. *The New York Times*, 3 May 1878, p. 3.
The transit of Venus. *The Chautauquan: A Weekly Newspaper*, *3*(4), 239 (January 1883).
The transit of Venus. *Christian Advocate*, *45*(19), 149 (7 May 1874).
The transit of Venus. *Christian Advocate*, *49*(53), 417 (31 December 1874).
The transit of Venus. *Christian Union*, *10*(23), 463 (9 December 1874).
The transit of Venus. *Littell's Living Age*, *156*(2013), 130 (20 January 1883).
The transit of Venus. *Friends' Intelligencer*, *39*(44), 696 (16 December 1882).
The transit of Venus. *Friends' Review; A Religious, Literary and Miscellaneous Journal*, *28*(18), 286–287 (19 December 1874).
The transit of Venus. *Friends' Review; A Religious, Literary and Miscellaneous Journal*, *28*(21), 326 (9 January 1875).
The transit of Venus. *Journal of the Aiken County Historical Society*, *11*(2), 1, 3–4 (September 1995).
The transit of Venus. *Maine Farmer*, *43*(4), 2 (26 December 1874)
The transit of Venus. *New York Observer and Chronicle*, *53*(1), 6 (7 January 1875).
The transit of Venus. *The New York Times*, 4 June 1874, p. 5.
The transit of Venus. *The New York Times*, 15 July 1874, p. 5.
The transit of Venus. *The New York Times*, 27 July 1874, p. 3.
The transit of Venus. *The New York Times*, 8 December 1874, p. 3.
The transit of Venus. *The New York Times*, 1 February 1875, p. 4.
The transit of Venus. *The New York Times*, 3 March 1875.
The transit of Venus. *The New York Times*, 21 September 1882, p. 8.
The transit of Venus. *The New York Times*, 12 November 1882, p. 9.
The transit of Venus. *The New York Times*, 14 November 1882, p. 1.
The transit of Venus. *The New York Times*, 27 November 1882, p. 2.
The transit of Venus. *The New York Times*, 3 December 1882, p. 1.
The transit of Venus. *The New York Times*, 4 December 1882, p. 5.
The transit of Venus. *The New York Times*, 10 December 1882, p. 1.
The transit of Venus. *The New York Times*, 13 December 1882, p. 2.
The transit of Venus. *The New York Times*, 14 December 1882, p. 2.
The transit of Venus. *The New York Times*, 29 December 1882, p. 2.
The transit of Venus (from *The Spectator*). *The New York Times*, 31 December 1882, p. 4.
Transit of Venus. *Ohio Farmer*, *45*(24), 376 (13 June 1874).

The transit of Venus. *Ohio Farmer, 46*(25), 395–396 (December 19, 1874).
The transit of Venus. *The Phrenological Journal and Science of Health, 59*(4), 258–260 (October 1874).
The transit of Venus. *Scientific American, 31*(8), 114 (22 August 1874).
The transit of Venus. *Scientific American, 31*(26), 400–401 (26 December 1874).
The transit of Venus. *Scientific American, 47*(21), 320 (18 November 1882).
The transit of Venus. *Scientific American, 47*(25), 384 (16 December 1882).
The transit of Venus. *Scribner's Monthly, 8*(6), 759 (October 1874).
The transit of Venus. *Western Christian Advocate, 41*(29), 228 (22 July 1874).
The transit of Venus. *The Youth's Companion, 47*(46), 379–380 (12 November 1874).
The transit of Mercury and the inter-Mercurial body. *Scientific American, 38*(21), 322 (25 May 1878).
The transit of Venus as seen at the Seagrave Observatory. *Scientific American, 47*(25), 385 (16 December 1882).
The transit of Venus; An account of the United States expedition by Prof. Harkness. *The New York Times,* 29 February 1876, p. 1.
The transit of Venus; The American party at Cape Town. *The New York Times*, 5 September 1874, p. 1.
The transit of Venus; Arrival at Bahia. *The New York Times*, 19 September 1874, p. 3.
The transit of Venus; Description of the scientific instruments. *The New York Times,* 5 June 1874, p. 5.
The transit of Venus; Entertainment of Captain Chandler. *The New York Times*, 22 September 1874, p. 4.
The transit of Venus; The expedition at the Cape of Good Hope. *The New York Times*, 6 December 1874, p. 1.
The transit of Venus; Experiences of the observing parties. *The New York Times*, 23 February 1875, p. 2.
The transit of Venus; Government preparations for its observance. *The New York Times*, 18 May 1874, p. 5.
The transit of Venus; Homeward bound. *The New York Times*, 29 March 1875, p. 3.
The transit of Venus; Le Gentil's observation. *The New York Times*, 25 July 1874, p. 2.
The transit of Venus; Lecture by Prof. C.A. Young, in Hoboken. *The New York Times*, 3 March 1875, p. 7.
The transit of Venus; A letter from Prof. Watson. *The New York Times*, 25 February 1875, p. 2.
The transit of Venus; Letter from Prof. Young at Peking. *The New York Times*, 31 December 1874, p. 1.
The transit of Venus; List of the persons designated for the observation. *New York Times*, 22 May 1874, p. 1.
The transit of Venus; Observation in Japan successful. *The New York Times,* 10 December 1874, p. 5.
The transit of Venus; Observations at Hobart Town. *The New York Times*, 9 February 1875, p. 8.
Transit of Venus; Preliminary account by Prof. Davidson. *The New York Times*, 11 February 1875, p. 5.
The transit of Venus; Preparations made for its observation. *The New York Times*, 21 May 1874, p. 3.
The transit of Venus; Report of Dr. Peters. *The New York Times*, 12 February 1875, p. 1.
The transit of Venus; Results of the observations at Chatham Island. *The New York Times*, 2 May 1875, p. 9.
The transit of Venus; Successful observations at Beyrout, Syria. *The New York Times*, 13 December 1874, p. 1.
The transit of Venus; Successful observations in Japan. *The New York Times,* 31 January 1875, p. 9.
The transit of Venus; The *Swatara* on her way from Capetown. *The New York Times*. 7 December 1874, p. 5.
The transit of Venus; United States observing expeditions. *The New York Times*, 30 April 1874, p. 3.

The transit of Venus; Voyage of the *Swatara* from Crozet. *The New York Times*, 30 December 1874, p. 1.
The transit of Venus expedition. *The New York Times*, 22 September 1874, p. 4.
The transit of Venus in Chili[sic]. *The New York Times*, 3 February 1883, p. 8.
The transit of Venus telegrams received at the Greenwich Observatory. *New York Times*, 10 December 1874, p. 1.
Transits of Venus. *Frank Leslie's Popular Monthly*, *15*(5), 628–638 (May 1883).
Transits of Venus. *Scientific American*, *47*(22), 336 (November 25, 1882).
Transits of Venus – The solar parallax. *Astronomical Register*, *6*, 219–222 (1868).
Transits of Venus behind the Sun. *Scientific American*, *33*(9), 131 (28 August 1875).
Trowbridge, D. (1867). Words from the people. *Beadle's Monthly, a Magazine of Today, 3*, 475–477.
The truth about the Sun. *The New York Times*, 3 August 1878, p. 4.
Tucker, L. L. (1975, Winter). Cincinnati, Athens of the West, 1830–1861. *Ohio History*, 10–25.
Tucker, L. L. (1964, Winter). The semi-colon club of Cincinnati. *Ohio History*, 73, 13–26.
Tupman, G. L. (1878). On the mean solar parallax as derived from the observation of the transit of Venus, 1874. *Monthly Notices of the Royal Astronomical Society, 38*(8), 429–457.
Turner, H. H. (1889). Notes on some total solar eclipses. *The Observatory, 12*, 65–70.
Turner, G. L. (1976). *The patronage of science in the nineteenth century*. Leyden: Noordhoff International Publishing.
Turner, S. (2010). Henry Fitz's telescope shop. National Museum of American History 12 May podcast: americanhistory.si.edu/connect/podcasts/history-explorer-henry-fitz-telescope-shop 2004, Transit of Venus, 2004. *Puck*, 12(301), 227 (12 December 1882).
Uniformity in time. *The New York Times*, 2 May 1880, p. 2.
United States Naval Observatory. (1878). *Instructions for observing the total solar eclipse of July 29, 1878*. Washington, DC: Government Printing Office.
United States Transit of Venus Commission. (1882). *Instructions for observing the transit of Venus, December 6, 1882. Prepared by the Commission, Authorized by Congress, and Printed for the Use of the Observing Parties by Authority of the Hon. Secretary of the Navy*. Washington, DC: Government Printing Office.
U.S. Congress. *New York Observer and Chronicle*, *47*(5), 38 (4 February 1869).
University of Cincinnati Archives, Box 21, Five Memoranda Notebooks of Participants in the Solar Eclipse Expedition of 1869 to the Dakota Territory.
An unscientific view of the transit of Venus (from the *Pall Mall Gazette). New York Times*, 23 January 1875, p. 2.
[Untitled]. *The New York Times*, 21 July 1878, p. 7.
[Untitled] *Zion's Herald* 51(11), 85 (12 March 1874).
Van Helden, A. (1976). The importance of the transit of Mercury of 1631. *Journal for the History of Astronomy, 7*, 1–10.
Van Loan, W. (1884). *Van Loan's Catskill Mountain Guide of 1884*. Catskill: Van Loan & Van Gordern.
Vander Weyde, P. H. (1874). The relative attraction of the Earth and the Sun (letter to the editor). *Scientific American, 30*(6), 84.
Vanity Fair, 13 November 1875.
Vanity Fair, 11 May 1878.
Vanity Fair, 3 March 1883.
Vanity Fair, 21 May 1903.
Vanity Fair, 13 April 1905.
Venus crossing the Sun's face. *The New York Times*, 5 December 1882, p. 12.
Veron, P., & Tammann, G. A. (1979). Astronomical broadsheets and their scientific significance. *Endeavour, 3*(4), 163–170.
Viewing Venus. *San Antonio Express*, 7 December 1882.
Views from the rostrum; Mr. Proctor's second lecture. *The New York Times*, 14 November 1879, p. 2.
Vulcan and the corona. *The New York Times*, 16 August 1878, p. 5.

Waldo, L. (1879). In L. Waldo (Ed.), *Report of the observations of the total solar eclipse, July 29, 1878, made at Forth Worth, Texas*. Cambridge, MA: John Wilson and Son.
Walker, S. Letter to the editor. *Boston Courier*, 5 April 1843.
Walker, S. C., & Kendall, E. O. (1843). On the Great Comet of 1843. *The American Journal of Science and Arts, 45*, 188–208.
Walker, S. C. (1842). In Sir J.F.W. Herschel's *Treatise on Astronomy (American edition)*, Philadelphia: Tea & Blanchard.
Walters, A. N. (1999). Ephemeral events: English broadsides of early eighteenth-century solar eclipses. *History of Science, 37*, 1–43.
Walters, R. (1943). The centenary of the Cincinnati observatory. *Science, 98*(2556), 551–555.
Warner, D. J. (1968). *Alvan Clark & Sons. Artists in optics*. Washington: Smithsonian Institution.
Warner, D. J. (1978). Science education for women in antebellum America. *Isis, 69*, 58–67.
Warner, D. J. (1979). Astronomy in antebellum America. In N. Reingold (Ed.), *Science in the American context* (pp. 55–75). Washington, DC: Smithsonian Institution Press.
Warner, D. J., & Ariail, R. B. (1995). *Alvan Clark & Sons – Artists in OPTICS* (2nd ed.). Richmond, VA: Willmann-Bell, Inc.
Warren, B. (1883), Discrepancies in astronomy. *The Chautauquan: A Weekly Newspaper,* 3(8), 458.
Washington. *The New York Times*, 10 December 1874, p. 1.
Washington – Astronomical observations. *The New York Times*, 7 October 1870, p. 4.
Washington – The coming solar eclipse. *The New York Times*, 7 July 1869, p. 1.
Washington – Congressmen and senators in consultation. *The New York Times*, 10 December 1874, p. 1.
Washington – The last great eclipse. *The New York Times*, 22 August 1870, p. 4.
Washington – The solar eclipse in August. *The New York Times*, 17 June 1869, p. 5.
Washington – The transit of Venus observers. *The New York Times*, 24 March 1874, p. 1.
Watching the transit. *The New York Times*, 8 December 1882, p. 5.
The weather – Distance of the Sun. *The New York Times*, 24 July 1875, p. 5.
Webb, T. W. (1869). American photographs of total solar eclipse of August 7, 1869. *Monthly Notices of the Royal Astronomical Society, 30*, 4–5.
Weiss, E. (1888). *Bilderatlas der Sternenwelt: Eine Astronomie fur Jedermann*. Stuttgart: Schreiber.
Wells, W. (1868). The recent solar eclipse. *Christian Advocate, 63*(43), 337.
Werrett, S. (2006). Transits and transitions: Astronomy, topography, and politics in Russian expeditions to view the transit of Venus in 1874. In D. Aubin (Ed.), *L'événement Astronomique du Siècle? Histoire Sociale des Passages de Vénus, 1874–1882* (pp. 147–176). Nantes: Université de Nantes.
What has been discovered by spectrum analysis. *The Ladies' Repository; A Monthly Periodical, Devoted to Literature, Art and Religion*, 252–256 (October 1869).
What the 'transit' Means. *Friends' Intelligencer, 31*(43), 687 (19 December 1874).
What we hope to learn from the approaching solar eclipse. *The Eclectic Magazine of Foreign Literature, 12*(5), 551–558 (November 1870).
Where to observe the solar eclipse of July 29th. *Scientific American, 39*(2), 24–25 (13 July 1878).
Whitesell, P. S. (2000). Nineteenth-century longitude determinations in the Great Lakes region: Government-university collaborations. *Journal of Astronomical History and Heritage, 3*(2), 131–157.
Wiggins. *The New York Times*, 10 February 1883, p. 4.
Wilczynski, E. J. (1898). On the depth of the reversing layer. *Astrophysical Journal, 7*, 213–214.
Willach, R. (2004). The heliometer: Instrument for gauging distances in space. *Journal of the Antique Telescope Society, 26*, 5–16.
William Henry Wesley. (1922a). *Monthly Notices of the Royal Astronomical Society, 83*, 255–259.
William Henry Wesley. (1922b) *The Observatory, 45*, 354–355.
Williams, T. R. (1984). Nathaniel Bowditch, early American amateur astronomer. *Journal of the American Association of Variable Star Observers, 13*, 64–66.

Williams, T. R. (1987). Criteria for classifying an astronomer as an amateur. Paper presented at IAU Colloquium 98, Paris (Note: the paper published in the Proceedings is merely a precis of the full paper).
Williams, T. R. (1988). A galaxy of amateur astronomers. *Sky & Telescope*, 484–486.
Williams, T. R. (1996). The development of astronomy in the Southern United States, 1840–1914. *Journal for the History of Astronomy, 27*, 13–44.
Williams, T. R. (2000, May). *Getting organized: A history of amateur astronomy in the United States* (Ph.D. Thesis), Chapters 1, 2, 4.
Wilson, J. M. (1869). The corona. *Nature,* 1, 139.
Wines, F. H. (1869). The eclipse – The eclipse in Illinois. *The Independent, 21*(1081), 2.
Wonderful appearances. *Saturday Evening Post*, 13 November 1869, p. 6.
The wonders of astronomy. *The New York Times*, 28 March 1874.
Wonders of the star depths. *The New York Times*, 21 January 1874.
Woolf, H. (1959). *The transit of Venus. A study of eighteenth science*. Princeton: Princeton University Press.
Wood, H. (1869). The late King of Siam. *New York Observer and Chronicle, 47*(10), 77.
The world of science. *The New York Times*, 3 October 1874, p. 1.
Worlds beyond the skies. *The New York Times*, 21 November 1879, p. 2.
Wright, B. (1878). Astronomical notes. *Scientific American, 39*(18), 277.
The Yale astronomers busy. *The New York Times*, 5 December 1882, p. 7.
Yearbook of Brooklyn Institute of Arts and Sciences, 1891. Brooklyn, Press of Eagle Book Printing Department, 85–90.
A year's discoveries. *The New York Times*, 26 February 1875, p. 2.
Yeomans, D. K. (1981). Comet Tempel-Tuttle and the Leonid meteors. *Icarus, 47*, 492–499.
Yeomans, D. K., Yau, K. K., & Weisman, P. R. (1996). The impending appearance of Comet Tempel-Tuttle and the Leonid meteors. *Icarus, 124*, 407–413.
Young, C.A. (1875a). The transit at Peking. *The Independent, 27*(1370), 2–3.
Young, C. A. (1875b, February 25). The transit of Venus. *The New York Times*, p. 2.
Young, C. A. (1876). Science: American contributions to astronomy. *The Independent, 28*(1442), 8.
Young, C. A. (1878a). The recent solar eclipse. *Princeton Review, 1878*, 865–888.
Young, C. A. (1878b). Observations upon the solar eclipse of July 29, 1878, by the Princeton eclipse expedition. *American Journal of Science and Arts, 16*(94), 279–290.
Young, C. A. (1880). Recent progress in solar astronomy. *Princeton Review, 1880*, 88–104.
Young, C. A. (1882a, October 7). The coming transit of Venus. *The Critic, 46*, 263–264.
Young, C. A. (1882b, 14 November). The preparations for photographing the transit. *New York Times*, p. 4.
Young, C. A. (1882c, December 2). The coming transit of Venus. *Scientific American, 47*(23), 357.
Young, C. A. (1882d, 16 December). Science: The transit of Venus. *The Critic, 51*, 345.
Young, C. A. (1896). *The Sun* (2nd ed.). New York: D. Appleton and Co.
Young, C. A. (Ed.). (1898). *A text-Book of general astronomy for colleges and scientific schools* (pp. 172–227, 237). Boston: Ginn.
Zochert, D. (1974). Science and the common man in ante-bellum America. *Isis, 65*, 448–473.

Index

A
Abbe, C., 65, 66, 69, 75, 80
Abbe, R., 69, 70, 72, 73
Abbot, E., 177
Abney, W., 99
Academy of Science (Paris), 97
Actinometer, 70
Active, 46, 63, 70, 82, 191, 246, 262, 274, 278, 281
Adams, J.Q., 27, 29, 139
Aden (Yemen), 41, 58, 61, 63, 143, 145, 169, 185
Agassiz, L., 27, 257
Aiken (South Carolina), 111–114, 235, 244
Airy, G., 32, 207, 212
Alaska, 63, 144, 159, 187, 190, 191, 215
Albion, 130
Alexander, J., 21
Alexander, S., 196, 197, 254
Allegheny Observatory, 192, 193, 199
Almagest, 17
Almanacs, 25, 27, 29
Amateurs, 67, 75, 128, 255–262, 272–275, 284, 285, 287
American Academy of Arts and Sciences, 127
American Association for the Advancement of Science (AAAS), 128, 156, 198, 241, 255, 274
American Astronomical Society, 274, 275
American Catholic Quarterly Review, 153, 154, 235
American Journal of Science, 32, 175
American Journal of Science and Arts, 10, 19, 20, 175–178, 218
American Literary Gazette and Publlishers' Circular, 131
Anderson, H.J., 230
Ann Arbor Observatory, 85
Appel, T., 173
Appleton's Journal of Literature, Science and Art, 201
Apprentice Mechanics' Association, 27
Arago, F.J.D., 11, 157
Archer, F.S., 98
Aristarchus of Samos, 88, 207
Arrhenius, S.A., 82, 83
Ashe, E.D., 75, 190
Assyrians, 161, 199
Astronomia àDesigned for the Amusement and Instruction of Young Ladies and Gentlemen, 30
Astronomical and Astrophysical Society of America, 275
Astronomical Journal, 25, 33, 271, 284, 287, 289
Astronomical Society of the Pacific (ASP), 272, 274, 277–279, 283, 289
Astronomical unit, 5, 87–128, 290
Astronomy and Astrophysics, 257, 272
Astrophysical Journal, 61, 272
Auckland (New Zealand), 125, 250, 251
Aurora borealis, 13, 139, 176
Auroral theory, 46, 168, 176

B
Bache, A.D., 257
Baily, F., 48–50
Baily's beads, 48, 70, 190
Ball, R.S., 233, 234
Barker, G.F., 178, 179, 193, 198
Barnard, E.E., 284, 285

S. Cottam, W. Orchiston, *Eclipses, Transits, and Comets of the Nineteenth Century*, Astrophysics and Space Science Library 406,
DOI 10.1007/978-3-319-08341-4

Bassat, Commandant, 116, 117
Beadle's Monthly, a Magazine of Today, 134
Beejapur (India), 59
Behring Strait, 130, 167, 191
Bell, T., 255, 267, 269–271, 274, 288
Berkowski, 51, 150, 151
Berlin Observatory, 97
Bigelow, F.H., 82
Black drop, 94–96, 109, 128, 211, 214, 219, 247
Black ligament, 94
Blake, F. Jr., 67
Bond, G.P., 177
Bond, W.C., 20, 21, 255
Bonn Observatory, 111
Bonpland, A., 11, 12, 14, 15
Borneo, 62, 136
Boss, L., 124, 126, 237, 250, 251
Boston Scientific Society, 272, 274
Boston Society for the Diffusion of Useful Knowledge, 27
Bowditch, N., 3, 48
Brackett, C.F., 158, 192
Brashear, J., 255, 256, 262, 267, 268, 274
Brattle, T., 7
British Scientific Association, 131
Broadsides, 34
Broken-tube transit telescope, 109
Brooklyn, 168, 195, 207, 212, 250, 266, 274
Brooks, W.R., 30, 242
Brown, A., 17, 47, 48
Bullis House Inn, 116, 118
Bull, R.H., 235
Bunsen, R., 53
Burckhalter, C., 282, 283
Burlington (Iowa), 145, 159
Burnet, J., 275, 276
Burnham, S.W., 111, 278
Burr, E.F., 175
Burritt, E., 30
Burr, T.W., 48

C

Cadiz (Spain), 192
Campbell Town, Tasmania (American transit expedition site), 105
Campbell, W.W., 43, 44, 82, 84, 86, 274
Cape of Good Hope, 19, 133, 183, 209, 225, 226, 237, 250, 251
Cape Sidmouth (Australia), 131
Carleton College Observatory, 272
Carnatic, 62
Casey, Jr., T.L., 250
Cassini, G.D., 162
Catania (Sicily), 143, 171
Catholic World, a Monthly Magazine of General Literature and Science, 154, 211
Cedar Keys (Florida), 117, 120, 246
Central City (Colorado), 80
Central High School (Philadelphia), 19, 250
Central High School Observatory, 21, 250
Central Park (New York), 265
Chabot Space and Science Center, 273, 280
Chandler, Captain, 225, 226, 230
Channing, W.E., 34
Chase, P.E., 161–162
Chatham Island (American transit expedition site), 105, 226–227
Chautauquan: A Weekly Magazine, 204, 231
Cherry Creek (Colorado), 76, 77, 178
Chicago Astronomical Society, 273, 277, 278, 289
Chilcat (Alaska), 191
Christian Advocate, 155, 174, 211, 216, 235, 239
Christian Union, 155, 156, 212, 236
Chromosphere, 139, 147, 170, 177, 178, 182, 183
Cincinnati Astronomical Society, 273, 275, 277, 278, 289
Cincinnati Observatory, 28, 33, 65, 69, 73, 80, 255, 265, 271, 289
Circular, 38, 70, 84, 131, 134, 284–286
Clark, A., 40, 41, 67, 105, 108, 159, 232, 240, 243, 255, 267–269, 277, 278, 288
Clark, A.G., 67, 75
Clark, G., 67
Clerke, A.M., 5, 19, 51, 74, 82, 87, 152
Clous, J.W., 119
Coakley, G.W., 235
Coffin, I.H.C., 74
Coffin, S.J., 247, 248
Collodion plates, 51
Columbia University Observatory, 231
Comet 109P/Swift-Tuttle, 17
Comet P55/Tempel-Tuttle, 13
Comet X/1106 C1, 23
Compton, A., 69, 70, 195, 236
Contacts, 5, 76, 90, 91, 93, 94, 96, 97, 100, 104, 111, 112, 115, 120, 123–125, 127, 214, 215, 219, 227, 229, 235, 238, 239, 244, 245, 247, 250–252, 275, 280
Cooper, J.F., 152, 187, 188
Córdoba (Chile), 250
Coronagraph, 86–87
Coronium, 74, 163, 173
Crabtree, W., 93
Creston (Wyoming Territory), 75, 80
Crookes, W., 186
Cros, É.-H.C., 201, 202
Croswell, W., 38
Cumana (Venezuela), 11
Curtis, E., 187, 241

D
Daguerreotype, 51, 98, 150, 186
Dallas (Texas), 80
d'Auteroche, Abbé, 96
Davidson, G., 123–125, 187, 190, 209, 210, 212–214, 228, 229, 246
Dean, G.W., 67
Defforges, G., 116
Deichmüller, F., 111
De la Rue, W., 7, 51, 63, 75, 98, 131, 138, 145, 165, 166, 175, 177, 182
Delisle, J.-N., 91, 92, 94, 212, 235
Delisle's method, 92, 103, 210
Denver (Colorado), 75, 77, 85
Des Moines (Iowa), 167–168, 174, 176, 187
DeWitt, S., 48, 50
Diamilla-Muller, D.E., 160
Donati, G.B., 57, 58
Draper, D., 189, 190
Draper, H., 76, 77, 152, 159, 170, 178, 192, 195, 197, 198, 204, 205, 209, 210, 220
Dry plate process, 51, 99
Dudley Observatory, 124, 170, 255
Dumas, J.-B. A., 242
Dumichen, J., 152
Dun Echt Observatory, 102
Durban, 244, 251

E
Easterday, L.F.M., 170
Eastman, J.R., 75, 120, 122, 168, 171, 246
Eclectic Magazine of Foreign Literature, 135, 141, 206
Edison, T., 77, 78, 156, 164, 178, 179, 193, 196
Egress, 5, 90, 91, 93–95, 97, 100, 103, 104, 244, 251
Egypt (transit expedition site), 219
Einstein, A., 148, 149
Electromagnetic theory, 82
Elementary Treatise on Astronomy, 29
Elements of the Four Inner Planets and the Fundamental Constants of Astronomy, 128
Encke, J.F., 97, 98, 219
Every Saturday, 141, 142

F
Familiar Astronomy, 31, 32
Fayé, H., 137, 138
Field, C.W., 133
Fireballs, 11
Fitz, H., 39, 40, 255, 267
Flag of Our Union, 141
Flammarion, C., 18, 19, 162
Florence Observatory, 57
Floyd, R.S., 125
Forest and Stream, 206
Fort Marion (Saint Augustine, Florida), 116
Fort Sam Houston, 114, 119–121
Fort Seldon (New Mexico), 117, 123, 246
Fort Worth (Texas), 78, 277, 281, 289
Frank Leslie's Popular Monthly, 112, 232, 233
Franklin, B., 1
Franklin Institute of Philadelphia, 74
Franz, J.H., 112, 113
Fraunhofer lines, 134, 145, 147, 153, 162–165, 170, 173, 178, 183
French Academy of Science, 201, 242
Friend. *A Religious and Literary Journal*, 156, 212
Friends' Intelligencer, 158–160, 212, 236, 238
Friends' Normal School, 160, 238
Friends' Review; a Religious, Literary and Miscellaneous Journal, 160, 213, 238

G
Gaillot, J.B.A., 196
Galaxy. A Magazine of Entertaining Reading, 143, 206
Gardner, J., 159
Gassendi, P., 89, 90, 92
Geography of the Heavens, 30, 182
Gibraltar, 141, 142, 168
Gill, D., 102, 274
Gould, B.A., 25, 26, 176, 257, 271, 289
Great Comet of 1843, 8, 17–25
Great Horn, 58, 59, 61, 63
Greenwich Observatory, 211, 216, 222, 236, 244
Gummere, J., 29, 30
Gummere, S.J., 158, 160
Guntoor (India), 57–59, 63

H
Haig, C.T., 59
Haines, J., 69
Hale, G.E., 86, 87, 269, 272, 274, 277, 278, 290
Hall, A., 52, 63, 80, 119, 177, 209, 245–246, 274
Halley, E., 34, 90, 91, 93, 116
Halley's method, 90, 91, 103, 211
Halsted Observatory, 232
Harkness, W., 73, 74, 105, 123, 128, 241, 245
Harper's Bazaar, 144, 207
Harrison, J., 100
Hartford (Connecticut), 111, 112, 235, 236, 244
Harvard College Observatory, 18, 20, 65, 99, 127, 177, 248, 255, 281

Hastings-on-the-Hudson, 204
Havana (Cuba), 80
Hawes, J.J., 150, 151
Heliometer, 51, 52, 99, 102, 104, 111, 115, 116, 237, 248, 249
Heliostat, 99, 100, 105, 119, 127, 162, 209, 217, 236, 238
Helium, 54, 61, 154
Hennessey, J.P., 62
Hennessy, J.B.W., 54, 222
Henry, James, 145
Henry, Joseph, 11, 105, 106, 133, 257
Herodotus, 47
Herschel, J., 7, 29, 32, 54, 59, 139, 140, 226
Herschel, W., 7, 139, 140
Hilgard, J., 123, 125, 208
Himes, C.F., 218
Hind, J.R., 130, 131, 133
Hinode Satellite, 87
Hipparchus of Nicaea, 88
Hitchcock, E., 11
Hobart College Observatory, 247
Hobart, Tasmania (American transit expedition site), 105, 108, 226, 227, 229, 247, 261, 277
Holden, E.S., 80, 177, 272
Horizontal photographic telescope, 99
Horrocks, J., 93, 210, 215, 234, 236
Horrox, J., 93
Hough, G.W., 170
Hours at Home, 144
Houzeau, J.-C., 114, 115
Huggins, W., 44, 45, 86
Huguenot Seminary for Girls (South Africa), 123–124
Humboldt, A., 11

I
Independent, 162–164, 214, 238
Ingress, 5, 90, 91, 93–95, 100, 103, 111, 244
Instructions for Observing the Total Solar Eclipse of July 29th, 1878, 75, 280, 281, 289
International Astronomical Union, 95
International Commission (France), 242
International Review, 208, 216
Intra-Mercurial planets, 67, 69, 82, 85–86, 130, 144, 145, 148, 152, 153, 155, 156, 162, 164–167, 173, 180, 183, 185, 194, 195, 197, 288
Isfahan (Persia), 104

J
Jamkandi (India), 59
Janssen, J., 54, 59, 104, 131, 136, 214
Jefferson (Iowa), 75

K
Keay, C., 13
Kendall, E., 19
Kepler, J., 89
Kerguelen (American transit expedition site), 103–105, 202, 204, 226
Kerguelen Island (foreign transit expedition site), 104, 105, 108, 226
King Rama IV, 169
Kirchhoff, G., 53, 132
Kirkwood, D., 13, 14, 241
Königsberg heliometer, 51
Kreutz Group of Sun-grazing comets, 23

L
Lacaille, A., 226
Lackawanna, 214, 228
Ladies' Repository; a Monthly Periodical Devoted to Literature, Art and Religion, 165
Lafayette College Observatory, 248
La Junta (Colorado), 80, 195
Lancaster, A., 115
Langley, S.P., 80, 81, 170
Lankford, J., 255, 258
Lardner, D., 22
La Sylphide, 97
Lazzaroni, 257, 271
Leach, J.H., 182
Lectures, 2, 6, 15, 22, 23, 25, 27, 34, 44, 173, 175, 197, 202, 206, 207, 220, 229, 236, 259, 262, 265–267, 288, 290
Lefaucher, J., 60
Le Gentil de La Galasière, 97, 220
Leonid Meteor Storm of 1833, 8–17
Leonids, 8–17, 237
Letters on Astronomy Addressed to a Lady, 30
LeVerrier, J.J., 196
Library of Congress, 272, 273, 289, 290
Lick Observatory, 8, 43, 46, 82, 125, 178, 270, 271, 274, 282, 284, 288
Lindsay, J.L., 104, 171, 172, 212, 213, 224
Lippincott's Magazine of Literature, Science and Education, 145
Litchfield Observatory, 167, 168, 187, 188, 247

Littell's Living Age, 145–147, 208, 233
Livermore, W.R., 119
Lockett, S.H., 282
Lockwood, J., 197
Lockyer, J. N., 131, 143, 147, 148, 164
Loewy, B., 175
Logan, J., 1
Logarithmick Arithmetick, 30
Longworth, N., 275, 276
Loomis, E., 15, 16
Loud, F., 281
Lowell Institute, 27
Lydians, 47, 135, 180
Lyot, B.F., 86

M
Maclear, J.P., 161
Madison, J., 2, 3
Madras Observatory, 57, 59, 135, 159
Magic lantern, 220, 265, 288
Magnetism, 13, 43, 63, 82, 160, 242
Maine Farmer, 147, 208
Masoorie (India), 222
Massachusetts Ploughman and New England Journal of Agriculture, 147, 209, 233
Massachusetts Teacher, 209
Masulipatam (India), 57
Ma-twan-lin, 144, 153
Maunder, E.W., 84
Mauritius Islands (transit expedition site), 104, 213
Mayer, A., 75, 206
McGill University Observatory, 243
McLeod, C.H., 243
Mechanical theory, 46, 82
Medes, 47, 135, 180
Medical and Surgical Reporter, 209
Medical News, 148
Meigs, J., 25, 26
Methodist Quarterly Review, 165
Method of difference of absolute time, 91, 103
Method of duration, 103
Meudon Observatory, 86
Meunier, S., 207
Mitchell, M., 77, 79, 127, 159, 174, 212, 249, 265, 274
Mitchel, O.M., 1, 27, 28, 33, 139, 175, 255, 265, 266, 271, 275, 277, 289
Mohican, 149
Monthly Religious Magazine, 165
Morse, S., 57, 100
Morton, H.J., 193, 194
Mount Pleasant (Iowa), 74–75
Mount Wilson Observatory, 277
Müller, F., 111
Myer, A.J., 183, 184

N
Nagasaki (American transit expedition site), 101, 105, 209, 212, 213, 228–229
Nagasaki (Japan), 101, 105, 109, 209, 212, 214, 222, 228–229
Nassau Literary Magazine, 148
National Academy of Science, 105, 196, 219, 241
National Archives (Washington, D.C.), 122, 124, 125, 128, 281, 284, 288
National Nautical Almanac, 74, 251
Nebular hypothesis, 196
Newcomb, S., 76, 97, 99, 101, 105, 107, 111, 123, 124, 128, 148, 155, 168, 171, 187, 193, 207, 212, 237, 239, 250, 251, 274, 275
New England Farmer, 280–281
New England Farmer and Horticultural Register, 148
Newton, H., 7, 12
Newton, I., 145, 165
New York Academy of Science, 198, 231, 241
New York Evangelist, 166–168, 214, 215, 238, 239
New York Observer and Chronicle, 168–170, 216
New York Teacher and American Educational Monthly, 148
New York Times, 6, 15, 16, 23, 34, 184–199, 210, 213, 220 230, 240–254, 263–266, 275, 287, 288, 290
New Zealand (transit expedition site), 104, 226, 229, 239
Norton, W.A., 46, 168, 176, 177
Noumea (New Caledonia), 202, 204
Noyes, J.H., 134

O
Observatories, 3, 7, 43, 135, 204, 255, 288
Ohio Farmer, 150, 210
Old and New, 170
Oliver, H.E., 194
Olmsted, D., 8–14, 25, 30
Orreries, 34, 38
Oskaloosa (Iowa), 160
Ottumwa (Iowa), 74–75, 148

Our Own Correspondent, 190, 193, 224, 226, 227, 264
Overland Monthly and Out West Magazine, 148, 210
Oxygen, 12, 84, 159, 161, 196, 198

P
Paine, R.T., 48, 63
Palermo Observatory, 82
Palomar Observatory, 277
Paris Observatory, 11, 60, 89, 111, 137, 196, 236
Parker, G., 10
Parkhurst, H.M., 100, 220, 274
Parthia, 251
Payne, H.V., 243
Payne, W., 272, 289
Peirce, B., 20, 22, 23, 65, 105, 107, 143, 155, 157, 163, 171, 177, 257
Peking (American transit expedition site), 105, 214, 224, 229
Penrose, F.C., 75, 76
Perrier, F., 116–118, 239
Perry, S., 104, 146, 192
Perseids, 16, 17
Peters, C.F.W., 13, 167
Peters, C.H.F., 135, 168, 187, 190, 215, 216
Peterson, H.M.B., 31, 32
Phosphoric lines, 11
Photography, 33, 44, 129, 204, 257
Phrenological Journal and Science of Health, 150, 210
Pickering, E.C., 127, 171, 274
Pickering, W.H., 77, 79, 86
Pike's Peak (Colorado), 76, 80, 199
Planisphere, 38, 39, 259
Pogson, N.R., 54, 59–61
Polariscopy, 44, 55–57, 76, 77, 82, 84, 130, 141, 146, 147, 154, 156, 171–173, 176, 177, 181, 182, 193, 194, 196
Polarization, 43, 55, 57–59, 63, 76, 77, 84–85, 147, 154, 164, 171, 172, 177, 182
Pondicherry (India), 97
Pope, J., 38, 62, 135, 136
Popular Astronomy, 129–199, 201–254, 272, 273, 287
Potsdam Astrophysical Observatory, 111
Potter´s American Monthly, 233
Preston, E.D., 116–117
Princeton Review, 172
Pritchard, C., 136, 137, 206
Proctor, R., 44–46, 82, 85, 90–95, 97, 103, 110, 131, 132, 136, 138, 139, 141, 146, 156, 197, 198, 202, 203, 206, 207, 210, 211, 219–222, 233–234, 253, 254, 265–267, 273, 288
Professionals, 1, 6, 8, 10, 20, 34, 39, 128–130, 148, 175, 209, 212, 254–262, 265, 267–269, 272–277, 279, 282, 284–290
Prominences, 8, 32, 43–46, 48, 54, 55, 57–60, 63, 67, 69, 70, 77, 82, 130–132, 135–139, 141, 143, 145, 147, 148, 150–157, 159–161, 163, 165, 167, 168, 170, 172, 177, 180–182, 185, 189, 190, 192, 264, 271
Protuberances, 43, 60, 131, 135, 152, 158, 189
Psyche, 171
Public, 4–6, 14, 18, 21, 22, 25, 27, 29, 30, 33–35, 38, 112, 129–130, 136, 139, 143, 144, 148, 153, 160, 175, 183–185, 190, 192, 196–197, 199, 201, 204, 206, 210, 218, 220, 225, 227–229, 231, 232, 235, 236, 240, 242, 246, 253, 254, 257–267, 269–271, 273–275, 277, 279–290
Publications of the Astronomical Society of the Pacific, 272, 274, 277–279, 283, 289
Puck, 234
Pueblo (Colorado), 75–77, 80, 85, 144, 195, 196, 199, 228, 231, 253, 273, 281, 285–286
Pulsifer, W.H., 78, 281
Punta Arenas (Chile), 19, 111, 114, 115, 123, 124, 235, 245, 250, 251, 253
Putnam´s Monthly Magazine, 152

Q
Quebec (Canada), 75, 80, 111, 190, 195, 243, 252
Quebec Observatory, 75
Queenstown, New Zealand (American transit expedition site), 105
Queen´s University Observatory, 243

R
Radau, R., 99, 204
Radiant, 13
Railroad, 4, 75, 112, 226
Rangoon, 62
Rawlins (Wyoming Territory), 75, 76, 80, 144, 148, 159, 170, 177, 178, 192, 193, 195
Rayet, G., 60, 61
Raymond, C.W., 212, 226
Red House Observatory, 242
Rees, J., 231, 281
Reformed Church Messenger, 173
Respighi, L., 141, 143
Revolving photographic device, 104

Riles, J.K., 78
Ritchie, J., 272
Rittenhouse Astronomical Society, 273
Rittenhouse, D., 38, 37, 96
Rivabellosa (Spain), 145
Robie, T., 7–8
Rodriguez (transit expedition site), 104, 202
Rogers, S., 15, 191, 195
Round Table, 152
Rowan, S., 119, 120, 122–124
Royal Astronomical Society (England), 133, 197
Royal Observatory (Brussels), 144
Royal Observatory (Greenwich), 84, 214, 216
Royal Observatory (Königsberg), 112, 150
Rudolphine Tables, 89, 92, 93
Russell, J., 73, 130
Rutherfurd, L.M., 255, 256

S
Saigon (Cochin-China), 202
Saint Joseph (Baja California), 96
San Antonio (Texas), 114–118, 245–246, 253
Sands, B.F., 63, 105, 106, 176, 187
Sandwich Islands, 104, 206, 215, 216
Santa Cruz (Patagonia), 124, 250
Santiago (Chile), 114, 124, 250, 253
Saturday Evening Post, 234
Savannah (Georgia), 117, 122
Sayre Observatory, 246
Schaeberle, J.M., 46, 82
Schiaparelli, G., 13
Schmidt, J.F., 161
Schuster, A., 55, 56, 76, 84, 85
Science Observer, 127–129, 139, 153, 154, 163, 164, 168, 175–178, 185, 190, 196, 198, 204, 210, 212, 214, 219, 224, 231, 235, 236, 241, 242, 255, 271–274, 277, 284, 286
Scientific American, 144, 175, 178, 180–183, 219, 239, 240
Scribner´s Monthly, 144, 153, 210
SDO. *See* Solar Dynamics Observatory (SDO)
Seagrave, F., 78, 240, 281
Seagrave Observatory, 78, 80, 240, 281
Searle, G., 67
Secchi, A., 55, 56, 135, 138, 165
Separation Point (Wyoming Territory), 85, 148
Sestini, B., 154
Seven Years War, 94, 96, 97
Seward, W.H., 190, 191
Shaker Manifesto, 173
Sharpless, I., 237, 238
Shelbyville (Kentucky), 65, 130, 143, 152, 155, 159, 165, 172, 183, 187
Shunk, W.A., 119
Sidereal Messenger, 33, 175, 271, 272, 289
Silliman, B., 32, 175
Silliman´s Journal, 175
Sioux Falls City, 69, 70
Skinner, A.N., 75
Smith, E., 124, 126, 250, 251
Smithson, J., 27
Society of Jesus, 153
SOHO, 87
Solar Dynamics Observatory (SDO), 87
Solar eclipses, 5, 6, 34, 41, 43–88, 129–199, 201, 206, 238, 258, 259, 261, 264, 277, 281, 286, 287, 290
Solar parallax, 88, 91, 94, 95, 97, 102, 104, 110, 115, 123, 128, 173, 204, 211, 213, 219, 241
Southworth, A.S., 150
Spectroheliograph, 277
Spectroscopy, 41, 44, 52–55, 57–60, 67, 69, 70, 75, 76, 82, 100, 129–132, 134–139, 141, 143–148, 152–159, 165, 168, 171–173, 175–178, 180, 182, 183, 185, 187–189, 192–199, 211, 247, 250, 252, 264, 287
Springfield (Illinois), 147, 163, 186, 187, 189, 190
Star maps, 34, 38, 139
Stebbins, R., 260–262, 265, 275, 290
Steele, J.D., 132, 155
Stewart, B., 175
St. Nicholas: An Illustrated Magazine for Young Folks, 210
Stokes, G., 131, 132
Stone, E.J., 97, 133, 134, 183
St. Paul (Campbell Island), 202
Sturdy, E.W., 250
Suggestions for Observing the Total Eclipse of the Sun on January 1, 1889, 282
Sundials, 34
Sunspots, 43, 46, 69, 82, 84, 90, 133, 134, 139, 159, 161, 162, 175, 194–196
Swatara, S.S, 108, 207, 209, 211, 212, 214, 225–228, 230
Swift, L.A., 85, 86, 162, 164, 183, 195–197

T
Tacchini, P., 82, 83, 100, 183
Tasimeter, 77, 78, 156, 159, 164, 170, 173, 178, 179, 193, 195, 196, 198
Taylor, W.C., 69, 70
Tebbutt, J., 61, 62, 259, 274

Telegraph, 4, 57, 100–102, 104, 144, 168, 180, 184, 189, 194, 209, 214, 229, 234, 241, 243
Telescopes, 1, 23, 25, 29, 34, 39, 40, 50, 51, 59, 67, 69, 75, 77, 87, 91, 99, 105, 109, 116, 128, 145, 156, 159, 161, 164, 169, 175, 176, 180, 187, 193, 204, 213, 217, 235, 236, 238, 247, 255, 258, 262, 267–269, 271, 275, 278, 280, 281, 284, 288, 290
Tempel, E., 13
Tennant, J.F., 54, 57
Tevis, R., 67
Thatcher, A.E., 189
Thompson, J.F., 206
Thomson, W., 219
Todd, D.P., 57, 80, 81, 110, 125, 281
Todd, M., 134
Transition Region and Coronal Explorer Satellite (TRACE), 87, 128
Transit of Mercury, 2, 89, 92, 180, 183, 230, 258
Transit of Venus, 2, 5, 89, 90, 92–96, 101–110, 112–115, 117, 118, 121, 123, 128, 201–252, 264–267, 269, 277–279, 284–289
Transit of Venus Commission, 119, 123, 127, 128, 238, 269, 283, 284, 289
Trigonometric Survey of India, 57–59, 222, 223
Trinity College, 111, 236, 244
Trowbridge, D., 134
Tschifu, China (transit expedition site), 104
Tupman, G.L., 215, 218
Tuttle, H., 13
Twining, A.C., 10
Twin-objective heliometer with unequal focal lengths, 115
Twitchell, H., 73, 255

U
U.S. Coast and Geodetic Survey, 65, 116–117, 123, 124, 163, 187, 208, 251, 257
U.S. Naval Observatory, 13, 19, 52, 63, 73, 75, 80, 105, 108, 109, 119, 120, 123, 157, 167, 168, 171, 187, 191, 192, 230, 236, 245, 246, 270, 280, 281, 289
U.S. Signal Service Station, 199
U.S. Weather Bureau, 80, 82

V
Vassar College Observatory, 249
Vassenius, B., 48
Vatican Observatory, 54
Very, S., 124, 250
Vladivostok (American transit expedition site), 101, 105, 109, 209, 212, 214, 219, 229
Von Fraunhofer, J., 54
Von Oppolzer, T., 13, 164
Vulcan, 85, 86, 141, 147, 148, 156, 159, 162, 165, 170, 183, 195, 196, 230, 280

W
Wah-koa (Siam), 60, 169
Waldo, L., 78, 220, 277, 281, 289
Walker, S., 19–21, 29
Warder, R., 69, 72
Watson, J.C., 85, 148, 152, 156, 162, 164, 165, 170, 183, 195, 196, 215, 216, 229
Webb, T.W., 75
Wells, W., 155
Wesley, W.H., 51
Western Christian Advocate, 174, 216, 239
West Las Animas (Colorado), 77, 80, 85, 144, 195, 196, 199, 228, 231, 273, 281, 285
Whipple, J.A., 67, 152, 187
Williams, T., 262, 267
Willson, R.W., 78, 281
Wilson, J., 34, 36
Winchester Observatory (Yale University), 248, 249
Winder, D.K., 73
Wines, F.H., 163
Winlock, J., 65–67, 99, 130, 143, 159, 183
Winthrop, J., 1, 2
Wollaston, W.H., 132, 165
Wright, F.E., 76, 183
Wyberd, T., 137

Y
Yerkes Observatory, 86, 275
Yohkoh X-Ray Satellite, 87
Yokohama (Japan), 202
Young, C.A., 73, 74, 125, 134, 172, 177, 192, 196, 198, 214, 224, 229, 232, 236, 239, 246, 266, 274, 288
Youth's Companion, 174, 216

Z
Zion's Herald, 174, 216

MIX
Papier aus verantwortungsvollen Quellen
Paper from responsible sources
FSC® C105338

If you have any concerns about our products,
you can contact us on
ProductSafety@springernature.com

In case Publisher is established outside the EU,
the EU authorized representative is:
Springer Nature Customer Service Center GmbH
Europaplatz 3, 69115 Heidelberg, Germany

Printed by Libri Plureos GmbH
in Hamburg, Germany